"Mathesis of the Mind"

A Study of Fichte's *Wissenschaftslehre* and Geometry

Fichte-Studien-Supplementa

Band 29

David W. Wood

"Mathesis of the Mind"

A Study of Fichte's *Wissenschaftslehre* and Geometry

Rodopi

Amsterdam - New York, NY 2012

Fichte-Studien-Supplementa

Die Supplementa zu den *Fichte-Studien* präsentieren Forschungen zur Geschichte und Systematik der Transzendentalphilosophie. Es werden in diesem Rahmen umfangreichere Untersuchungen veröffentlicht, z.B. Monographien, Dissertationen und Habilitationsschriften, die dem Verständnis der Transzendentalphilosophie dienen oder ihre Erneuerung und Weiterentwicklung voranbringen können.

Typographie und Satz: Christoph Asmuth (Berlin)

The paper on which this book is printed meets the requirements of "ISO 9706:1994, Information and documentation - Paper for documents - Requirements for permanence".

ISBN: 978-90-420-3491-4
E-Book ISBN: 978-94-012-0768-3

Printed in The Netherlands

Contents

Acknowledgements

The following study is a revised version of my PhD dissertation. It examines Johann Gottlieb Fichte's conception of geometry, theory of mathematics and the role of geometry in the *Wissenschaftslehre*. Carried out under the joint supervision of Professor Dr. Jean-François Courtine of the Sorbonne (Université Paris IV) and Professor Dr. Günter Zöller of the Ludwig-Maximilians-Universität (Munich), the dissertation was defended in Munich in July 2009.

I would like to thank the following people for their support during the writing of this dissertation. Firstly, to Professor Jean-François Courtine at the Sorbonne for accepting me into his doctoral programme and for his philosophical encouragement and expert help. To Professor Günter Zöller, for facilitating the co-tutelle transfer to Munich, and for his valuable advice and suggestions concerning the transcendental philosophies of Kant and Fichte. And to the other members of my doctoral committee, Professor Emmanuel Cattin (Université Blaise Pascal, Clermont-Ferrand II) and PD Dr. Ives Radrizzani (LMU/Schelling Kommission) – their close readings and constructive feedback have greatly improved my text.

My warmest thanks also go to Dr. Erich Fuchs, the editor-in-chief of the *Johann Gottlieb Fichte Gesamtausgabe*, for kindly providing me with hitherto unpublished manuscripts, elucidating the nuances of Fichte's German, and for hosting me at the Fichte Kommission for a 12-month post-doctoral fellowship during the period of the revision of this dissertation, a fellowship generously funded by the Fritz Thyssen Stiftung.

I would also like to thank the following people for their scholarly encouragement and philosophical comments: Professor Daniel Breazeale, Professor Marco Ivaldo, Dr. Rodrigue Nzameyo, Dr. Norman Sieroka, PD Dr. Andreas Schmidt, Dr. Peter K. Schneider, Professor Michael G. Vater, Paul S.-J. Wood, and Dr. Renatus Ziegler.

A special mention is due to Mrs. Fabienne Mailé (doctoral office Sorbonne, Paris), Dr. Cathleen Muehleck and Mrs. Doris Jacobson (Promotionsausschuß LMU), for their excellent administrative assistance regarding the co-tutelle doctoral process between Paris and Munich.

I am also grateful to my dear family and friends in Australia – to Loraine, Nick, Andrew, Aidan, Katrina and the rest of the Wood and Psarros families, as well as the Maurel and Cahen families in France and America. And to the memory of my dear departed parents – anything of value in this text is due to them.

Finally, my greatest debt and thanks are to Laure Cahen-Maurel. *L'amour ne rend pas aveugle aux faiblesses d'une bien-aimée; au contraire, il ouvre l'œil précisément à ses qualités les plus profondes.* This book is dedicated to her.

Introduction: Fichte as the Euclid of Philosophy?

> However, even a fleeting perusal of the work is sufficient to arouse the conviction in every informed reader that if any mortal were granted *to become to philosophy what Euclid was to mathematics*, it would be Herr *Fichte*.[1]

a. Fichte and Mathematics

According to standard histories of philosophy and mathematics, the German idealistic philosopher Johann Gottlieb Fichte (1762–1814) never wrote any work on mathematics or the philosophy of mathematics, nor apparently had any influence on their evolution and development. Hence, the question might be immediately asked: Why undertake a study of Fichte's *Wissenschaftslehre* and geometry?

Well, we only have to examine the above quote from October 1794 to be confronted with a mystery: here Fichte's system of philosophy is explicitly brought into connection with Euclid's system of geometry. And this is not an isolated statement. Just consider Fichte's pronouncements from around the same time in letters to three different colleagues: »I am convinced that philosophy can only become a science through the development out of a single *Grundsatz* (first principle), but that it must then receive the *self-evidence* of geometry [...] In short, in a few years I believe we will have a philosophy to rival the self-evidence of geometry. [...] this project concerns nothing less than a scientific philosophy that could be measured against mathematics itself.«[2]

1 Announcement for the *Grundlage der gesammten Wissenschaftslehre* (GWL), in the *Intelligenzblatt der Jenaer Allgemeine Literatur Zeitung*, no. 113, October 1, 1794 (Quoted in: ed. Erich Fuchs, *Fichte im Gespräch* (FG), vol. 1, footnote 1, p. 209; my emphasis). Unless otherwise indicated, all translations are my own.

2 See letters to J. F. Flatt, November or December 1793 (draft) (GA III/2: 18); H. Stephani, December 1793 (GA III/2: 28); F. V. Reinhard, January 15, 1794 (GA III/2:

We could add to this the enigmatic account of Fichte's discovery of the *Wissenschaftslehre* in late 1793. In a scene recalling none other than Descartes' discovery of the Cartesian philosophy[3], and likewise immersed in his »own *meditations*«[4] by a warm winter stove, Fichte divined the outlines for a brand new system of philosophy. – This philosophy would be capable of isolating the first act of knowledge with »the same certitude possessed by mathematics«, yet able to avoid the latter's »rigid and lifeless formalism«, because it is generated and grounded in our own »living and active« self-consciousness.[5]

The genesis for the present study is to be attributed to encountering these early remarks on the relationship between the *Wissenschaftslehre* and mathematics. These remarks gave rise to a number of questions: was Fichte really considered by his contemporaries to be the Euclid of philosophy? If so, by whom and for what reason? What is Fichte's conception of geometry? Does Fichte have a philosophy of mathematics? If this proves to be the case, how does it compare with those of his contemporaries (e.g. Kant, Schelling, and Maimon), and how could it be classified in line with modern philosophies of mathematics? Finally: what is the *philosophical* relationship between Fichte's *Wissenschaftslehre* and geometry; that is to say, what is the role of geometry within the Fichtean system of philoso-

40); and to Karl August Böttiger, February 4, 1794 (GA III/2: 55). [GA = *J.G. Fichte – Gesamtausgabe der Bayerischen Akademie der Wissenschaften*, eds. Reinhard Lauth, Hans Jacobs, Hans Gliwitzky, Erich Fuchs, Peter K. Schneider, Ives Radrizzani, Hans Georg von Manz, Martin Siegel, Günter Zöller *et al.* (Stuttgart/Bad Cannstatt: Frommann-Holzboog, 1962-...). Henceforth: GA, followed by series, volume and page numbers. All further abbreviations of Fichte's texts and the works of others are indicated in the bibliography].

3 This parallel between Descartes' philosophical discovery in Neuburg an der Donau and Fichte's in Zurich has also been noted by Andreas Schmidt. See: *Der Grund des Wissens: Fichtes Wissenschaftslehre in den Versionen 1794/95, 1804/II und 1812* (Paderborn: Schöningh, 2004), pp. 7–8, footnote.

4 See Fichte, *Eigne Meditationen über ElementarPhilosophie* (GA II/3: 21–177).

5 Report on Fichte by Henrik Steffens from 1841 (in: FG, volume 1, pp. 63 f.). See too the similar report by Eduard Fichte from 1863. This report contains echoes of Descartes' discovery when he says of Johann Gottlieb Fichte: »after long and incessantly meditating on the highest principle of philosophy, [Fichte] was seized in sudden self-evidence, while standing near a warm winter stove, by the thought that the highest principle could only be the *I* (*Ich*), the concept of pure subject-objectivity.« (ibid. p. 63).

phy? Simply put, the aim of this study is to provide a contribution towards answering these important questions.

I will show that Fichte placed a huge importance on geometry because of its strict deductive and axiomatic form, its clear and lucid process of reasoning, its analysis of construction and indemonstrable proofs, and its reliance on non-sensible intuitions and universally valid first principles. On account of these aspects, mathematics – and Euclidean geometry in particular – became a model of certitude and scientific thinking to him. Although Fichte was not a professional geometer, it did not stop him from harbouring a lifelong fascination for the topic of geometry. Indeed, Fichte's reflections on the need for a philosophy of mathematics, and the influence of mathematical method and reasoning on his thought can be discerned throughout his career as a whole, from the origins of the *Wissenschaftslehre* in the winter of 1793/1794 to the final expositions of 1813/1814.[6] With regard to the external or formal presentation of his philosophy, the mathematical influence is perhaps most visible in late 1800 in Fichte's attempt at a so-called »geometric« exposition of his system – the unfinished *Neue Bearbeitung der Wissenschaftslehre*.[7]

In general, however, Fichte was not so much interested in the technical and *quantitative* aspects of mathematics and geometry, i.e. in the solution and measurement of specific geometric figures and problems, as in their *qualitative* aspects, in geometrical thinking or *esprit géométrique*, to borrow a title from Pascal.[8] Hence, I will above all argue – as Joachim Widmann aptly puts it in his excellent study of the *Wissenschaftslehre 1804/II* – that the relationship between Fichte's philosophy and geometry should be understood »not with regard to the technical externalities and the presentational conventions of mathematicians, but must concern the *internal, abstract cognitive structure* of the *Wissenschaftslehre*.«[9] Or again, in the words of Andreas Speiser, the following study particularly aims to

6 Among others, see the lecture of February 27, 1794, in the *Zurich Wissenschaftslehre* (ZV), GA IV/3: 32–37); the *Announcement* dated November 4, 1800 (GA I/7: 153–164); the *Erlanger Logik* of 1805 (GA II/9: 124–137); and the *Einleitung in die Wissenschaftslehre*, from November/December 1813 (GA II/17: esp. 253–268).

7 GA II/5: 331–401.

8 Cf. Blaise Pascal, »De l'esprit géométrique«, essay from 1657/58, reprinted in: *De l'esprit géométrique – Ecrits sur la Grâce et autres textes* (Paris: Flammarion, 1999), pp. 67–96.

determine Fichte's »*mathematische Denkweise*«[10] – that is, his inherently mathematical style of thinking and philosophizing. In fact, the self-evidence, intuitions and constructions of traditional geometry and mathematics will ultimately be seen to constitute particular and limited forms of the more universal self-evidence, intuitions and constructions of the Fichtean system of philosophy. And although Fichte's model of an »original geometry« and reflections on the infinite polygon and mathematical infinity should be viewed in this context, they are also worth investigating in their own right as independent and innovative contributions to the development and history of geometric thought.

In terms of its geometric presentation, Fichte's system of philosophy naturally does not stand in a vacuum, but fits into a long tradition of philosophers in 18th century Germany who were convinced that philosophy could only be raised to the level of a science by following the lead of Euclidean geometry. This philosophical current contains such distinguished names as Christian Wolff (1679–1754), Immanuel Kant (1724–1804) and Moses Mendelssohn (1729–1786), whose endeavour – for at least some portion of their careers – was for »metaphysics to attain the same clarity and rigorousness in its proof as mathematics.«[11] Fichte's *Neue Bearbeitung der Wissenschaftslehre* from 1800 can therefore be said to rival other paradigmatic attempts in this tradition – such as Wolff's *Anfangsgründe aller mathematischen Wissenschaften* (1710), Kant's *Physical Monadology* (1756) or Schelling's *Darstellung meines Systems der Philosophie* (1801), all of which are set out *ordine geometrico demonstrata* and embellished with a wealth of theorems, postulates, corollaries, proofs and scholia.[12] Ultimately, all these German attempts drew their inspiration from the rigorous and self-evident geometric style of reasoning in Spinoza's

9 Joachim Widmann, *Die Grundstruktur des transzendentalen Wissens nach Joh. Gottl. Fichtes Wissenschaftslehre 1804*[2] (Hamburg: Felix Meiner Verlag, 1977), p. 16 (henceforth: *Grundstruktur)*.

10 Cf. Andreas Speiser, *Ein Parmenideskommentar – Studien zur platonischen Dialektik*, 2nd ed. (Leipzig: K.F. Koehler, 1959), esp. pp. 73–78.

11 Giorgio Tonelli, »Der Streit über die mathematische Methode in der Philosophie in der ersten Hälfte des 18. Jahrhunderts und die Entstehung von Kants Schrift über die ›Deutlichkeit‹,« *Archiv für Philosophie* 9 (1959): 64. Tonelli's words here refer to Kant's earliest writings. Tonelli's seminal article provides a nice overview of this tradition; unfortunately, his study does not extend as far as Fichte.

Ethica of 1677[13] – as Fichte implicitly admits in the first and most famous presentation of his philosophy – the *Grundlage der gesammten Wissenschaftslehre* of 1794/95 – when asserting that there have only been two »completely consistent systems« in the history of philosophy – the critical and Spinozistic.[14]

Like many other thinkers at the time, Fichte still considered the science of geometry to constitute the inner core of mathematics – to be its ground and foundation, and hence he often employs the terms »mathematics« and »geometry« interchangeably. Unlike the other main figures of this tradition though, particularly Kant and Wolff, where the relationship of their philosophy to the mathematical sciences has been treated in depth[15], the connection between the *Wissenschaftslehre* and mathematics has only been occasionally touched upon by Fichte scholars and scholarship. It is therefore perhaps both surprising and encouraging to note the reception of Fichte's thought on a group of leading 20th century philosophers of mathematics. As we shall see, distinguished philosophers of the calibre of Hermann Weyl, Andreas Speiser and Jules Vuillemin, all under-

12 Cf. Christian Wolff, *Anfangsgründe aller mathematischen Wissenschaften (Halle: Rieger,* 1710); Immanuel Kant, *Metaphysicae cum geometria junctae usus in philosophia naturali, cuius specimen I. continet monadologiam physicam* (AA I: 473–487); the full title in English of Kant's text is: *The Employment in Natural Philosophy of Metaphysics Combined with Geometry, of Which Sample I Contains the Physical Monadology* (1756); a translation is available in: Kant, *Theoretical Philosophy (1755–1770)*, ed. David Walford, with Ralf Meerbote (Cambridge: Cambridge University Press, 1992), pp. 47–66; and Schelling, *Darstellung meines Systems der Philosophie*, in: *Zeitschrift für spekulative Physik*, vol. 2, 1801 (repr. Meiner, 2001, pp. 329–436), where Schelling says: »[Spinoza's] form permits the greatest brevity of presentation and most distinctly allows us to judge the self-evidence of the proofs« (ibid. p. 335).

13 Written in Latin, Spinoza's *Ethica* was published posthumously in 1677. See Benedictus de Spinoza, *Die Ethik*, Latin-German edition, trans. Jakob Stern (Stuttgart: Reclam, 1977); English translation in: *The Ethics and Selected Letters*, trans. Samuel Shirley (Indianapolis, IN: Hackett, 1982).

14 GA I/2: 264.

15 For Kant, see the following influential studies listed in the bibliography: M. Friedman (1985, 1992, 2000); E. Carson (1997), J. Hintikka (1981, 1984), and Charles Parsons (1980). For Wolff, see Frängsmyr (1975); Shabel (2003, pp. 49–90; 2006, pp. 95 ff.); and Tutor (2004). See part d. below of my introduction for details of the Fichtean reception.

scored the fruitfulness of Fichte's *philosophical* ideas for understanding certain complex mathematical structures in group theory, set theory and abstract algebra. Interestingly, their analyses are primarily based on Fichte's philosophical writings and his theory of intuition (*Anschauung*) and not on any of his discussions of mathematical topics or issues *per se*. The results of their studies are only now beginning to receive the critical attention they deserve.[16]

To be fair, this paucity of attention to Fichte's mathematical ideas may be partly explained by the fact that some of his most important texts have only been published in the *J.G. Fichte Gesamtausgabe* in the last twenty years.[17] Nevertheless, in other available texts Fichte repeatedly drew attention to the similarities and links between mathematics and the *Wissenschaftslehre*. – And this was not just in private letters to colleagues or in unpublished manuscripts. In at least three significant public announcements, for the *Grundlage der gesammten Wissenschaftslehre* (1794), the *Neue Bearbeitung der Wissenschaftslehre 1800*, and the second series of lectures on the *Wissenschaftslehre 1804*, Fichte clearly highlighted the conceptual affinity between the science of mathematics and his own philosophical enterprise.[18] In fact, the second of these announcements – with its declaration that »the *Wissenschaftslehre* is mathematics and has the distinctions of mathematics« – is a veritable manifesto for the mathematical pretensions and underpinnings of the *Wissenschaftslehre*.[19] The underlying claim of all three announcements may be expressed as follows: just

16 For instance, see Norman Sieroka's recently published detailed study on the Fichte-Weyl relationship in: *Umgebungen: Symbolischer Konstruktivismus im Anschluss an Hermann Weyl und Fritz Medicus* (Zurich: Chronos, 2010), esp. pp. 57-59, 102-111, 142-152, 223-226.

17 For example, the *Erlanger Logik* of 1805 was only published in 1993 (see GA II/9: 57–171; especially pages 124–137; English translation of these pages in the Appendix below); the *Zurich Wissenschaftslehre* in 1996 (separate edition, then in 2000 in the *J. G. Fichte Gesamtausgabe*); and the *Ankündigung* was only reprinted for the first time in 1988.

18 Cf. the Announcement for the *Grundlage*, reprinted: FG vol. 1, footnote 1, p. 209; *Ankündigung* (also titled: *Seit sechs Jahren*) of 1801 (GA I/7: 153–164); and the »Ankündigungen« (announcements) for *Die Wissenschaftslehre. Zweiter Vortrag im Jahre 1804* in the newspapers: *Ungersche Zeitung*, Jan. 3, 1804; *Haude-Spenersche Zeitung*, 5 Jan. 1804; *Oberdeutsche Allgemeine Litteraturzeitung* Jan. 14, 1804 (reprinted in: »Ankündigungen zu den Vorlesungen in Berlin«, GA I/8: 17).

as Greek geometry reached its scientific summit and definitive formulation in Euclid's *Elements*, so the scientification and mathematization of metaphysics had attained its apotheosis in Fichte's *Wissenschaftslehre*, to finally become a »*mathesis of the mind*.« But does this claim stand up to scrutiny?

b. Theses and Methodology

This study has two principal aims: 1. To present and reconstruct the main outlines of Fichte's philosophy of mathematics and theory of geometry, and 2. To demonstrate the critical role played by geometrical reasoning and method in the presentation and arguments of the *Wissenschaftslehre*.

My fundamental theses may be stated as follows:

I will argue that Fichte not only had a philosophy of mathematics, but that his conception of mathematics as a whole may be classified as a brand of mathematical Platonism or neo-Platonism. My classification of Fichte as a mathematical Platonist is primarily due to his own model of a pure and original geometry. – A key but fragmentary text in this respect is Fichte's *Erlanger Logik* 1805.[20] Here Fichte outlines his own conception of an »original« (*ursprüngliche*) geometry that is grounded in ideal archetypal elements, and he contrasts it with »ordinary« (*gewöhnliche*) or Euclidean geometry. Following Kant, Fichte's theory of geometry may be additionally termed »synthetic«, since it is also based on synthetic constructions and intuitions. Nevertheless, Fichte departs from both the Kantian and classical Euclidean conceptions in a number of significant ways, particularly with regard to the starting point, the treatment of the parallel problem, and its positing of elements at infinity.

Furthermore, I maintain that in terms of their *genesis*, Fichte's views on scientific methodology, *Grundsätze* (first principles), intellectual intuition and construction, share distinct internal parallels with Euclidean ge-

19 *Ankündigung* (GA I/7: 160); it bears the date: Berlin, November 4, 1800, and was first published January 24, 1801 in the *Allgemeine Zeitung* (reprinted in: GA I/7: 153–164). English translation, see »Public Announcement« in: Fichte, *Introductions to the Wissenschaftslehre and Other Writings*, trans. Daniel Breazeale (Indianapolis: Hackett, 1994), pp. 186–201.

20 The *Erlanger Logik* (GA II/9: 57–171) These lectures on logic have the heading: »Logik. Erlangen, 1805.« I will refer to them simply as *Erlanger Logik*.

ometry and mathematical method. This is not to say that Fichte's philosophy has its origin in geometry, but only that he repeatedly appeals to geometric intuitions, postulates and constructions as a confirmation of the scientific nature of his own enterprise. This is because in a *structural* or foundational sense Fichte considers the intuitions, constructions and axioms of geometry to be in fact *particular instances* of the *more universal* intuitions, constructions and first principles (axioms) of the *Wissenschaftslehre*. In other words, although Fichte's transcendental idealism is directly inspired by his *geometrical idealism*, the *Wissenschaftslehre* structurally enjoys a higher status in his hierarchy of the sciences.[21]

With regard to the relationship of mathematics to the *Wissenschaftslehre*, my study is twofold: I will examine the *external* similarities between mathematics and the *Wissenschaftslehre*, e.g. the striking procedure of *more geometrico*, but more importantly, I will analyse the *internal* structure and mathematical conceptions employed by Fichte in his philosophical expositions. Or to use the distinction frequently employed by Fichte himself, I will look at both the *letter*, i.e. the formal presentation, and the *spirit*, or mathematical reasoning animating his system. I will do this by a treatment of Fichte's own words, texts, letters and writings (both published and unpublished) on this topic, showing that despite the ever-changing and revised versions of the *Wissenschaftslehre*, his views on geometry and mathematics possess an internal consistency and remain relatively constant throughout his career.

Thus, in terms of methodology, I will generally proceed in two stages. 1) Firstly, I will look at one particular aspect or branch of mathematics and geometry, such as the nature of the Euclidean postulates, for example, taking into consideration the state of mathematical knowledge at Fichte's time. My aim here is to *reconstruct* Fichte's principal views on this particular aspect of mathematics. I will of course note the mathematical conceptions of other philosophers of the period. Chiefly, the crucial influence of Kant's ideas on Fichte, but occasionally too those of F.W.J. Schelling (1775–1854) and Salomon Maimon (1753–1800) if they shed light on Fichte's views. This section will especially take into account contemporary

21 See Chapter Three below for more details on the reciprocal relationship between genesis and structure. I am indebted here to Jules Vuillemin's presentation in: *La philosophie de l'algèbre* (Paris: PUF, 1st ed. 1962, 2nd ed. 1993), esp. pp. 112–120.

Fichte scholarship and research. 2) And secondly, I will consider Fichte's *philosophical interpretation* of this geometrical or mathematical element, i.e. what specific role and application it had within the framework and arguments of the *Wissenschaftslehre*.

Looking at each chapter in more detail:

Introduction: After this statement of the theses and methodology of my study, the remainder of the introduction will furnish an overview of the primary Fichte texts I have consulted, followed by an aperçu of the contemporary reception of Fichte's ideas on geometry and a review of the current state of research in this domain.

Chapter One presents an overview of Fichte's philosophy of mathematics, which I find most fully articulated, though in a fragmentary manner, in the *Erlanger Logik* (1805). I contend that a certain 14-page excursus in this text may be read as an exposition of Fichte's *philosophy of mathematics in nuce*. This chapter will also outline nine different elements belonging to a Fichtean theory of mathematics, including among others: geometry, arithmetic, algebra, mathematical infinity, and the theory of the dimensions of space.

Chapter Two reconstructs and attempts to classify Fichte's theory of geometry. My principal thesis is that Fichte possessed his own definite model of geometry that he called »*ursprüngliche Geometrie*« (original geometry). Fichte's conception of geometry is neither a logical theory, as in the current dominant interpretation of Kant, nor a simple reproduction of the finite Euclidean conception. Instead, I argue that Fichte's model of original geometry is a *constructivist and synthetic* model founded on the immediate self-evidence of geometrical intuition and grounded in an ontology of ideal or archetypal (or *Ur-*) elements. In this sense it is best classified in line with traditional synthetic or projective geometry on the one hand, and may be interpreted on the other according to modern-day classifications in the philosophy of mathematics as a species of mathematical Platonism.

Chapter Three provides a theoretical framework for situating and interpreting intuition, axiomatic first principles and construction in Fichte's system of philosophy. The first section determines the extent to which Fichte's *Grundsatz* is equivalent to a geometric axiom or postulate. Section two ascertains the connection between intellectual and mathematical intuition, with the latter classified by Fichte as a mode of intelligible intuition. I conclude this chapter with a study of the role of construction in Fichte's theory of mathematical cognition. Here I outline four main modes of construction in the Fichtean system: technical (schematic), sensible, free, and infinite construction. The first three are applicable to geometry, whereas the latter is the construction of philosophical knowledge itself.

Chapter Four provides an examination of Fichte's conception of the basic elements of geometry (including points, lines and planes) and 3-dimensional space, especially contrasting it with Kant's views. It shows how Fichte extrapolates a philosophical interpretation or image (*Bild*) from these fundamental geometrical elements that helps shed light on the activity and genesis of the I (*das Ich)* in the *Wissenschaftslehre*. This is most visible in Fichte's theory of the mental drawing of a geometrical line, which serves as a philosophical preparation for students who are unable to directly proceed to intellectually intuiting the I.

Chapter Five investigates the relationship between geometry and the *Wissenschaftslehre*, focusing on a number of further specific similarities and differences. These similarities include those of self-evidence, irrefutability and determinacy, whereas their differences lie in their unequal relations to freedom and the formal structure of their languages. On the whole, Fichte considers geometry and mathematics to be an intellectual propedeutic for the *Wissenschaftslehre*, and I therefore again situate him in the spirit of the Platonic tradition.

My *conclusion* is that the *Wissenschaftslehre* may be read as a form of philosophical mathesis, or as a »mathesis of the mind«; that is to say, as a mathematized theory of the structure and activity of the rational mind or spirit. Geometry is not simply a rigorous inspiration to Fichte, but serves as a structural and scientific touchstone for his own methodology and philosophical system.

Principal Texts

My study is largely based on the following texts. It should be repeated that Fichte's philosophy of geometry and mathematics is fragmentary and incomplete, and though he left behind quite a number of specific indications, they are scattered and dispersed throughout his writings and necessitate a »reconstruction« of his principal theories. According to my research there are four primary clusters of texts in which Fichte discusses mathematics and geometry, with the most concentrated period spanning the years 1800–1805. These four clusters may be summarized as follows:

(1) *1793–1795*: The *Zurich Wissenschaftslehre* of February 24–28, 1794 (GA IV/3: 19–41), especially the 4th Lecture of February 27, 1794 entitled: »Verhältniß der Wissenschaftslehre zur Geometrie« (pp. 32–37). Also important from this same time are the unpublished texts *Eigne Meditationen/Practische Philosophie* from 1793/94 (GA II/3: 21–266) – first available in the Fichte GA in 1971 – as well as the essential published works: *Über den Begriff der Wissenschaftslehre* of 1794 (GA I/2: 107–172) and *Grundlage der gesammten Wissenschaftslehre* of 1794/95 (GA I/2: 249–451). The latter text is the only complete presentation of the *Wissenschaftslehre* published during Fichte's lifetime. All these early writings implicitly champion the necessity of philosophy attaining the scientific status of geometry by following the same self-evident and axiomatic method with an unprovable *Grundsatz* grasped in pure intuition. On the other hand, we already glimpse here a key Fichtean theme: that of demarcating the limits between geometry and the *Wissenschaftslehre*. The text of the *Zurich Wissenschaftslehre* is in fact a lecture transcript (or *Kollegnachschrift*) made by the Swiss thinker Johann Kaspar Lavater (1741–1801), and first published in 1996.[22] As it is not from Fichte's hand we are aware of the dangers of over-interpreting it. For our purposes what is important and even unique about the *Zurich Wissenschaftslehre* is that it contains an entire lecture devoted to the relationship between geometry and the *Wissenschaftslehre*, and

22 See J. G. Fichte, *Züricher Vorlesungen über den Wissenschaftslehre Februar 1794. Nachschrift Lavater. Beilage aus Jens Baggesens Nachlass: Exzerptseite aus der Abschrift von Fichtes Züricher Vorlesungen über die Wissenschaftslehre*, ed. Erich Fuchs (Neuried/Munich: Ars Una, 1996). Subsequently published in the Fichte *Gesamtausgabe* in the year 2000.

the ideas announced here do not contradict Fichte's other writings and extant manuscripts.

(2) *1800–1802*: The *Neue Bearbeitung der Wissenschaftslehre* from October-December 1800, first published in the Fichte *Gesamtausgabe* in 1979 (GA II/5: 331–401), its accompanying *Ankündigung*, published in January 1801 (GA I/7: 153–164); the so-called popular version: *Sonnenklarer Bericht* (1801, GA I/7: 183–268),[23] as well as the *Wissenschaftslehre 1801/02*. On the one hand, this group of texts especially inquires into how philosophical understanding is related to a formal or axiomatic exposition. – On the other hand, Fichte seeks to clarify further the function of intuition within a system of philosophical mathesis.

(3) *1804–1805*: *Wissenschaftslehre 1804/II* (in: GA II/8: 2–421), above all lecture three, April 19, 1804 (pp. GA II/8: 37–53). And the *Erlanger Logik 1805* – first published in the Fichte GA in 1993 (GA II/9: 57–171) – especially pages 124–137 (English translation in the Appendix below). Group three has renounced any further attempt at a presentation of the *Wissenschaftslehre more geometrico*, but deals with the *foundations* of geometry, and culminates in a *philosophy of mathematics in nuce* and an innovative model of geometry in the *Erlanger Logik*. Here Fichte makes a distinction between two species of geometry – *gewöhnliche* or ordinary (Euclidean) geometry, and *ursprüngliche* or original (archetypal) geometry. The latter Fichtean model of geometry also tackles specific problems of pure geometry, including: the correct starting point of geometry, the nature of ideal absolute elements, the determination of angles and directionality, and the problem of parallel lines.

(4) *1813*: *Einleitung in die Wissenschaftslehre* (November-December 1813). This constitutes the last complete introductory course on the *Wissenschaftslehre*, and is noteworthy for its treatment of geometric construction as a template for philosophical construction. Versions of the

23 Daniel Breazeale has also grouped this cluster of texts together, labelling them as a whole, ›the *Wissenschaftslehre* of 1800–1801‹. See his recent essay: »Toward a *Wissenschaftslehre more geometrico* (1800–1801)«, in: *After Jena: New Essays on Fichte's Later Philosophy*, eds. D. Breazeale and T. Rockmore (Evanston, Illinois: Northwestern University Press, 2008), p. 5.

text are present in *Fichtes nachgelassene Werke* (SW IX: 1–102) and in *›Ultima Inquirenda‹. J.G. Fichtes letzte Bearbeitungen der Wissenschaftslehre Ende 1813/ Anfang 1814*, ed. R. Lauth (Stuttgart/Bad Cannstatt: Frommann-Holzboog, 2001), pp. 1–129. However, I have made use of the recently revised manuscript of Fichte (forthcoming in GA II: 17: 233–314), the parallel *Kollegnachschrift* versions entitled Halle 6c (in GA IV/6: 351–472) and Berlin Ms.germ.qrt. 1964b, since they contain additional material with concrete examples of the nature of mathematical construction and cognition that is not included in the earlier published versions.[24]

This brief overview does not purport to argue for a rigid classification of Fichte's conception of mathematics into distinct phases, but simply to draw attention to Fichte's key texts and treatment of central mathematical topics. Because as mentioned at the outset, Fichte's interest in issues such as intuition and construction and their analysis in the light of geometry extends throughout his career.

There are a large number of other texts and letters in which Fichte talks about geometry, and we will refer to them in the discussion where appropriate. These texts include the *Grundriss des Eigenthümlichen der Wissenschaftslehre* (1795), the *Grundlage des Naturrechts* (1796/97), and the *Thatsachen des Bewußtseyns* (1810). However, my principal focus will be the above four clusters of texts because they offer Fichte's clearest and most detailed thoughts on the relationship between geometry and the *Wissenschaftslehre*.

24 I here express my thanks to Dr. Erich Fuchs for drawing my attention to this material and making it available to me prior to its publication in the *Gesamtausgabe*.

c. Contemporary Reception and Influences on Fichte

> Fichte est dans les idées abstraites une tête mathématique comme Euler ou Lagrange.
>
> Madame de Staël[25]

It is interesting to note that the proclamation of Fichte as the Euclid of philosophy found a *positive* resonance even during Fichte's lifetime, particularly in a couple of figures associated with the romantic movement. In her 1813 book *De l'Allemagne* Madame de Staël (1766–1817) sums up Fichte's style of philosophizing by equating his »mathematical mind« with the thoughts of the renowned 18th century mathematicians Leonhard Euler (1707–1783) and Joseph-Louis Lagrange (1736–1813) (as the above quote shows). Her words appear to be a translation or update using modern mathematicians of the Euclid claim from the *Intelligenzblatt der Jenaer Allgemeine Literatur Zeitung* in 1794.[26] Furthermore, a serious student of Fichtean philosophy – the poet-philosopher Novalis – compares Fichte's thought with another French scientist, Georges-Louis Le Sage (1724–1803). In this respect Novalis was perhaps the first thinker to make concrete references to the mathematical nature of Fichte's system. In philosophical notes and fragments from 1798–1800 he put forward the view that Fichte's systematic approach to philosophy was akin to the problem-solving power of the methods in calculus and algebra. Comparing it to the »speculative« mathematical methods utilised by Le Sage in physics,

25 Germaine de Staël, *De l'Allemagne*, vol. 2, p. 148 (Paris: Flammarion, 1968). Translation: »In abstract ideas Fichte has a mathematical mind like Euler or Lagrange.« Madame de Staël wrote *De l'Allemagne* during the years 1807–1810 but it was first published in France in 1813.

26 In addition to her remarks in *De l'Allemagne*, in connection with Fichte, Madame de Staël is also remembered for her comic faux pas in a personal meeting with him in Berlin in 1804, where to Fichte's embarrassment (and in spite of herself it seems) she had the sudden inspiration to equate the Fichtean philosophical enterprise with the hapless adventurer Baron von Münchhausen (cf. FG vol. 3, pp. 240–244). Initially »Fichte's face looked like a tragedy« (ibid. p. 242), but he was not put off by this remark and subsequently composed a number of aphorisms for Madame de Staël stressing the scientific unity of his own philosophy in contrast to the Kantian system, which is »incomplete and bereft of systematic unity, and sins against scientific form, even though it is mostly correct in its main results.« *Aphorismen [Für Mme de Staël]* (GA II/7: 247).

Novalis asserted that Fichte's system too was »completely analogous to mathematics with regard to its discoveries, its rectifications – and to what it can achieve«, and elsewhere advanced the possibility of a »mathematical *Wissenschaftslehre*.«[27] Little did Madame de Staël and Novalis know, but their remarks recognizing a mathematical affinity between the thought of Fichte and that of Lagrange and Le Sage would prove extremely prescient, to be finally elaborated in detail by the French philosopher of mathematics Jules Vuillemin in the twentieth century (see section d. below).

In contrast to the favourable opinions of Madame de Staël and Novalis, the majority of the criticism during Fichte's career concerning his alleged mathematical philosophy was almost overwhelmingly *negative*. Friedrich Heinrich Jacobi (1743–1819) spoke for a whole class of prominent German intellectuals, when in his famous letter on »nihilism« or *Open Letter to Fichte* (1799) he termed the mathematical bent in the Fichtean philosophy as »inverted Spinozism« (*umgekehrter Spinozismus*) or »*Mathesis pura*«, because for him it was a form of »materialism without matter.«[28] Employing the term »mathesis« in an altogether pejorative sense, Jacobi argued that pure mathematics at least presupposes mathematical *bodies* with mass, figures and movement, unlike the Fichtean philosophy, which for him lacked any real content except an abstracted and »empty« form of consciousness.[29] Fichte himself would publicly discuss the relationship between »mathesis« and the *Wissenschaftslehre* one year later in his *Ankündigung*. However, here Fichte turned Jacobi's criticism

27 See Novalis, *Fragmente und Studien 1799–1800*, no. 25 (HKA III: 559), and *Das Allgemeine Brouillon* (1798/99), nos. 429 and 464 (HKA III: 321, 336). For more details on the relationship between the mathematical thought of Fichte and Novalis, see Werner Volker Csech, »Von der Reichweite mathematischen Denkens bei Fichte und Novalis« *Theologie und Philosophie. Vierteljahrschrift* 70 (1995): 41–61; as well as my forthcoming article: »›*Mathematische Wissenschaftslehre*‹: On a Late Fichtean Reflection of Novalis.«

28 (GA III/3: 227). Letter of Friedrich Heinrich Jacobi to Fichte, March 3–21, 1799, reproduced in the Fichte *Gesamtausgabe* (GA III/3: 224–281).

29 Frustrated after reading this *Ankündigung*, in a letter to Jean Paul from April 1801 Jacobi declared that he intended to challenge Fichte to demonstrate the realist potential of his supposedly mathematical philosophy: »My topic will be the comparison between mathematics and philosophy. If Fichte were able to philosophically draw a line or to even simply place a point into empty space, he would have won hands down.« (Jacobi to Jean Paul, April 30, 1801; in: FG, vol. 3, p. 32).

on its head, using the example of mathematics as proof of the rational core and scientific claims of his system.

To Jacobi's public dismissal one could add Immanuel Kant's (1724–1804) own rejection of the *Wissenschaftslehre* in 1799 as nothing more than an abstraction from »mere logic« – for where Fichte had targeted a scientific philosophical method *à la* Euclid and Spinoza, Kant too only perceived empty abstractions and the absence of any »real object.« As a result, Kant claimed the *Wissenschaftslehre* to be »entirely untenable« and lacked the capability for advancing to the »material of cognition.« [30]

Notwithstanding these highly public rejections in 1799, harsh criticisms of Fichte's mathematical views and knowledge had already commenced immediately after the publication of the Announcement (for the *Grundlage*) in October 1794 that had declared Fichte to be a new philosophical Euclid.[31] Both Henrik Steffens (1773–1845) and Christoph Friedrich Nicolai (1733–1811) were of the conviction that Fichte had probably penned the announcement himself,[32] something Fichte was ambiguous about in the 1797 *Annalen des philosophischen Tons*, saying that his endeavour to raise »philosophy to a level of a self-evident science« would still be valid regardless whether he had or had not written the earlier announcement.[33] However, the criticism reached fever pitch when Fichte posed a seemingly innocuous question to the mathematical community in a footnote to the 1794 text *Über den Begriff der Wissenschaftslehre*. The

30 I. Kant, *Erklärung in Beziehung auf Fichtes Wissenschaftslehre*, dated August 7, 1799 (in: AA XII: 370f.).

31 Cf. FG vol. 1, footnote 1, p. 209.

32 In a retrospective from 1841 Henrik Steffens believed that the original announcement correctly expressed a latent kernel of Fichte's philosophy: »In the bookseller announcement for [Fichte's early] texts it was stated that the *Wissenschaftslehre* ought to become for philosophy, what *Euclid* was to mathematics. I do not believe that this statement, as spoken by him, ever became overly dominant: however, after I had learned about the history of the emergence of his philosophy, I am convinced that this thought articulated in the announcement was prompted by Fichte himself, and most purely expressed his hopes« (in: FG, vol. 1, p. 243). Nicolai's words about Fichte's remarks as the Euclid of philosophy stem from 1796: »That is surely a little strong, whether it be praise from the publisher or praise from the author. And it can't be praise from the publisher, for what does the publisher know about Euclid!« (Ibid., p. 325, footnote).

33 Cf. GA I/4: 295.

hostile reaction surrounding this footnote is not simply symptomatic of the polemical debates in which Fichte was frequently mired in Jena, but in my opinion plays a substantial role in the adverse contemporary reception (or ignorance) of Fichte's conceptions of geometry and mathematics. Hence I will discuss it here at relative length.[34]

In *Über den Begriff der Wissenschaftslehre* Fichte articulated the following query: »A question for the mathematician: Is not the concept of straightness already included in the concept of a line? Is there any other sort of line except a straight one?»[35] Here Fichte essentially claimed that all lines, according to their internal constitution, are always straight. Or to put it another way, only straight lines are *genuine* lines, because the very concept of a line implies the qualities of straightness and direction. As a result, the finite curve or curved line (the so-called *krumme Linie*) is not a true line at all for Fichte, but simply a stringing together of infinitely many points that are infinitely close to each other. Fichte did not view the straight line in the conventional manner as a set of discrete points, but as uniform and continuous whole, in which even the smallest part is still a line. And as we shall see in the following pages, there do exist circles for Fichte – but all are to be derived from the original circle at infinity grasped in inner intuition. Thus, though Fichte's contentions on the line and circle are perfectly legitimate in certain branches of geometry, and essentially had already been noted by Nicholas of Cusa, Kepler and Girard Desargues two to three hundred years before[36], his remarks put him in direct confrontation with a number of his contemporaries. Worst of all, he finished the footnote with an assertion that seemed to many to be contra-

34 See Paul Ziche's valuable discussion of this episode: *Mathematische und naturwissenschaftliche Modelle in der Philosophie Schellings und Hegels*. (Stuttgart-Bad Cannstatt: Frommann-Holzboog, 1996), pp. 39–71.

35 *Über den Begriff der Wissenschaftslehre* (BWL) (GA I/2: 135, footnote); English translation by Daniel Breazeale in: *Fichte: Early Philosophical Writings*, footnote, pp. 120 f.

36 Concerning Nicholas of Cusa (Cusanus) and Kepler on the opposition between straight and curved lines, see: F. Hallyn, »Symbolic Order in Kepler's World« in: J. Hintikka and F. Vandamme, eds. *Logic of Discovery and Logic Discourse* (New York: Plenum Press, 1985), pp. 114–115. On Kepler and Desargues, see: »Kepler's Invention of Points at Infinity«, in: *The Geometrical Work of Girard Desargues*, pp. 185–188.

dictory and unsubstantiated, when he declared the *reductio ad absurdum* procedure used to measure polygons to be an infinite task and the definition of a straight line to be impossible.[37]

The reactions to this ›question‹ were undoubtedly not what Fichte expected. A group of leading mathematicians and philosophers quickly ridiculed his apparent lack of geometric understanding. The philosopher Friedrich Karl Forberg (1770–1848), Fichte's colleague at the University of Jena (who was later to play a key role in the atheism dispute), was not alone in remarking in 1795: »What Fichte understands about mathematics is for the most part false.«[38] Moreover, in a letter to Friedrich Immanuel Niethammer (1766–1848), Johann Benjamin Erhard (1766–1827) would soon complain of Fichte's »idiotic and ignorant arrogance«[39] in the sphere of mathematics, while one of Fichte's own students, Johann Friedrich Herbart (1776–1841) (later successor to Kant in Königsberg), wrote a long letter to Fichte in October 1795 in which he assiduously listed the theories of a dozen mathematical thinkers ranging from Archimedes to Christian Wolff in an attempt to disabuse his teacher Fichte of his »obvious« errors about the geometric line.[40]

Jakob Sigismund Beck (1761–1840), who was not only a respected philosopher but possessed a sound mathematical mind, continued the criticism in a review of Fichte's writings, stating that Fichte's mathematical ideas were not even worth discussing, and implied that the same was therefore true of his philosophy, when he finished with the Latin flourish: *Ansas philosophiae non habes!*[41] However, as Günter Buhl has already

37 »This makes it clear *that* and *why* the task of measuring the circumference of a circle by a straight line is an infinite task, one which could only be accomplished if an approximation to infinity could be completed. At the same time it becomes clear why the straight line cannot be defined.« BWL (GA I/2: 135, footnote); English translation by Daniel Breazeale in: *Fichte: Early Philosophical Writings*, footnote, p. 121.

38 FG vol. 1, p. 241. Forberg continues: »and Fichte hasn't fared any better in this field than Maimon, who likewise wanted to deal with mathematics, but without any greater success.« (ibid.)

39 Erhard to Niethammer (January 31, 1797) in: FG vol. 1, p. 400.

40 See Herbart to Fichte October 1, 1795 (GA III/2: 411–415).

41 In English: »You have no handle on philosophy.« In: *Annalen der Philosophie und des philosoph. Geist*, Feb. 6, 1795 (Quoted in: Fichte, *Grundlage des Naturrechts*

shown thirty years ago,[42] Beck had misinterpreted Fichte's remarks to refer to the problem of *incommensurability*, i.e. to the relationship between two lengths, the length of the circumference of the circle and that of the diameter, and implied that Fichte conceived the curve to be a straight line.[43] Fichte too had noticed Beck's misunderstanding, and in a footnote to the *Grundlage des Naturrechts* of 1796/97 he rejected Beck's criticisms, saying that he was not talking about incommensurability but measuring the surface of a circle, i.e. about the quadrature of the circle. Fichte then stated that he had already required the requisite *knowledge* about this problem from his student days at Schulpforta, yet no mathematician had hitherto provided him with an *understanding* of how such a quadature procedure could be carried out and terminated, since for him it rested on inherent contradictions between the notions of straight and curved and the idea of a completed infinity.[44] In any event, he dropped this ›question for the mathematician‹ in the second edition of *Über den Begriff der Wissenschaftslehre* in 1798. – Despite all these contemporary criticisms and misunderstandings, I hope to show in this study that Fichte's views on geometry and mathematics are perfectly rational and coherent.

At this point we might ask the question: What was Fichte's mathematical education? Paul Ziche has confirmed that Fichte's education in mathematics only went as far as the high school level, and remarks that the geometry books at Schulpforta did not teach according to the Euclidean method but in line with practical needs.[45] And perhaps lending weight to some of Erhard's and Beck's criticisms, Ziche furthermore argues that Fichte made a number of errors when discussing certain algebraic formu-

(GNR), GA I/3: 318. For an English translation of this criticism, see Fichte, *FNR*, pp. 7–8).

42 See Günter Buhl, »Über eine Bemerkung Fichtes zur Quadratur des Kreises«, *Archiv für Geschichte der Philosophie* 59 (1977): 281–288.

43 Cf. Beck: »The incommensurability of the circle by means of the diameter (*Durchmesser*) is for Herr Fichte an infinite task! And how is this clearly incommensurable? Because the circle (*Kreislinie*) is a curved and not a straight line! It is not worth even speaking about this, and does not correspond to the respect that the reviewer has for the public.« Quoted in: Fichte, GNR (GA I/3: 317 f.).

44 See Fichte, GNR (GA I/3: 318).

45 See Paul Ziche, *Mathematische und naturwissenschaftliche Modelle in der Philosophie Schellings und Hegels*, op. cit. pp. 42–45.

lae in his *Eigne Meditationen* and *Grundlage der gesammten Wissenschaftslehre.*[46] In the 1805 *Erlanger Logik* Fichte too seems to confirm his ignorance of certain aspects of higher level geometry in a remark about his school years and the nature of geometrical self-evidence.[47] Nevertheless, this should *not* be read as an ignorance of higher geometry and geometrical problems *on the whole*, because as we will see, in the very same text the opposite seems to be the case, where Fichte proposes a new starting point for geometry in order to avoid the circularity of the so-called parallel problem. Moreover, from the available documentation, it now appears that Fichte returned to a more intensive study of mathematics after his school and university years, and even provided instruction in mathematics to students in his charge. Fichte's new-found »taste for mathematics« coincided with his acquaintance of the Kantian critical philosophy in 1790.[48]

What were the further mathematical influences on Fichte? It is obviously beyond the scope of our study to provide an exhaustive account of all the possible influences on Fichte, but we will mention the most important. Fichte was influenced by many mathematicians and philosophers on many different levels. On the one hand, we have classical ancient geometrical sources like Euclid (fl. 300 BC); Fichte's enormous debt to Euclid's masterpiece *The Elements* will hopefully be visible in the pages of this study. The same holds for Plato (428–347 BC) with his theory of ideal elements, and naturally for more modern philosophers such as Immanuel Kant. It would be hard to overestimate the influence of Kant's mathematical ideas on Fichte, especially his geometrical arguments and examples

46 Ibid., pp. 45f.

47 »Geometry in the form known and taught to me has factual self-evidence [...] (I said with good reason ›known to me‹, as I learned it in the past. Because I am not sure if this is the case in higher domains.« *Erlanger Logik* (GA II/9: 128f.).

48 See for example the letter from Leipzig dated November 2, 1790 to Johanna Rahn in Zurich: »Apart from the Kantian philosophy (to which I continue to devote all my time of the hours remaining over to me. I am working on part one of an explanation, which I will try to publish in time for the new year book fair). I have *particularly acquired a taste for mathematics*, and which I am *likewise teaching*. Hence, you see that I am mostly occupied with abstract things and give little scope to my imagination; and that is the reason for my calmness.« (GA III/1: 184) My emphasis.

from the *Critique of Pure Reason*. Likewise for Spinoza (1632–1677), whose *Ethics* is famously written *more geometrico* and furnishes the paradigm for all mathematized philosophical systems, and which Fichte considered as a »spiritual precursor« to the *Wissenschaftslehre*. Fichte's *Neue Bearbeitung der Wissenschaftslehre*, which constitutes one of our key texts, may in part be traced back to Fichte's own reading of Spinoza's Latin text in 1800, as well as to Leibniz's conception for a *characteristica universalis* and the general idea of a philosophical mathesis. And as already mentioned, we will occasionally take note of the mathematical theories of Maimon and Schelling if they shed light on Fichte's endeavours.

One further figure should be singled out, a figure who is often neglected in Fichte studies but who in my opinion had a decisive influence on Fichte's mathematical theories and early intellectual direction: Johann Friedrich Schultz (1739–1805). Schultz was a professor of mathematics and *Hofprediger* in Königsberg, an old and trusted friend of Kant, and thoroughly versed in philosophical matters. His *Prüfung der Kantischen Critik der reinen Vernunft* (1792) was one of the first expositions of Kant's *Critique of Pure Reason*, and Schultz was the thinker that Kant considered had best understood his philosophical ideas.[49] Significantly, Fichte and Schultz were also acquainted, and became friends during Fichte's first sojourn in Königsberg (July to October 1791), when Schultz helped Fichte find a teaching position near Danzig.[50] In light of Fichte's slim mathematical education at school, and his new found enthusiasm for mathematics in 1790, it is highly probable that Fichte's knowledge and views on geometry and mathematics, especially their technical aspects, were not only stimulated by Schultz's mathematical works such as the *Anfangsgründe der reinen Mathesis* (Königsberg: Hartung, 1790), but also personally by Schultz himself.[51] That Fichte knew of Schultz's mathematical writings is apparent from the 1805 *Erlanger Logik*, where Fichte rejects Schultz's solu-

49 Schultz's *Prüfung der Kantischen Critik der reinen Vernunft* was a relative success and Kant seemed grateful for the help in defending the critical philosophy, especially with regard to its treatment of mathematical principles. See Henry Allison, *The Kant-Eberhard Controversy* (Baltimore: Johns Hopkins University Press, 1973), p. 9; and Darius Koriako, *Kants Philosophie der Mathematik* (Hamburg: Meiner, 1999) pp. 279f.

50 See Fichte's outlines of letters to Schultz, dated: August 25, 1791; end of 1791 (GA III/1: 255, 274), and Schultz's reply dated January 18, 1792 (GA III/1: 284f.)

tion to the parallel problem.[52] Schultz's influence is all the more significant when we recall the pivotal (and frequently underestimated) role that Schultz played in the evolution of Fichte's *philosophical* ideas. Despite Fichte's criticisms of Schultz's »orthodoxy« – especially in matters of the philosophy of religion[53] – according to Fichte's account in the *Versuch einer neuen Darstellung der Wissenschaftslehre* (1797) it was in personal discussions with Schultz that he first outlined the idea to base his entire philosophy on a theory of the self-intuiting I.[54] Philosophically, though an ally of Kant, Schultz was still an independent enough thinker to criticize the Kantian theory of sensible intuition. In both his early review of Kant's inaugural dissertation and in a review of the first Critique, Schultz furthermore posited his own theory of intellectual intuition or *intuitus intellectualius*, on which Fichte favourably comments in the *Versuch einer neuen Darstellung der Wissenschaftslehre*.[55] Yet his interpretation differs from the Fichtean one, insofar as Schultz considers it to be a faculty in which the I is entirely passive and is simply acted upon by inner objects.[56]

51 Heinrich Theodor Schön (1773–1856) gives a brief account of the personal relationship between Fichte and Schultz in Königsberg in his *Selbstbiographie* (cf. *Fichte in Gespräch*, vol. 1, pp. 30–33). Schön's correspondence with Fichte also occasionally touches on mathematical and philosophical issues. See Schön's August 1792 sketch of an *Abhandlung* to Fichte, where he discusses love and marriage in the light of mathematics (GA III/1: 327 ff.).

52 Cf. *Erlanger Logik* (GA II/9: 131).

53 See Fichte's letter from Königsberg to Friedrich August Weißhuhn, October 11, 1791 (GA III/1: 268).

54 »My admirable friend, the court chaplain Schultz, to whom I confided my still as yet indefinite idea to construct the whole of philosophy upon the pure I long before it had become clear to me, and who found this same idea less repellent than anyone else, has a remarkable passage on this topic in his *Prüfung der Kantischen Kritik der reinen Vernunft*, part 2, p. 159: ›The pure active self-consciousness, of which in fact every single I consists, can and must directly instruct us, and should neither be confused with the *Anschauungsvermögen* (faculty of intuition) nor inferred that we possess a non-sensible, intellectual *Anschauungsvermögen*‹ (entirely like the *Wissenschaftslehre* has since then asserted).« (GA I/4: 225f.).

55 Cf. GA I/4: 225–230.

56 Concerning the details of these reviews see Reinhard Brandt, »Materialien zur Entstehung der Kritik der reinen Vernunft (John Locke und Johann Schultz).« In: *Beiträge zur Kritik der reinen Vernunft*, 1781–1981, eds. Ingeborg Heidemann and Wolfgang Ritzel (Berlin: Walter de Gruyter, 1981), pp. 59–66; and Schultz's review

Thus, Schultz's ideas on mathematics and philosophy will at times be referred to in this study.

The above early contemporary criticisms of Fichte's mathematical conceptions seem to have cast a long shadow over the historical reception of his ideas on mathematics on the one hand, and the connection between geometry and the *Wissenschaftslehre* on the other. There are references to Fichte among a number of professional mathematicians and philosophers of mathematics, but more often than not Fichte is used as a negative foil to bolster the supposed strengths of other systems. For example, in a 1937 article entitled »Die Wissenschaftslehre Bolzanos: Eine Jahrhundert-Betrachtung« the philosopher of mathematics Heinrich Scholz (1884–1956) compared Fichte's *Wissenschaftslehre* with the identically named *Wissenschaftslehre* (1837) of the logician Bernard Bolzano (1781–1848). However, Scholz rejected Fichte's attempt as superficial in comparison with what he viewed as Bolzano's logically superior conception of a science.[57]

Nevertheless, as noted above, we do find a wholly positive but relatively unexplored Fichtean influence on three highly distinguished figures in twentieth-century philosophy of mathematics – the German mathematician Hermann Weyl (1885–1955), the Swiss mathematician-philosopher Andreas Speiser (1885–1970) and the French philosopher of mathematics Jules Vuillemin (1920–2001). We will refer to their findings and readings

of J.A.H. Ulrich's »Institutiones Logicae et Metaphysicae« in ALZ of 13 December 1785 in: *Kant's Early Critics*, ed. and trans. Brigitte Sassen (Cambridge: Cambridge University Press, 2000), p. 212.

57 »We only want to say here that for the realization of this unified science (*Einheitswissenschaft*) in the sense that a mathematician connects with the word ›realization‹ (*Realisierung*) Fichte accomplished nothing at all [...] In this respect there only remains for Fichte the honour of creating the name ›*Wissenschaftslehre*‹, nothing more. It is entirely different with Bolzano. Here we encounter for the first time a *Wissenschaftslehre* that is worthy of the name.« Heinrich Scholz, »Die Wissenschaftslehre Bolzanos: Eine Jahrhundert-Betrachtung«, article reprinted in: *Mathesis Universalis: Abhandlungen zur Philosophie als Strenger Wissenschaft*, eds. H. Hermes, F. Kambartel & J. Richter (Basel: Benno Schwabe & Co., 1961, pp. 219–267), quote: p. 224. According to Scholz, Bolzano does not refer to Fichte in his works (ibid., p. 223). Steve Russ also remarks that Bolzano's *Wissenschaftslehre* signifies a rigorous system of logic and is unrelated to Fichte's endeavour, see: Bernard Bolzano, *The Mathematical Works of Bernard Bolzano*, trans. and ed. Steve Russ (Oxford: Oxford University Press, 2005), p. xxvii.

throughout this study. I will conclude this introduction with a brief overview of the present state of research on the topic of Fichte and mathematics.

d. Current State of Research

Max Wundt and Joachim Widmann, in 1929 and 1977 respectively, bemoaned a lack of Fichte scholarship on the relationship between the *Wissenschaftslehre* and mathematics.[58] Although no monographs have as yet appeared, a number of more recent articles and book chapters have touched on this subject.

In his important dissertation of 1902, *Fichtes Idealismus und die Geschichte*, Emil Lask (1875–1915) drew attention to the »mathematical method« of Fichte by saying that it is »geometrical in origin«, but he did not elaborate any further.[59] This is unfortunately true for a large majority of scholarly studies, which often highlight Fichte's apparently ›mathematical‹ method in the *Wissenschaftslehre*, but draw the conclusion without any further investigation that it must be derived from Spinoza's *more geometrico*.[60] In contrast, studies of Spinoza's influence *in general* on Fichte have long received their due, as testified by countless older and more recent studies. These studies demonstrate both the positive reception of Spinoza in Fichte, as well as the extent to which Fichte criticized, among

58 See M. Wundt, *Fichte-Forschungen* (Stuttgart: Frommann, 1929), pp. 306ff., and J. Widmann, *Grundstruktur*, pp. 17f.

59 Emil Lask, *Fichtes Idealismus und die Geschichte* (Tübingen/Leipzig: Mohr, 1902), p. 39. Cf. Widmann, *Grundstruktur*, pp. 17f.

60 For example, in volume 7 of his widely read history on German philosophy, Frederick Copleston says of Fichte's definition of a science in *Über den Begriff der Wissenschaftslehre*: »This model of science is obviously inspired by a mathematical model. Indeed, *Fichte* takes *geometry* as an example of a science. But it is, of course, a particular science, whereas philosophy is for Fichte the science of science, that is, the knowledge of knowledge or doctrine of knowledge (*Wissenschaftslehre*). In other words, philosophy is the basic science. Hence, the fundamental proposition of philosophy must be indemonstrable and self-evidently true.« Frederick Copleston, *A History of Philosophy* (Vol. 7: *Eighteenth and Nineteenth German Philosophy*, first published 1963, reprint, New York: Continuum, 2003) p. 37. An interesting exception is Marino Gentile, »Il matematismo spinoziano nella formazione della filosofia di Fichte«, *Giornale di metafisica* 12 (1957), pp. 330–340.

other things, the rigid »dogmatism«, understanding of being (*Sein*) and lack of freedom in Spinoza's system.[61]

As mentioned above, things begin to fundamentally change with respect to the topic of Fichte and mathematics when his philosophical ideas start to have an impact on a number of 20th century philosophers of mathematics. As early as 1918 Hermann Weyl began to privately and publicly acknowledge the positive contribution of Fichte's theories for understanding the phenomena of causality, continuity and constructivism.[62] Moreover, Weyl believed that the Fichtean theory of intuition had helped him to further clarify a number of his mathematical ideas on the foundations of set theory. On the negative side, Weyl harboured reservations about a wholesale appropriation of Fichte's theory, because in his eyes it was ultimately guilty of »lapsing into mysticism« instead of remaining in »symbolism.«[63] Nevertheless, Weyl seems to have retained a firm belief in the primacy of intuition. This is due in part to Fichte's rigorous »metaphysical« speculations, whose potential for mathematical logic Weyl had already noted in his 1918 book *Das Kontinuum*.[64] Weyl was above all critical of the tendency of modern mathematicians to formalize the foundations of mathematics without understanding what it is they were formalizing.

61 Among others, see R. Lauth, »Fichtes Sicht der Philosophie Spinozas« (1989); Marco Ivaldo, »Transzendentalphilosophie und ›Realistische‹ Metaphysik. Das Fichtesche Spinoza-Verständnis« (1992); W. E. Wright, »The Shadow of Spinoza in Fichte's WL 1804²« (2003); K. Hammacher »Fichte und Spinoza« (1992), »Spinoza und Fichte« (2003); G. Zöller, »Fichte als Spinoza, Spinoza als Fichte. Jacobi über den Spinozismus der Wissenschaftslehre« (2004); and Daniel Breazeale, »Fichte und Spinoza: ›Gemeinschaftlicher Standpunkt – wesentlicher Gegensatz‹« (unpublished manuscript from 2005).

62 For more details, see the recent paper of Norman Sieroka: »Weyl's ›Agens Theory‹ of Matter and the Zurich Fichte«, *Studies in History and Philosophy of Science* 38 (2007), pp. 84–107.

63 Cf. Erhard Scholz, »Hermann Weyl's ›purely infinitesimal geometry‹«, in: *Proceedings of the International Congress of Mathematicians Zurich, Switzerland 1994* (Basel: Birkhäuser, 1995), Vol. 1,2, pp. 1592–1603.

64 See Hermann Weyl, *Das Kontinuum* (Leipzig: Veit & Comp., 1918): »We cannot here proceed to the final explanation of the essence of Sachverhalt, judgment, object (Gegenstand), and quality (Eigenschaft); this task leads into metaphysical depths. Here one has to receive advice from men whose names cannot be mentioned among mathematicians without eliciting a compassionate smile – Fichte, for example.« (Chapter one, p. 2).

Through the mediation of the Fichte scholar and editor Fritz Medicus, Weyl came to especially appreciate the transcendental logic of Fichte, as Norman Sieroka's valuable recent study has shown.[65]

The Swiss mathematician Andreas Speiser – a former student of the great German geometers David Hilbert and Felix Klein – proposed a reading of Fichte in 1946 that situated him within reflections on the foundations of mathematics in a line leading directly back to Plato.[66] According to Speiser, Fichte's system was founded on abstract universal principles and archetypal elements in a manner reminiscent of Plato's philosophy, leading Speiser to assert an intellectual affinity between the *Wissenschaftslehre* and mathematics. Furthermore, because of its mathematical Platonism and understanding of ›facticity‹, the Fichtean system provides an excellent point of entry to Plato's own philosophy, as Speiser sought to demonstrate in 1959 by advocating a study of the *Wissenschaftslehre 1804/II* as an introduction to Plato's *Parmenides*.[67]

In my opinion, the most innovative study of Fichte and mathematics is Jules Vuillemin's groundbreaking but unjustly neglected study from 1962, *La philosophie de l'algèbre*.[68] Here Vuillemin compares Fichte's genetic method in philosophy with the algebraic method of the French mathematician Joseph-Louis Lagrange. We will refer to Vuillemin's volume throughout this study. Suffice to say that one of Vuillemin's principal contentions is that many recent mathematical breakthroughs have not been sufficiently appreciated *philosophically* to the extent that profound methodological developments in mathematics are only subsequently understood by philosophers.[69] In the *La philosophie de l'algèbre* Vuillemin

65 Weyl especially appreciated Fichte's *Transcendental Logik* of 1812; see Norman Sieroka, *Umgebungen: Symbolischer Konstruktivismus im Anschluss an Hermann Weyl und Fritz Medicus* (Zurich: Chronos, 2010), pp. 57-59.

66 Cf. Andreas Speiser: »Die Grundlagen der Mathematik von Plato bis Fichte«, *Eranos Jahrbuch* 14 (1946): 11–38.

67 See Andreas Speiser, *Ein Parmenideskommentar: Studien zur Platonischen Dialektik*, 2nd ed. (Stuttgart: K.F. Koehler Verlag, 1959), pp. 6–7, 73–106.

68 Jules Vuillemin, *La philosophie de l'algèbre* (Paris: PUF, 1962, repr. 1993). [All references refer to the reprint edition of 1993]. I am very grateful to Professors Courtine and Cattin for drawing my attention to the work of Vuillemin.

69 »[...] il existe un rapport plus intime quoique moins apparent et plus incertain entre les Mathématiques pures et la Philosophie théorique. L'histoire des Mathématiques et de la Philosophie montre qu'un renouvellement des méthodes de

proposes a new appreciation of the revolution that occurred in abstract algebra and group theory in the late eighteenth century. For him, there is a distinct parallel between the disciplines of mathematics and philosophy and Fichte's discovery of the genetic method, insofar as the latter reproduces and adapts the primary stages of the Kantian system, just as the new a priori method of Lagrange reproduces and elevates the chief a posteriori moments of the resolution of the algebraic equations of his predecessors. Both the philosopher and algebraist raise their methods to the highest level of abstraction by dispensing with particular entities or specific equations. With regard to Fichte, this concerns the philosophic transition from sensible to intellectual intuition, whereas in Lagrange it is the transition from general solutions of algebraic equations to the first elements of a theory of groups. In both thinkers the hallmark of their new critical method is that it becomes liberated from all earlier limited, contingent and necessary series, to finally become completely a priori and generalized.[70]

Another serious attempt to examine the relationship between Fichte's *Wissenschaftslehre* and mathematics is Joachim Widmann's monograph: *Die Grundstruktur des Transzendentalen Wissens nach Joh. Gottl. Fichtes Wissenschaftslehre 1804*^{2} (1977).[71] Though Widmann's volume is principally devoted to the cognitive structure of the *Wissenschaftslehre 1804/II*, it does include a rather detailed discussion of »Das Verfahren *more* geometrico«. After a brief preliminary survey of prior Fichte scholarship on the geometric method, Widmann's primary focus is on the properties of intuitability and irrefutability in a *logical* system of thought. Above all, his study is distinguished by its investigation into the nature of Fichte's *Grundsatz* and the definitions, postulates and theorems of geometry. He concludes by arguing that one may justly compare the logical structure of

celles-là a, chaque fois, des répercussions sur celle-ci. […] j'utiliserai les analogies de la connaissance mathématique pour critiquer, réformer et définir, autant qu'il se pourra la méthode propre à la Philosophie théorique.« Ibid., pp. 4–5.

70 Cf. Jules Vuillemin, *La philosophie de l'algèbre*, p. 116. On Fichte and Lagrange, see especially pp. 102–122.

71 *Die Grundstruktur des Transzendentalen Wissens nach Joh. Gottl. Fichtes Wissenschaftslehre 1804*^{2} (Hamburg: Felix Meiner Verlag, 1977), see especially pp. 21–30.

Fichte's starting point with the first principles and axioms of Euclid, Descartes and David Hilbert.[72]

Four significant studies appeared in the 1990s. Firstly, in 1990 Hans Niel Jahnke published a large historical investigation into the state of mathematical education, teaching and knowledge during the founding of the Humboldt university period, and included Fichte's philosophy and educational views among his discussions.[73] In 1995 Claude Piché highlighted a vital aspect of Fichte's thoughts on geometry with a study of the notion of self-evidence (*Evidenz*), and the reactions to Fichte's claim to be a new Euclid.[74] Piché maintains that Fichte's belief in the moral certitude of the practical sphere is based on the theoretical certitude of geometric self-evidence.[75] In 1996 Paul Ziche furnished an important study for understanding Fichte's mathematical ideas in the context of the emergence of Schelling's and Hegel's ideas for a philosophy of nature. Ziche particularly focuses on the role of Fichte in the 1794–95 debate on the so-called infinite polygon.[76] Finally, an impressive contribution to scholarship on Fichte's theories of geometry and space, and one that was a valuable stimulus in the preparation of the present work, is Werner Volker Csech's comprehensive text from 1999: *Die Raumlehre Johann Gottlieb Fichtes: Mit Berücksichtigung philosophiegeschichtlicher Konstellationen*.[77] Csech's study is chiefly devoted to explicating the Fichtean conception of space, yet his early chapters not only provide an extensive investigation of the roles of construction, intuition and limitation, but also of Fichte's thoughts on the geometrical point, line and plane. Csech emphasizes the

72 Ibid. pp. 21ff.

73 Hans Niel Jahnke, *Mathematik und Bildung in der Humboldtschen Reform* (Göttingen: Vandenhoeck & Ruprecht, 1990) esp. pp. 95–118.

74 See Claude Piché's book chapter entitled: »Évidence géométrique et certitude morale chez Fichte: le problème de la vérité en philosophie« from his volume: *Kant et ses épigones, le jugement critique en appel* (Paris: Vrin, 1995), pp. 132–189.

75 Ibid., pp. 144–150.

76 See Paul Ziche, *Mathematische und naturwissenschaftliche Modelle in der Philosophie Schellings und Hegels*, esp. pp. 40–132.

77 Werner Csech, *Die Raumlehre Johann Gottlieb Fichtes: Mit Berücksichtigung philosophiegeschichtlicher Konstellationen* (Frankfurt am Main: Peter Lang, 1999). (Henceforth: *Raumlehre*).

great extent to which Fichte appreciated the benefits of geometry in intellectually training the mind.[78]

A number of smaller articles and studies appeared in the first decade of the 21st century.[79] Knut Radbruch in particular furnished in 2003 a valuable article on the significance that mathematics held for Fichte's conception of philosophy.[80] He especially highlights the higher role that Fichte attributes to the philosopher of geometry, compared to the technical geometer, insofar as the former transcendentally reflects upon the nature and conditions of the geometer's activities when carrying out mathematical exercises. Radbruch additionally brings into relief (pp. 255ff.) the Euclidean examples hovering in the background of Fichte's *Zurich Wissenschaftslehre*, and concludes with the assessment that mathematics for Fichte played a propedeutic role in the study of philosophy. His article ends with a reference to Hermann Weyl's interest in Fichte's philosophical idea of intuition.[81] However, he voices his »regret« that he could not find

78 Ibid. Csech's analysis of the Fichtean conceptions of the fundamental elements of geometry and space, including the point, line and solid, is to be found on pages 91–149.

79 Among others, Violetta Waibel provides a short chapter on »Das Verhältnis von Philosophie und Mathematik bei Kant und Fichte« in her volume *Hölderlin und Fichte: 1794–1800* (Paderborn: Schöningh, 2000), pp. 269–277. She especially considers the mathematical references in Fichte's *early* texts and letters, including the *Grundlage* of 1794, the *Eigne Meditationen*, with a concluding focus on Fichte's *Zurich Wissenschaftslehre*. Here she explicates Fichte's argument about the inverted methods of geometry and philosophy, where he asserts that the »cognitive goal« of geometry is equivalent to its products, insofar as it posits its own purposes through free spontaneity. (Ibid., p. 276). In the course of his discussion on Fichte's conception of intellectual intuition and construction, Paul W. Franks makes some interesting remarks about Fichte and geometry in his 2005 monograph: *All or Nothing. Systematicity, Transcendental Arguments, and Skepticism in German Idealism* (Cambridge Mass: Harvard University Press, 2005). On the one hand he contends that Fichte's method of construction was inspired by Maimon, and on the other that Fichte started using the term »postulate« in a geometrical sense in 1796 in the *Wissenschaftslehre nova methodo* after Jakob Sigismund Beck's criticism of the Fichtean *Grundsatz*. (Ibid., p. 350).

80 Knut Radbruch, »Die Bedeutung der Mathematik für die Philosophie bei Fichte« *Fichte-Studien* 22 (2003), pp. 251–263.

81 Ibid. 262f.

anywhere in Fichte's writings a realization of Fichte's wish to carry out a study of the philosophy of mathematics.[82]

Fichte's unpublished text of the *Neue Bearbeitung der Wissenschaftslehre* (1800) appeared in the *Gesamtausgabe* in 1979. The present investigation has greatly profited from a recent 2008 article on Fichte's text by Daniel Breazeale: »Toward a *Wissenschaftslehre more geometrico* (1800–1801).«[83] Breazeale's essay simultaneously examines the circumstances of the composition of the *Neue Bearbeitung*, its continuity with the Jena versions of the *Wissenschaftslehre*, as well as its geometrical presentation. Like Widmann's text, Breazeale supplies a decisive contribution to Fichte scholarship with his interpretation of Fichte's employment of the mathematical terms »theorem«, »postulate«, »demonstration« and »corollary« [84], which I discuss below in Chapter Three.

This concludes our overview of the current state of research on the topic of Fichte and mathematics. To summarize: Fichte apparently had a modest early mathematical education, but seemed to continue and deepen his readings on mathematical subjects throughout his career. And despite the hostile criticisms of his philosophical contemporaries, and the relative lack of scholarly research in this domain, I will argue that Fichte possessed a thoroughly consistent and coherent philosophy of mathematics and original theory of geometry.

82 Ibid.

83 »Toward a *Wissenschaftslehre more geometrico* (1800–1801)«, in: *After Jena: New Essays on Fichte's Later Philosophy*, eds. D. Breazeale and T. Rockmore (Evanston, Illinois: Northwestern University Press, 2008), pp. 3–40.

84 Ibid., see in particular pages 18–25 of Breazeale's essay.

Chapter One: Fichte's Philosophy of Mathematics

1. A Philosophy of Mathematics *in nuce*: Fichte's *Erlanger Logik*

I will begin this chapter by evaluating and comparing some of Fichte's most prominent announcements on geometry and mathematics, such as those in the early Zurich lecture of February 27, 1794 on the relationship between geometry and the *Wissenschaftslehre*, the *Ankündigung* from 1800, and certain remarks in the second series of lectures on the *Wissenschaftslehre 1804*.[1] After such an examination, however, it will become evident that these texts do not furnish Fichte's most detailed reflections on the foundations of mathematics. For that we must turn instead to a little-studied work from 1805: the *Erlanger Logik* (GA II/9: 57–171). I will argue that the *Erlanger Logik 1805* is unique in the Fichtean oeuvre insofar as it offers a short but sustained exposition of what I will term: Fichte's »philosophy of mathematics *in nuce*.« That is to say, we find here a compact summary of Fichte's principal thoughts on the nature of mathematics. I will compare these reflections in the *Erlanger Logik* with his other pronouncements from the texts mentioned above and in other texts in order to show the essential continuity of Fichte's thought in this domain.

a. The Nature of the Erlangen Text

Fichte arrived in Erlangen in May 1805 to give lectures on his philosophy to the university.[2] Fichte prefaced his lectures on the *Wissenschaftslehre* as such with lectures on logic and metaphysics, or »*Institutiones omnis philosophiae*«, to use the official title (cf. GA II/9: 25–27). These introduc-

1 See the lecture Feb. 27th 1794, in the *Zurich Wissenschaftslehre* (GA IV/3: 32–37); the *Ankündigung* (GA I/7: 153–164); and the *Wissenschaftslehre 1804* (GA II/8: 2–421).

2 Fichte arrived in Erlangen on May 4, 1805, and began his introductory lectures on May 13, 1805. See the editors' introduction in GA II/9: 27ff.

tory lectures deal with such topics as the »logical function«, »syllogisms« and »higher logic.«[3] However, Fichte also makes a number of interesting digressions in the lectures. These include remarks on aesthetic feeling and philosophical understanding with regard to Raphael's masterpiece, the Sistine Madonna (GA II/9: 98–100) – which Fichte had presumably viewed in Dresden on his way to Erlangen – a comparison between the Pythagorean theorem and philosophical insight (pp. 101–103; 108–9), not to mention notoriously voicing his displeasure at »the declining attendance« of his lectures (GA II/9: 86–97).

However, our attention will above all focus on pages 124–137 of this printed *Erlanger Logik* text. I will argue that this excursus of 14 pages supplies the materials for a genuine Fichtean philosophy of mathematics. – Yet as noted above, *in nuce*, or in a highly concentrated form. Hence my reading of Fichte is at odds with certain scholars who consider that Fichte did not possess or execute a theory of mathematics.[4] My approach shares instead a certain affinity with that of Andreas Speiser, who in an article from 1946 included Fichte in a short survey of philosophies of mathematics from Plato to modern times.[5]

One final point must be kept in mind: Fichte did not publish any complete article, chapter or monograph on the topic of mathematics or geometry, and his thoughts on this subject are scattered among his many works. Hence, they are mostly unrevised, terse, sometimes cryptic, and require a certain amount of *reconstruction* and interpretation. This also holds for the text of the *Erlanger Logik*. As a consequence, Fichte's expositions cannot be compared with the technical sophistication of Kant's or Descartes' presentations for example. Nevertheless, after drawing together Fichte's texts and indications on this subject, I contend that it is still possible to determine a genuinely coherent philosophy of mathematics that remains consistent throughout his career.

3 Cf. GA II/9: 59–85, 116–123.

4 See K. Radbruch, »Die Bedeutung der Mathematik für die Philosophie bei Fichte« *Fichte-Studien* 22 (2003), p. 260; and W. Csech »Von der Reichweite mathematischen Denkens bei Fichte und Novalis« *Theologie und Philosophie. Vierteljahrschrift* 70 (1995), p. 58.

5 Cf. Andreas Speiser, »Die Grundlagen der Mathematik von Plato bis Fichte«, *Eranos Jahrbuch* 14 (1946): 11–38.

Turning to the text of the *Erlanger Logik* we find Fichte making a 14-page digression on geometry. Fichte explicitly states *twice* that his digression forms a part of a philosophy of mathematics. – He asserts this once at the beginning and once at the end of this excursus.[6] Fichte prefaces his difficult digression by telling his students that they should consider his remarks as an important »fragment« from his theory of metaphysics. His reflections might be completely new to his listeners, but they still ought to be considered as philosophy, or more exactly, as »philosophy of mathematics.« Thus, on page 124 of the *Erlanger Logik 1805* we read: »*Geometry*, as the ground of all mathematics. [...] Moreover, it is philosophy. – *Ph. [ilosophy] of mathematics* [*Ph. der Mathematik*], and new things might emerge; it is not assumed that you know them already, but they should be first learned here.«[7] –

As we can see, Fichte's philosophy of mathematics has geometry at its centre, because in his eyes geometry forms the foundation or »ground« (*Grund*) of all the mathematical sciences. This focus on geometry was not unusual, because at that time many thinkers still considered geometry to be not just the queen of mathematics, but the queen of *all the sciences*. – The secure and »royal road« to science undergone by geometry over the centuries, and its revolutionary advance during the time of the Greeks, was a model from which philosophy could learn, as Kant declared in the Preface to the *Critique of Pure Reason*.[8] Fichte mentions other branches of mathematics in this excursus, including arithmetic, algebra and mathematical physics, yet it is vital to see that geometry undeniably forms the core of his examination, just as it overwhelmingly does for Kant.[9] In this

6 See GA II/9: 124, 136.

7 GA II/9: 124; my emphasis. See the Appendix below for a complete translation of this excursus.

8 Cf. I. Kant, *Critique of Pure Reason*, trans. and eds. Paul Guyer and Allen Wood (Cambridge: Cambridge University Press, 1998) (Bxi -Bxii) pp. 107f. Here Kant traces the origin of geometry as a science back to Thales and the ancient Greeks. His reference to the Greek revolution in understanding a priori truths in geometry was of course an ancient precedent of his own so-called »Copernican Revolution« in philosophy (cf. B xvi; p. 110) (All references to Kant's first Critique in English will be to this edition).

9 See Lisa Shabel's view: »Indeed, Kant himself chooses almost all of his illustrations of the source of mathematical cognition throughout the Critique from among the examples of Euclidean geometry«, *Mathematics in Kant's Critical Philosophy* (New

sense, Fichte's philosophy of mathematics in the *Erlanger Logik* is primarily a *philosophy of geometry*.

Turning to the end of this 14-page sequence, we see that Fichte again clearly remarks that his exposition belongs within a philosophy of mathematics. Noting that he would now like to move on to a subject that he calls a »philosophy of physics«, he reaffirms what he has just said about the content of the preceding 14 pages: »As preliminary remarks I might add (1) That this subject will contain the most essential elements of the philosophy of physics and its history; *just as the immediately preceding contains the most essential aspects of the philosophy of mathematics*: [they are] extremely essential parts of metaphysics, whose latter foundations coincide with the plan of this course.«[10]

In my view it is clear – Fichte explicitly states and invites us to consider that his (albeit brief and fragmentary) exposition contains the essentials of his philosophy of mathematics. Although only first published in 1993, it is still surprising to learn of the almost total absence of reference to these passages and to this text in Fichte scholarship, not just in studies on Fichte's ideas in this domain, but even in studies of the Erlangen period.[11] Using concrete examples from geometry to illustrate his views, this 14-page sequence is a minor *tour-de-force* of Fichte's views on mathematics, ranging over specific topics such as the problem of parallel lines, his conception of pure and infinite space and ideal elements, the essence of geometry as the theory of space, and the nature of arithmetic.

York and London: Routledge, 2003), p. 115. And compare Henry Allison's similar remarks in *The Kant-Eberhard Controversy*, p. 91.

10 GA II/9: 136, my emphasis.

11 For example, neither Radbruch (2003), Franks (2005 & 2006), nor Widmann (1977) mention the *Erlanger Logik 1805*; the exception is Csech (1995 & 1999), who frequently cites its reflections on geometry, but does not view it as containing Fichte's conception for a philosophy of mathematics. Rebecca Paimann, in her large recent study of Fichte's different courses on logic: *Die Logik und das Absolute. Fichtes Wissenschaftslehre zwischen Wort, Begriff und Unbegreiflichkeit* (Würzburg: Könighausen & Neumann, 2006) does indeed discuss the *Erlanger Logik* at some length (pp. 242–277), but skips over precisely this excursus on the philosophy of mathematics (cf. p. 274).

b. What is a Philosophy of Mathematics?

We will analyse in detail the contents of Fichte's excursus in the next section, but first we should ask: what exactly constitutes a philosophy of mathematics? According to modern scholars of this subject a philosophy of mathematics is not easy to define, primarily due to the difficulties of articulating what exactly »mathematics« itself is. Simply put, it is the goal of the former to precisely define and determine the latter, as Shapiro and Benacerraf have remarked.[12] That is to say, the philosophy of mathematics is not concerned with specific mathematical problems and examples, but with a rigorous examination of its foundations, epistemological principles, assumptions and the ontological determination of mathematical entities, such as numbers and figures.[13] If it wishes to provide a complete philosophical theory of mathematics, then one should also have a theory that can explain how mathematics relates to the real (physical) world and to the other sciences.[14] As a consequence, to the extent that Fichte philosophically investigates in his writings the underpinnings of mathematics and its assumptions, ontological nature and methods of reasoning, his attempts may therefore be classified as belonging to the philosophy of mathematics.[15]

12 See *Oxford Handbook to the Philosophy of Mathematics*, ed. S. Shapiro (Oxford: Oxford University Press, 2005), pp. 5f., as well as P. Benacerraf & H. Putnam, eds., *Philosophy of Mathematics: Selected Readings* (Cambridge: Cambridge University Press, 2nd edition 1983), p. 1. Apart from these two standard texts, for this study I have additionally found the following works useful: L. Magnani, *Philosophy and Geometry: Theoretical and Historical Issues* (2001); R. Torretti, *Philosophy of Geometry from Riemann to Poincaré* (Dordrecht: Reidel, 1978, corrected reprint, 1984).

13 Cf. P. Benacerraf & H. Putnam, eds., *Philosophy of Mathematics: Selected Readings*, pp. 1–18.

14 See Shapiro: »The burden of a complete philosophy of mathematics is to show how mathematics is related to the material world and to show how the methodology of mathematics (whatever it may be) fits into the methodology of the sciences (whatever *it* may be)« *Oxford Handbook to the Philosophy of Mathematics*, p. 4.

15 Cf. Lisa Shabel's similar rationale with Kant. In her article »Kant's Philosophy of Mathematics« she asserts that an examination of Kant's philosophy of mathematics is equivalent to an »analysis of his reasoning and mathematical practice.« In: ed. Paul Guyer, *The Cambridge Companion to Kant and Modern Philosophy* (Cambridge: Cambridge University Press, 2006) p. 94.

Five specific aspects of a philosophy of mathematics may be outlined. They concern its scientific status, the questions it poses and tries to answer, and the method, knowledge and ontology of mathematics. We will now briefly see if they apply to Fichte's presentation in the *Erlanger Logik* before explicating them in greater detail in the course of this chapter:

(1) Scientific status: It might seem obvious, but one of the assumptions we must make is that mathematics and geometry are to be considered as *sciences*.[16] Fichte absolutely agrees with this statement, saying: »We consider, however, the whole of geometry and history as a science.«[17] As already mentioned, this view of mathematics as a rigorous science par excellence was current coinage at the time, and the philosophers Schelling and Novalis (both one-time followers of Fichte's thought, with the latter trained in higher mathematics) also lauded the precision, scientific certainty and lawfulness of geometry.[18] However, the assumption that mathematics is a science obviously entails another question: What is a science?

In his first published programmatic text on his new philosophy – the 1794 text *Über den Begriff der Wissenschaftslehre* – Fichte had begun by listing the elements crucial for a body of knowledge to be acknowledged as a science. These include, having a systematic form, its contents must be knowable, all principles must be related to an absolutely certain and self-evident *Grundsatz* or grounding first principle, and this *Grundsatz* should have the ability to communicate its certainty to the other propositions in the science.[19] In order to avoid an infinite regress, this first principle itself cannot be proved or demonstrated but should be so simple that it can be

16 See Magnani, *Philosophy and Geometry: Theoretical and Historical Issues* (2001), pp. 99ff.

17 BWL (1794) (GA I/2: 114). See too Fichte's letter to Reinhard on the 15th January 1794, where he takes people to task who doubt the scientific nature of (his own) philosophy, and compares them to those who »thought that geometry would never become a science because it had not been one up to that point.« (GA III/2: 40).

18 For Novalis, see, among others, the fragment from 1799: »The concept of mathematics is the concept of science in general. Thus all the sciences should become mathematics« (HKA III: 593). For Schelling, see his remarks on mathematics and geometry in *Fernere Darstellungen* (*Sämmtliche Werke* I, 4: 345–350).

19 Cf. BWL (GA I/2: 117ff.).

grasped in intuition.[20] – Fichte was especially critical of how the contingent and inessential aspects of science had become conflated, resulting in imprecise notions of what actually constitutes the nature of a science.[21] In the Nov./Dec. 1813 *Einleitung in die Wissenschaftslehre*, i.e. towards the end of his career, Fichte explicates that mathematics is to be considered a science not because of external factors, but because it is based on self-evidence and the free construction of an inner intuition, and these are the hallmarks of an authentic deductive science: »If we take mathematics, for instance, mathematics has a different kind of relation to philosophy. Why is this? [...] Mathematics is differently related not because it is mathematics but because it is a science. This is obviously not true of the philosophy of my opponents. The essence of science does not consist in the external form and course of the demonstration, but in the free construction of an inner intuition, where self-evidence and free construction locate and provide the shape.«[22]

The topics of self-evidence, intuition and construction will interest us in due course, suffice to say that for Fichte any definition of science must take into consideration these two points: its internal content and external form. In the *Erlanger Logik* this is precisely the reason why he introduced the topic of geometry and began his discussion on the philosophy of mathematics: to explicate the general rules and laws of genuine scientific content and form, i.e. to further explain their relationship and interconnectedness.[23] Moreover, here in the *Erlanger Logik* Fichte gives us a con-

20 »Every science has a *grounding principle* (*Grundsatz*), on the basis of which all the propositions are proven, and which itself stands in need of no proof« *Zurich Wissenschaftslehre* (1794), lecture one (GA IV/3: 19).

21 See Fichte's remarks in lecture one of the *Zurich Wissenschaftslehre* (1794): »On this point everyone is agreement: philosophy is a science. The dispute concerns only the object of this science. The concept of a science itself is indeterminate and undeveloped. This is the source of the conflict: essential identifying marks of this concept have been omitted and others that are contingent have been taken to be essential.« (GA IV/3: 19f.).

22 (*Kollegnachschrift Halle*) in: GA IV/6: 398 (Cf. SW IX: 40f. and the text from Fichte's own hand forthcoming in GA II/17: 268).

23 With taking geometry as an example in the *Erlanger Logik* Fichte says: »It is all the more significant as an example, in that I will not only indicate the rules of the external form, but also the internal penetration of the concept. This is the only

cise and elegant definition of a science: »Science = bringing together the accidents under the unity and universality of a law, etc.«[24]

(2) Questions: The philosopher of mathematics poses questions that the mathematician does not pose or is not interested in posing. For example: what is the nature or origin of mathematical objects? Do they belong to space and time, or are these entities simply mental abstractions? If so, how do we have knowledge of them, and what is the status of our knowledge? In Fichte's eyes, to provide a satisfying answer to these questions requires the transition to a »higher« science, such as the *Wissenschaftslehre* as an interpreter of mathematics. This higher (transcendental) science attempts to provide the philosophical framework and explain the conditions for the possibility of mathematics and geometry that are frequently ignored by the geometer.[25] We here see the boundary limit between geometry and philosophy: as Fichte states in the *Erlanger Logik*, the ordinary geometer is content with the »facts« of his science, whereas the philosopher of the *Wissenschaftslehre* seeks to ascertain their nature and justification: »[Geometry carries it out] absolutely through the act of its existence: fact; without justification, and without wondering about it: (and it is entirely right to do this: no geometry is possible without this unexplained fact). Question: does the *Wissenschaftslehre* explain this [fact]? Certainly. Why does the geometer not do this? Answer: It does not take place as geometry but as *Wissenschaftslehre*.«[26]

(3) Method: The methods of mathematics often differ from other sciences, especially the natural sciences, insofar as it does not use induction, but

way for science to receive its true scientific content and inner spirit« (GA II/9: 124).

24 GA II/9: 136. See too the *Zurich Wissenschaftslehre*, where Fichte remarks on the connection between form/content and part/whole: »Science as such has to be considered with respect to its *form*. Form is the relationship of the parts of a science to the whole of the same, and of its whole to the parts. / Then science has to be considered with respect to its *material* or *inner content* as well.« (GA IV/3: 19f.).

25 See his remarks from 1804: »The principles of the possibility of [mathematics] lie in another higher, science.« *Wissenschaftslehre 1804/II* (GA II/8: 47f.).

26 GA II/9: 125.

construction, intuition and deduction from basic principles.[27] Indeed, the inability of philosophy to construct its concepts is what separates for Kant the methods of mathematics and philosophy.[28] Fichte is aware of this Kantian distinction, however, he is not concerned so much with what separates the two sciences but with what they share in common. In the *Erlanger Logik* he investigates the nature of intelligible (geometrical) intuition, as well as the synthetic method in mathematics, that is, a science deducing itself as opposed to the analytic method, and from this tries to determine the appropriate rules. This leads him in turn to reflections on the extent to which the methods of philosophy are similar or even identical to those in geometry.[29]

(4) Knowledge: A philosophy of mathematics is not just concerned with the foundations of mathematical entities and structures, but with mathematical knowledge or cognition as such. This is because mathematical knowledge appears to be more fixed and secure and less subject to revision and overturning than the knowledge in other sciences. In the *Erlanger Logik* Fichte grants that geometry has factual evidence, yet says that the philosopher must ascend to a higher cognitive standpoint.[30] As we saw, this relates to the posing of questions, since the philosopher of mathematics must raise himself to the transcendental standpoint and actively reflect on the »unconscious« operations of the geometer, and explain the conditions for their possibility and their *genesis*. We shall see in Chapter Three that Fichte's account of mathematical knowledge is anchored in intuition and the method of construction.

27 See L. Magnani, *Philosophy and Geometry: Theoretical and Historical Issues* (2001), pp. 23f., 34–39.

28 Cf. I. Kant, *Critique of Pure Reason*, (A713/B741), p. 630.

29 Cf. Fichte, *Erlanger Logik* (GA II/9: 119f., 124); and *Ankündigung* (GA I/7: 153–164).

30 »Geometry, on account of its essence and the determination from its principle, *necessarily* only has factual self-evidence; because it is a fragment from the domain of knowledge. Therefore – as a fragment – it cannot have insight into its own possibility within this domain, but only realises its possibility factually. (By saying and wanting to prove this I have to go beyond [geometry] into philosophy: how deeply, depends on the present aim).« *Erlanger Logik* (GA II/9: 125).

(5) Ontology, or Tradition: According to modern studies in the philosophy of mathematics, any philosopher's view of mathematics can be classified as belonging to a main ontological stream or tradition: for example, to logicism, intuitionism, formalism, Platonism or some other stream.[31] Again, being conscious of the dangers of attributing modern classifications to older philosophical views, I will argue in Chapter Two that Fichte's theory of mathematics ought to be classified as a version of mathematical Platonism or neo-Platonism.[32]

In broad brush strokes these are the criteria for any presentation on the philosophy of mathematics, and Fichte's exposition in the *Erlanger Logik* largely satisfies these five criteria. As already stated, we should try to avoid the twin extremes of anachronism and prefiguration. We should not be anachronistic and try to find in Fichte the same principles animating modern twentieth and twenty-first century philosophies of mathematics, neither should our goal be to try to detect in Fichte's texts a prefiguring of these same modern tendencies. Our aim is to simply show that Fichte's concerns are the same universal concerns of philosophers since antiquity: a serious concern with not just the theorems and specific technical problems of mathematics, but with its foundations and with the nature of mathematical knowledge.

c. *Situating the Erlanger Logik among Fichte's other Texts on Mathematics*

We will now briefly examine Fichte's other courses on logic before comparing the *Erlanger Logik* with his other main texts on mathematics. This

31 Among others, see P. Benacerraf & H. Putnam, eds., *Philosophy of Mathematics: Selected Readings* (Cambridge: Cambridge University Press, 2nd edition 1983), p. 30ff.; *The Mathematical Experience* by R. Hersh and P. Davies, pp. 318–362; L. Magnani, *Philosophy and Geometry: Theoretical and Historical Issues* (2001), p. 82. These are standard classifications, and there are of course others. For example, Mark Balaguer adds in »conventionalism« and »deductivism« (cf. *Platonism and Anti-Platonism in Mathematics*, Oxford University Press, 1998, pp. 11–14, 103f.).

32 Indeed (and it is not meant as a proof), but at the beginning of his excursus on geometry in the *Erlanger Logik* 1805 it is the precisely the figure of Plato who is invoked: »Plato: we people of modern times are required to be more modest«, perhaps in the sense that despite our seeming intellectual progress our *philosophical* understanding of geometry has still scarcely evolved past that of the ancient Greeks (GA II/9: 124).

will be done in order to underscore why we consider the Erlangen course to be unique in Fichte's oeuvre.

The 1805 *Erlanger Logik* is different from Fichte's other courses on logic, for it is only here that he makes a detour from the announced topic to give a presentation on mathematics.[33] The other two major courses on logic – the ones in Jena (1794–98) and Berlin 1812 – do not at all undertake this foray into the philosophy of mathematics. The early Jena lectures on logic and metaphysics are based on a reading of Ernst Platner's *Philosophische Aphorismen* (Leipzig 1793).[34] Since Platner himself does not expound on the nature of mathematics and geometry, neither does Fichte, save for the odd remark on mathematical talent and the roles of the philosopher and mathematicians.[35] And when the 1812 *Logik* briefly mentions mathematics it is solely to refer to it as an example of a science that involves »construction« in its type of cognition.[36] Yet what is logic and how does it relate to mathematics for Fichte? In the strict philosophical sense Fichte does not consider logic to be a proper branch of philosophy but simply a rational theoretical tool employed by philosophers to express their arguments. – As Fichte says at the end of the *Wissenschaftslehre nova methodo*: »Logic cannot be considered a part of philosophy. It is merely an instrument of philosophy – and not simply of philosophy, but of every part of

33 Of course, other texts such as *Über den Begriff der Wissenschaftslehre*, 1794 (GA I/2: 107–172) and the *Zurich Wissenschaftslehre* (GA IV/3: 22–28) distinguish Fichte's system from other theoretical sciences, especially the disciplines of logic and geometry. However, these texts are not courses on logic as such.

34 See GA II/4; Platner's *Philosophische Aphorismen* are reprinted in a supplementary volume to the Fichte *Gesamtausgabe*: GA II/4 S.

35 For example: »One can be a good mathematician without possessing a single philosophical concept; but not without having the talent for it. – One can be a philosopher – although mathematics to a certain extent, i.e. according to its premises, also belongs to philosophy etc.« (GA II/4: 123), which is a comment by Fichte on Platner's aphorism §345 on the imagination (cf. GA II/4S: 86). In these lectures Fichte also makes use of the triangle, parallelogram and rectangle to clarify points about logic (cf. GA II/4: 134, 140, 143f., 149f.). In the *Kollegnachschriften* of Fichte's lectures on Platner's Aphorisms from 1797 entitled: *Vorlesungen über Logik und Metaphysik SS 1797* (GA IV/ 1: 175–450) Fichte also talks about the different procedures of science and philosophy and briefly mentions mathematics.

36 Cf. GA II/4: 224; and GA II/14: 170f.

knowledge, and hence of reason itself.«[37] Unlike mathematics, logic is unique with regard to its »content«, since it is actually devoid of content and has to borrow it from the other sciences: »What distinguishes logic from all the other sciences is this: that it has no proper *content* of its own in the philosophical sense.«[38] As Fichte stressed often enough, the goal of the *Wissenschaftslehre* was *not* to be a mere formal science like logic, a so-called *Formular-Philosophie* (formulaic philosophy), but a living or »*reelle Philosophie*«, a system of real and genuine thinking. A formulaic philosophy is only concerned with how one can think about an object or the »conditions for the necessity of this thinking«, neglecting the object as such. – The *Wissenschaftslehre*, on the other hand, has its origin in the living ›I‹ of the thinker carrying it out, and as a »reelle« (real) philosophy it presents the »concepts and object at the same time, and never treats one without the other.«[39] Though subsequently expressed in language and symbols that are necessarily formal, only someone performing its original act can see its truth, and hence this active practical aspect differentiates it from the finished abstract rules of modern logic.[40] To express this difference in another short Fichtean formula: »*I* am *I* is *philosophy. A* is *A* – is *logic.*«[41]

Fichte's references to a »Philosophy of Mathematics«

Fichte's ideas and calls for a proper »philosophy of mathematics« do not first arise in the year 1805 in Erlangen, but may be found much earlier in his career. In fact, the idea for a separate philosophical discipline for mathematics that would not only examine specific problems but also the theories and presuppositions behind them can already be discerned during the period of the discovery of the *Wissenschaftslehre* in Zurich. In the midst of an investigation into the nature of space in the *Eigne Meditationen* (1793–94) Fichte speculates on the contents of a philosophy of mathematics and says that such a discipline ought to include the theory of geo-

37 GA IV/2: 265. In the same vein Fichte calls logic in the *Zurich Wissenschaftslehre*, the »daughter of philosophy.« (GA IV/3: 28).

38 *Zurich Wissenschaftslehre*, lecture two (GA IV/3: 28).

39 Fichte, GNR (1796/97) (GA I/3: 316 ff.).

40 Ibid.

41 *Zurich Wissenschaftslehre* (GA IV/3: 24).

metrical entities that relate to the limiting of the dimensions of space: »A mathematical line is in space, in *quantitas positiva*, limitation of the second dimension. – This belongs to the philosophy of mathematics.«[42]

In 1796 Fichte had obviously been discussing the relationship between philosophy and mathematics with his students in Jena. One of them – Johann Rudolf Steck (1772–1805) – who later became a politician in Switzerland, relates from this time that as a result of »numerous hints« from Fichte, and from studying the works of the mathematician Schulze, as well as a close acquaintance with the Fichtean synthetic method, had allowed him to make further progress in understanding the »philosophy of mathematics.«[43] A few years later in December 1799, in the so-called *Entwurf zu einem Plan über ein zu errichtendes Kritisches Institut*, Fichte again made a call for a separate discipline bearing the name the philosophy of mathematics, and described it as imperative for the health of mathematics itself. Contained in a letter to August Wilhelm and Friedrich Schlegel, this *Entwurf* (sketch) lamented the »lack of a philosophy of mathematics and the *disadvantages* that arise for mathematics on account of this; this will continue throughout history until this lack is finally remedied.«[44]

As part of this plan for a so-called »Critical Institute« Fichte repeated his call in August 1800 for some philosophical work to already be carried out on the subject of a philosophy of mathematics, and thought that Schelling should be the person to begin this work. Concerning the journal that was to be the organ for the Critical Institute, Fichte made the following request to Schelling in a letter dated August 2, 1800: »For the first volume, could I ask you to especially do something on the *fundamentals of a philosophy of mathematics*, as well as a *philosophy of history*?»[45] Shortly after, in another philosophically significant letter to Schelling dated 15th November 1800, Fichte turns to a discussion of the dimensions of space and says that the theory of dimensions does not belong in the *Wissenschaftslehre* per se, but in a philosophy of mathematics. In words recalling his earlier remarks on dimensions in the *Eigne Meditationen*, Fichte says:

42 Fichte, *Eigne Meditationen* (GA II/3: 82).

43 See J.R. Steck's letter to J.S. Ith quoted in: FG, vol. 6.1, p. 200.

44 J.G. Fichte, letter to A.W. Schlegel and Friedrich Schlegel, December 23, 1799 (GA III/4: 171).

45 J.G. Fichte to Schelling, August 2, 1800 (GA III/4: 278).

»[T]he deduction of the three dimensions is not at all incumbent upon the pure *Wissenschaftslehre*, but initially upon the *philosophy of mathematics*; and the *philosophy of nature* presupposes this deduction from the latter.«[46] Schelling, for his part, could not envisage the philosophy of mathematics as a separate branch within the system of philosophy. Instead, Schelling saw the philosophy of mathematics *as an abstraction* – it was simply a discipline abstracted from his own philosophy of nature. In a reply to Fichte four days after receiving the above letter Schelling wrote: »In my opinion, the philosophy of mathematics is an abstraction from the *Naturphilosophie*, just as the philosophy of mere formal thought, i.e. logic, is an abstraction from the *Wissenschaftslehre*.«[47] Hence it is clear, well before his 1805 remarks in the *Erlanger Logik* Fichte had already been intermittently reflecting on a philosophy of mathematics.

Finally, if we look at certain statements from 1812, it seems that Fichte was even contemplating teaching a course in Berlin on the philosophy of mathematics. Unlike in Erlangen, this topic is absent from his lectures on logic at this time. However, in a letter to Count (*Graf*) Zichy-Vásonykeö, while Fichte was the rector at the University of Berlin, Fichte outlined the following plan mentioning a possible course on the philosophy of mathematics, which unfortunately did not come to pass: »My teaching plan is as follows: my principal course will be on the *Wissenschaftslehre* [...] I will apply the *Wissenschaftslehre* in order to generate the particular philosophical sciences. This semester I will lecture on the theory of right (*Rechtslehre*). Then the theory of ethics. Later on a philosophy of nature, perhaps even on a philosophy of mathematics.«[48]

A Comparison of the Erlanger Logik with Fichte's other Texts on Mathematics

We will now compare the *Erlanger Logik* with Fichte's other principal writings discussing mathematics. In February 1794 in Zurich Fichte gave his very first public talks on his new philosophy. This set of lectures, partly available under the title *Zurich Wissenschaftslehre*, is also notable for an entire lecture devoted to »The Relationship of the *Wissenschaftslehre* to

46 GA III/4: 361.

47 F.W.J. Schelling, letter to J.G. Fichte, November 19, 1800 (GA III/4: 367).

48 J.G. Fichte to Graf Zichy-Vásonykeö, beginning of May, 1812 (GA III/8: 6).

Geometry.«[49] Despite some highly interesting remarks about the nature of geometry, in my view this lecture 4 is not an exposition of Fichte's theory of mathematics, since it is mainly concerned with demonstrating the *superiority* of Fichte's *Wissenschaftslehre* over geometry. On the other hand, the beginning of this lecture 4 in the *Zurich Wissenschaftslehre* shares distinct parallels with the statement in the *Erlanger Logik* about geometry being the foundation of mathematics: »The external relationship between the *Wissenschaftslehre* and mathematics, and in particular between the *Wissenschaftslehre* and geometry, as the ground of all mathematics, still remains to be indicated (the inner relationship between them depends *per se* on the *Wissenschaftslehre*).«[50]

Again, in my opinion this early text (only available as a *Kollegnachschrift* in Lavater's hand) should not be classified as a text on the philosophy of mathematics, but viewed as a condensed argument for outlining the similarities and differences between two individual disciplines that aim at scientific certainty. According to Fichte, the main similarities between geometry and the *Wissenschaftslehre* concern their precision and exactness, whereas one of their salient differences is with regard to their methods of proof. This already hints at Fichte's classification of the sciences, in which geometry is relegated to a sub-science of the *Wissenschaftslehre*.[51] Fichte's own philosophical system, as he publicly declares after his arrival in Jena in *Über den Begriff der Wissenschaftslehre* (1794), is to form a »science of the sciences«, which is to ground every other scientific discipline, including mathematics.[52] This classification of the sciences is already implicit in the *Zurich Wissenschaftslehre*, and Fichte refers to it again when he says that even geometric elements such as points and space itself are grounded in the *Wissenschaftslehre*: »First of all, geometry needs a *point, by means of which* it draws; [secondly] *space within which* it draws [a line]; [and thirdly, it needs] the free self-activity of the power of the

49 German: »Verhältniß der Wissenschafts-Lehre zur Geometrie«, *Zurich Wissenschaftslehre* (GA IV/3: 32–37).

50 Ibid., p. 32.

51 Ibid.

52 See GA I/2: 117–119. That the *Wissenschaftslehre* was a »science of the sciences« is a belief that Fichte still harboured almost twenty years after his first lectures in Jena. See Fichte's comments in the *1812 Logik*, where »philosophy aims to be a science of the sciences.« (GA II/14: 170f.).

imagination, which does the drawing. It obtains all three of these from the *Wissenschaftslehre*, and hence the possibility of all construction (and hence the possibility of demonstration as well) is grounded in the *Wissenschaftslehre.*«[53]

The *Zurich Wissenschaftslehre* therefore foreshadows the identical remarks eleven years later in the *Erlanger Logik*, when Fichte states that a geometer is usually ignorant about the *origin* of geometric entities, or *why* and *how* they can occur within consciousness: »This is a question that geometry cannot answer. / Geometry presupposes the answer to this question, which comes from the *Wissenschaftslehre*.«[54] Already at this early stage of his career Fichte is inclined to think that the philosophy of mathematics – with geometry as the ground of mathematics – could be subsumed under the *Wissenschaftslehre*.

A valuable document for Fichte's views on mathematics is the public *Ankündigung* (Announcement) for a new version of the *Wissenschaftslehre*. The text is dated 4th November 1800, but was first published in January 1801.[55] Fichte here talks at length about the similarities between the *Wissenschaftslehre* and mathematics, chiefly seeking to argue that the *Wissenschaftslehre* is a science of the same rigour and self-evidence as mathematics. For Fichte, the *Wissenschaftslehre* and mathematics share a number of »advantages« in common, including their scientific traits of »determinacy« and »irrefutability«.[56] But like the lecture on geometry from the *Zurich Wissenschaftslehre*, the *Ankündigung* is more concerned with a comparison between philosophy and mathematics, with the *Wissenschaftslehre* as the ultimate science of the sciences. Thus, the *Ankündigung* too should not be considered as dealing with a philosophy of mathematics in the conventional sense, i.e. as determining what the nature of mathematics is, and its conditions and possibility.

Written at the same time as the November 1800 letter to Schelling mentioned above, the aim of the *Ankündigung* was to publicly announce a *Neue Bearbeitung der Wissenschaftslehre*, which was to be published in

53 *Zurich Wissenschaftslehre* (GA IV/3: 34).

54 Ibid., p. 35.

55 Reprinted in: GA I/7: 153–164.

56 GA I/7: 160–164. See Chapter Five below regarding these »advantages« in common.

1801, but remained a fragment and was never published during Fichte's lifetime.[57] The *Neue Bearbeitung* has been viewed as an attempt to present the *Wissenschaftslehre more geometrico*.[58] Whenever the *Neue Bearbeitung* broaches the topic of geometry and mathematics, it is chiefly to highlight the parallels between its procedure or method and that of the *Wissenschaftslehre*.[59] Thus it also does not discuss the contents of a philosophy of mathematics as such.

In 1804, in the 2nd series of lectures on the *Wissenschaftslehre*, Fichte reinforces his view that a philosophy, or at least a theory, of mathematics is needed. This is because a philosophy of mathematics would explain, for example, how and why our conception of space arises, and, for example, the nature of numbers. These are philosophical questions that belong in a philosophical setting, and are not required to be answered or explored in the field of pure mathematics. They belong in a »higher science« than ordinary mathematics, one which could provide answers to these questions and explain the *possibility* of mathematics itself. As Fichte says in a significant passage from lecture three: »Let the arithmetician as mere arithmetician, tell me how he is able to generate a solid and stationary number one; or let the geometer explain what holds and keeps his space stationary for him while he draws his continuous lines in it; [...] This is not in any way meant as a reproach to mathematics; because mathematics should and cannot be anything else. And it is certainly not our business to blur the boundaries of the sciences; but it should simply be recognized that this science [of mathematics], like all the other sciences, is not the first science, and it is not independent, but that the principles of its possibility lie in another, higher science.«[60]

This »higher science« is of course the *Wissenschaftslehre* – or at least that section of it devoted to the philosophical underpinnings of mathe-

57 *Neue Bearbeitung der Wissenschaftslehre* (GA II/5: 331–401).

58 See Daniel Breazeale's essay: »Toward a *Wissenschaftslehre more geometrico* (1800–1801)« in: *After Jena: New Essays on Fichte's Later Philosophy*, eds. D. Breazeale and T. Rockmore (Evanston, Illinois: Northwestern University Press, 2008), pp. 3–40.

59 »The procedure is entirely like in mathematics [...] Corollary. It is wholly like a kind of mathematical proof.« *Neue Bearbeitung der Wissenschaftslehre* (GA II/5: 335, 338).

60 *Wissenschaftslehre 1804/II* (GA II/8: 47f.).

matics. Thus, in 1804 Fichte again clearly articulates the need for a philosophy of mathematics, and implicitly views the *Wissenschaftslehre* as a system of philosophy able to explain the principles and possibility of mathematics.

In conclusion, we should mention the *Einleitung in die Wissenschaftslehre* from autumn 1813, where Fichte speaks in relative detail about the relationship between philosophy and mathematics one final time. Like in the *Zurich Wissenschaftslehre* and the *Ankündigung*, here Fichte again compares the two disciplines and emphasizes how the *Wissenschaftslehre* should learn from the constructive and intuitive methods of mathematics, as well as the necessity of philosophy becoming a precise and self-evident science like mathematics. According to Fichte, if we learn to understand what transpires in our minds when we carry out mathematics, it should cast light on what happens when we philosophize or carry out philosophy. In an echo of his lament to the Schlegel brothers in 1799 about the lack of a philosophy of mathematics, Fichte criticizes his contemporaries for failing to even notice that they have not correctly philosophized about mathematics, or as he terms it here, »mathematicized« about the discipline of mathematics itself: »If [my opponents] had not simply mathematicized, but also mathematicized about their mathematicizing itself [...] then the light of the essence of science in general and also that of philosophy would have dawned on them.«[61]

These late lectures reaffirm one final time the continuity of Fichte's thoughts on mathematics, and the necessity for a genuine philosophy of mathematics. Nevertheless, it would be fair to say that Fichte's observations here are not complete, but are only further fragments from his theory of geometry and mathematics. On the whole, these comments from 1813 also concern the similarities between the discipline of geometry and the *Wissenschaftslehre*, and therefore do not match the dense presentation given in the *Erlanger Logik* in 1805 under the rubric of a »philosophy of mathematics«.

At this point we could ask: where does a philosophy of mathematics fit into the Fichtean system? Apart from the pure *Wissenschaftslehre* as

61 Fichte, *Einleitung in die Wissenschaftslehre 1813* – forthcoming in GA II/17: 268 (cf. SW IX: 41). See too the similar remarks in Kollegnachschrift Halle version (GA IV/6: 398).

such, which is divided into theoretical and practical philosophy, with the latter having primacy in his system, Fichte further divides his system into four principal so-called applied sub-disciplines.[62] The latter include the *Rechtslehre* (theory of right), *Sittenlehre* (theory of ethics), *Naturlehre* (theory of nature), and *Religionslehre* (theory or doctrine of religion).[63] As already noted, the science of mathematics was above all the science of geometry for Fichte, and he used the terms synonymously. Since geometry was primarily the theory of *space* for Fichte, any philosophy of mathematics or geometry would ultimately have to relate to and describe the world of nature, for it is mathematics that has proved to be so effective in scientifically describing the natural world. Hence, it seems that geometry for Fichte would have to be a part of his *Naturlehre* or theory of nature. However, not simply in its material or physical sense, which is chiefly the domain of the physicist, biology or chemist for example, but especially in the ideal sense, as an ideal system of the laws of space. Unfortunately, compared to the other three sub-disciplines, which strongly highlight the primacy of practical reason, the *Naturlehre* was the least developed of Fichte's four sub-disciplines.[64] Hence, the necessity of not only reconstructing Fichte's theory of geometry, but also to some degree his theory of nature.[65]

Finally, both Radbruch and Csech have remarked that it is an unfortunate deficit that Fichte never completed his wish to write and lecture on

62 For a general overview of Fichte's classification of the *Wissenschaftslehre* and its sub-disciplines, see Reinhard Lauth: »J. G. Fichtes Gesamtidee der Philosophie«, *Philosophisches Jahrbuch* 71 (1964/65): 353–384.

63 Concerning Fichte's own classifications, see the remarks on his system in *Über den Begriff der Wissenschaftslehre* (GA I/2: 150ff.); *Wissenschaftslehre nova methodo*, which also highlights the intermediary role of aesthetics between the transcendental and common standpoints (GA IV/3: 520–523); *Die Anweisung zum seligen Leben, oder auch die Religionslehre* (GA I/9: 103–114); as well as the editor's introduction in the Fichte *Gesamtausgabe* to the 1813 *Staatslehre*, about Fichte's idea for an »Allgemeine wissenschaftliche Encyclopädie« (general scientific encyclopedia) (GA II/16: 3–4).

64 In this regard see Andreas Schmidt's remarks on Fichte's classifications, in: »Johann Gottlieb Fichte (1762–1814)« in: ed. Otfried Höffe, *Klassiker der Philosophie: Von Immanuel Kant bis John Rawls* (Munich: Beck, 2008), pp. 36–38.

65 Reinhard Lauth has attempted such a reconstruction of Fichte's transcendental theory of nature in: *Die Transzendentale Naturlehre Fichtes nach den Prinzipien der Wissenschaftslehre* (Hamburg: Meiner, 1984).

the philosophy of mathematics.[66] It is certainly a deficit to the extent that Fichte never published a fully worked out exposition of his ideas in this domain. Nevertheless, it is my contention that Fichte did at least lecture once on his philosophy of mathematics, and this was precisely in 1805 in the *Erlanger Logik*. As already noted, the Erlangen presentation is not an entire lecture course on the philosophy of mathematics, and Fichte's extant notes in volume II/9 of the *Gesamtausgabe* are certainly fragmentary. Yet this excursus from 1805 manages to marshal together Fichte's most important thoughts on the subject. Thus, unlike the other texts just mentioned, whose aim is mostly to compare or distinguish the respective qualities of the *Wissenschaftslehre* and geometry, Fichte's examination in the *Erlanger Logik* contains his most concrete reflections on the foundations of mathematics and his own model of geometry.

2. The Contents of a Fichtean Philosophy of Mathematics

a. Nine Elements of a Philosophy of Mathematics in the Erlanger Logik 1805

I will now analyse Fichte's *Erlanger Logik* to determine the specific elements of his philosophy of mathematics. The following presentation does not aim to be definitive or complete. On the contrary, this chapter solely seeks to summarize the main elements of Fichte's theory. Those elements specifically relating to Fichte's model of geometry will be outlined in more detail in the subsequent chapters. The author is also well aware that an exhaustive analysis of Fichte's philosophy of mathematics as a whole is still needed, as well as an in-depth study of his place in the history of the philosophy of mathematics. Our study simply aims at a basic reconstruction of Fichte's theory, and its chief focus is on the Fichtean theory of geometry and its relation to the *Wissenschaftslehre*. I will outline nine different aspects of Fichte's philosophy of mathematics in the Erlangen text and indicate where the same ideas are elaborated in more detail in other texts in the Fichtean corpus. It will be seen that the main essentials of Fichte's theory can already be found in the earliest Zurich and Jena texts on the *Wissenschaftslehre*, but become expanded and more precise in his later writings.

66 See Radbruch (2003), p. 260 and Csech (1995), p. 58.

Turning to the text of the *Erlanger Logik*, we can now ask: which elements exactly did Fichte think should belong in a philosophy of mathematics? For example, does Fichte have anything to say here about arithmetic or algebra? And how are the different branches of mathematics related to one another in Fichte's theory?

Fichte begins this excursus on mathematics in the *Erlanger Logik* with a preliminary discussion of the nature of a science and its methodology. As already noted, a scientific method for Fichte should furnish both the rules of its external form and internal content. It is a matter of intellectual insight into the two.[67] According to Fichte, a philosophy of mathematics therefore presupposes understanding a certain number of basic scientific rules or »propositions.« We will enumerate them as they appear in the text. What is a scientific rule? (1) It has a universal law of material being; (2) It has a universal law of form. They latter includes: (a) No leaps or gaps (*saltus*) in reasoning. (b) No circles. (c) No pauses (or »standing still«). d. Rigorous determinacy.[68] Our goal is not to go into details concerning these principles, but to briefly summarize them before moving on to the main task of determining the contents of Fichte's philosophy of mathematics. One thing to recall about these points is that they are the standard hallmarks of a science for Fichte – they are at once »elementary« and the »ideal of science«, and are characterized with regard to the synthetic and analytic methods.[69] The so-called »scientific« analytic method descends downwards from definitions and classifications to specific concrete facts, whereas the synthetic method ascends upwards to general abstractions. But each of the two methods should be continuous and without gaps in the argument: »In their manner of expression, these rules are bound to the analytic method as the actual scientific method. For example, they descend using definitions, classifications, sub-classifications: down to the concrete, i.e. ever further determining and characterizing. The synthetic method ascends upwards by means of combinations (already approaching from the classification, which has to be complete for the crafter of science otherwise he does not have any exhaustive combination) and abstractions. Likewise, in accordance with these rules, in the

67 See Fichte, *Erlanger Logik* (GA II/9: 124).

68 Cf. Ibid., pp. 120ff.

69 Ibid., p. 123

continuous path of science there are no gaps and circles or pauses while ascending.«[70]

These rules form the basis of all scientific practice for Fichte and they are applicable to both the synthetic and analytic methods. If we now ask: why did Fichte make this excursus on the philosophy of mathematics in the *Erlanger Logik*? – We can answer: the first function of Fichte's excursus is as an example to illustrate the scientific rules and methodology of his philosophy. A second important function seems to be to finally provide his listeners with a detailed overview of his theory of mathematics and geometry. Yet, however much philosophy is inspired by the transparency and rigorous methods of geometry, it ought to remain a distinct and independent science. – As with saw in the *Wissenschaftslehre 1804/II*, anyone wishing to *literally* join philosophy and geometry together would no longer be traversing the path of science for Fichte but blurring its boundaries: »It – the *Wissenschaftslehre* – can never become geometry as such. If the scientist of the *Wissenschaftslehre* wanted to be a geometer, [...] then he would be either (1) A pedant (2) Or something worse, a confused and confusing mind.«[71]

If we analyse this section of the *Erlanger Logik* we discover that Fichte's philosophy of mathematics contains nine principal elements. This number nine is of course approximate, and a different method of classification would no doubt yield a different number of elements. Though brief, Fichte's discussion covers a large amount of topics and territory. I will provide a commentary on these 14 pages of the *Erlanger Logik 1805* by means of a close reading of the text, as well as pointing to the principal passages in other works of Fichte where these thoughts are discussed in more detail.

i. Geometry

Fichte's excursus begins with the following assertion: »*Geometry*, as the ground of all mathematics.«[72] Fichte's theory and model of geometry will be treated in detail in Chapter Two below. For this reason our main question here is simply: what is the relationship of geometry to the other

70 Ibid., p. 122f.

71 Ibid., p. 125. Cf. *Wissenschaftslehre 1804/II* (GA II/8: 47f.).

72 Original German: »*Geometrie*, als den Grund aller Mathematik.« (GA II/9: 124)

branches of mathematics for Fichte? If we look again at the beginning of this excursus, it is clear from the very outset: in Fichte's philosophy of mathematics the science of geometry is considered to be the branch that forms the foundation or »*Grund*« for mathematics as a whole. That is say, in terms of a theoretical framework, all the other branches can be traced back and have their basis in geometry. As mentioned above, the statement that geometry is the *foundation* of mathematics is not surprising, since at Fichte's time geometry was still valued as a model of scientific excellence, and it was a widely held view that the majority of the other branches of mathematics could be reduced to geometry.[73] Thus, whenever Fichte speaks of mathematics, he generally has the science of geometry in mind. One should therefore bear this in mind: like many other contemporary philosophers, including Kant and Schelling, Fichte often employs the terms »Geometrie« and »Mathematik« synonymously. Geometry *is* mathematics for Fichte, its inner core or ground. We have seen that an identical view is already expressed in the lecture of February 1794 in the *Zurich Wissenschaftslehre.*[74] Yet what is surprising here in 1805 is Fichte's intention to discuss in more depth *foundational* issues, i.e. to investigate the *nature and foundations* of the science of geometry itself. Three other characteristics that Fichte attributes to geometry should also be mentioned. For Fichte, geometry is the most secure and certain science and its truths are indisputable and irrefutable.[75] – We acquire them by means of a rigorous process based on transparent inner construction and for this reason mathematics contains the »spirit of scientificity«, as Fichte could still say in 1813.[76] In the 1794/95 *Grundlage der gesammten Wissenschaftslehre* Fichte chose to emphasize the more abstract aspect of mathematics, classifying it as the science that »fully abstracts from all quality and solely takes quantity into consideration.«[77] Whereas in lesson six of the *Sonnenklarer Bericht* from 1801 his focus is more positive and pedagogical, terming geometry the branch of mathematics that excites intuition in the most im-

73 See, L. Magnani, *Philosophy and Geometry: Theoretical and Historical Issues* (2001), pp. 67ff.

74 Cf. GA IV/3: 32.

75 See *Erlanger Logik* (GA II/9: 125) and the *Ankündigung* (GA I/7: 161).

76 *Einleitung in die Wissenschaftslehre* (1813) (SW IX: 41).

77 Fichte, GWL (GA I/2: 292).

mediate manner, especially with respect to spatial figures.[78] The 1805 *Erlanger Logik* may therefore be read as a summary of all these views, and encapsulates the classical notion of geometry as the science of the determination and limitation of space. Yet like in the 1794/95 *Grundlage*, Fichte also takes into account the formal aspect of geometry: »The *conditio sine qua non* of this science [i.e. geometry] is to empty space of everything that is forming and limiting in itself, and to grasp space as empty and formal.«[79]

Thus, if the goal of certain philosophers of mathematics today is to classify mathematics as a science of patterns, shapes or as a logical or conventional game[80], Fichte interpreted the goal of geometry to be traditional and uncontroversial. On the one hand, the practical function of geometry is *the measurement and determination of space*. On the other, it is also possible to consider pure space or space as such. This pure science of geometry is a science that limits and »resides entirely in relations«, and where its purpose is to formally grasp space, i.e. as abstract.[81] In this conception, space can be infinitely extended: »The extension of space is infinite: this means only geometry can unconditionally enlarge this circle or semi-circle and must be able to do this (unconditionally enlarge it).«[82]

ii. Arithmetic

There are of course other branches of mathematics apart from geometry, such as arithmetic, algebra and analysis (infinitesimal calculus). What does Fichte say about these? Although in the *Erlanger Logik* Fichte almost exclusively confines his discussion to the discipline of pure geometry, he does briefly mention the subject of arithmetic. From other texts we know that Fichte considers analysis or infinitesimal calculus to be a more complex part of arithmetic, yet a more detailed consideration of calculus

78 See GA I/7: 253.

79 *Erlanger Logik* (GA II/9: 125).

80 See, among others, Magnani's characterization of geometry as »the science that studies shapes« – *Philosophy and Geometry: Theoretical and Historical Issues* (2001), pp. 99f. We shall see other interpretations in the concluding section of Chapter Two when discussing Fichte's Platonism.

81 Fichte, *Erlanger Logik* (GA II/9: 125, 135).

82 Ibid. p. 134. See Chapter Four below for a discussion of Fichte's conception of space.

seems to be completely absent in Fichte's writings[83] Like Kant, Fichte chiefly uses examples from Euclidean geometry whenever discussing mathematics. Unlike Kant, who employed examples of arithmetic in the first Critique to demonstrate the synthetic nature of mathematical cognition[84], Fichte did not place the same value on the contribution of arithmetic to mathematics. Schelling too underscored the importance of arithmetic, and following the lead of Kant, made a distinction between the cognitive modes of arithmetic and geometry, especially in their functions of seizing the particular and the universal.[85]

Arithmetic is generally not of prime consideration in Fichte's conception of mathematics, and he above all views it as an auxiliary tool to assist in illustrating specific problems.[86] Arithmetic (like geometry) is furthermore grounded in the *Wissenschaftslehre*, and this stance is already evident in the *Zurich Wissenschaftslehre*: »The *Wissenschaftslehre* is not merely the arithmetical rule – it is the calculation itself.«[87] This was to highlight the »infallibility« of the *Wissenschaftslehre* compared to the fallible nature of all counting and calculating. Despite their brevity, Fichte's pronouncements on arithmetic in the *Erlanger Logik* are some of his most

83 Cf. section § 51 *Deducirter Plan einer zu Berlin zu errichtenden höhern Lehranstalt* (1807) (GA II/11: 111, 145–148). Regarding this point, see J. Widmann's small study: »Dialektik more geometrico. Fichtes Prinzip des Systemgenesis« (1983), pp. 188 ff.

84 Cf. Kant, *Critique of Pure Reason*: »The arithmetical proposition is therefore always synthetic; one becomes all the more distinctly aware of that if one takes somewhat larger numbers« (B16, p. 144), as well Kant's long discussion in the section »The Discipline of Pure Reason in Dogmatic Use« (A713/B741-A738/B766), pp. 630–643.

85 This difference between geometry and arithmetic for Schelling concerns their presentations of the particular and universal: »If we understand by ›universal‹ the purely intellectual or discursive universal, then it is easy to see that these two kinds of intuition are really given in the two branches of mathematics, that arithmetic expresses a particular (relation of single quantities) in the universal, geometry expresses a universal (the concept of a figure) in the particular [...] the particular in geometry is not the empirical triangle that I draw on the paper, but, according to Kant, the triangle of pure intuition.« *Über die Construktion in der Philosophie* (*Sämmtliche Werke* I, 5, 130f.).

86 In his unpublished 1804 aphorisms on education (nos. 5 and 7), Fichte briefly mentions the benefits of arithmetic in helping to stimulate the imagination of children (GA II/7: 19f.).

87 Lecture Five (GA IV/3: 41).

significant. Taking a *reductive* or derivative view of arithmetic, Fichte claims in a »corollary« that: »all arithmetic is based on geometry« – or that arithmetic is nothing but an »exemplar« of geometry.[88] This is because in Fichte's eyes arithmetic is simply a more abstract manner of presenting geometric spatial relations.[89] Although arithmetic is able to posit unities, Fichte sees the difference between geometry and arithmetic with regard to the issue of *divisibility*. Unlike geometric space, the »space« of arithmetic is *not* infinitely divisible. The ultimate goal of arithmetic is to terminate the process of division and arrive at a final solution or »unity«: »The essence of arithmetic is to posit a unity (which incidentally has nothing to do with geometry and consists entirely in relations through which arithmetic becomes a mere exemplar of geometry). However, it only arrives at this unity by ceasing to divide. The space of arithmetic, therefore, in accordance with its own existence, is not inwardly infinite or infinitely divisible (as geometrical space is of course), but the division terminates and this is where arithmetic starts.«[90]

Fichte's reductive theory of arithmetic brings him in line with the Kantian view, which also values the primacy of geometric constructions over arithmetic ones.[91]

Finally, although Fichte does not devote much time to the branch of mathematics called arithmetic, it is striking to see that the concept of number still plays an importance role in his philosophy. This is especially the case in his famous theory of »Fünffachheit« – »quinticity«, or the five-fold synthesis.[92] – This fivefold numerical basis to Fichte's philosophy is

88 *Erlanger Logik* (GA II/9: 134).

89 Ibid. (GA II/9: 134–135).

90 Ibid. (GA II/9: 135).

91 See Shabel's discussion of this point: »It is plausible to suppose that Kant's philosophy of arithmetic is derivative on his philosophy of geometry, with arithmetic ›construction‹ entirely dependent on geometric construction« *Mathematics in Kant's Critical Philosophy*, p. 163.

92 The number three or a threefoldness/trichotomy (Dreifachheit/Trichotomie) also plays a role. See for example the *Wissenschaftslehre 1810*: »*Threefoldness*, which I have deduced. Synthetic member, subjectivity (not at all still, subject, as it were), *object*.« (GA II/11: 28); and *Wissenschaftslehre 1812* (GA II/13: 131). While in the *Wissenschaftslehre 1804/II* Fichte argues that the five standpoints each have five main aspects to result in »25 Hauptmomente«, a »splitting into 25 forms« or »original basic determinations of knowledge« (GA II/8: 418f.).

particularly visible in the Berlin presentations. For example, in *Die Anweisung zum seligen Leben* (1806), Fichte outlines in a popular manner five principal ways of viewing the world, or considering its essence. The first and lowest view is that of the senses, or what falls within the range of the senses. The second is viewing the world through the lenses of rational order, and the third view is that of »true and higher morality«. The fourth view is the standpoint of religion and the fifth is the standpoint of science: »That science, I say, of the One, absolute science that is perfected in itself.«[93] These fivefold philosophical schemas are not fixed in an absolute sense, as can be seen in the second chapter of the *Wissenschaftslehre 1812*. Here Fichte presents a deduction of the *Fünffachheit* in the form of the appearance, in which the first four divisions become united in a fifth: »Intuition and concept, yields themselves, of which each in turn is again split into subject and object; everything in absolute unity.«[94]

iii. Algebra

As for the mathematical discipline of algebra in the strict sense, Fichte scarcely comments on the subject, save to mention its abuse by mathematicians with »formal minds« who use algebraic methods to solve geometrical problems. In one sense, algebra constitutes a paradigm of a purely »formal science«, as opposed to Fichte's own attempt at a »living« and »real philosophical science.«[95] For instance, in the *Grundlage des Naturrechts* (1796/97) Fichte observes that the famous problem of the squaring of the circle is impossible, and that formal mathematicians who believe they have actually grasped infinity because they can express it using the symbolic notation of algebra are under a delusion. They are ignorant of »the foundations of science« because they have confused infinity itself with the formal expression of it, and only possess an »empty concept of infinity.«[96] (See the next section on Infinity). Despite this mistrust of alge-

93 Lecture Five, *Die Anweisung zum seligen Leben* (GA I/9: 105ff.).

94 *Wissenschaftslehre 1812* (GA II/13: 102, 105).

95 Fichte, GNR (1796/97) (GA I/3: 317).

96 Ibid. Fichte also mentions algebra in the Preface to the *Sonnenklarer Bericht*, but only to say that one should not talk about sciences like the *Wissenschaftslehre* if one has no knowledge of it, just as one should not speak about »trigonometry or algebra if one has not learnt these sciences.« (GA I/7: 187).

bra's formal and symbolic notation Fichte still manages to acknowledge some of its benefits in training one to think abstractly, especially if carried out correctly and not rote learnt. He suggests algebra as »an excellent exercise for learning to retain the most abstract concepts in the intellect.«[97] In the *Critique of Pure Reason* Kant likewise stresses the abstract nature of algebra, but is more favourable to it than Fichte, since he considers the symbolic constructions of algebra to be on a par with the constructions of geometry.[98] Though not explicitly mentioned in the *Erlanger Logik*, Fichte's view of the benefits of formal languages such as algebra for expressing abstract conceptions is implicitly present when considering the formal nature of geometry. For it is precisely the aim of geometry to grasp space as »empty and formal«; indeed »this formal perfection« of geometry is its »triumph.«[99]

Despite his mixed estimation for algebra as a mathematical discipline, Fichte nevertheless had a sound *philosophical* appreciation of the symbolism of algebra and of the formalism of mathematics as a whole. Like the philosophical employment of arithmetic in the theory of *Fünffachheit*, explicitly algebraic formulations of Fichte's philosophy can be found in various presentations of the *Wissenschaftslehre*. This is especially the case during the discovery of his system in 1793/94 in Zurich, in Fichte's first tentative and preliminary articulation of the *Wissenschaftslehre*, the so-called *Eigne Meditationen*.[100] Fichte's concern with algebraic or linguistic sym-

97 *Die am Ende des Winters [Sommers] 1794 von Fichten Gehaltene Schlussvorlesung als Beylage der Verantwortung* (1797) (GA I/4: 413).

98 »But mathematics does not merely construct magnitudes (*quanta*), as in geometry, but also mere magnitude (*quantitatem*), as in algebra, where it entirely abstracts from the constitution of the object that is to be thought in accordance with such a magnitude. [...] and thereby achieves by a symbolic construction equally well what geometry does by an ostensive or geometrical construction.« Kant, *Critique of Pure Reason* (A717/B745), p. 632. Kant's belief in the certainty and power of algebra is also visible in his early *Inquiry* essay: »Furthermore, in geometry the signs are similar to the things signified, so that the certainty of geometry is even greater, though the certainty of algebra is no less reliable.« In: Kant, *Theoretical Philosophy 1755–1770*, p. 265.

99 Fichte, *Erlanger Logik* (GA II/9: 125, 135).

100 For example: »Consequently the concept of quantity must still be applicable to A. Then it would be *unlimited*. Indeed, it would in any case be *applicable*; but in the

bolism is explicitly mentioned in the 1800 *Ankündigung*, where he raises the possibility of increasing the accessibility of the *Wissenschaftslehre* by further determining its »terminology using linguistic signs, as well as a symbolic system of pure concepts.«[101] Or again in the 1807 Königsberg presentation of the *Wissenschaftslehre*, for example, where we see that Fichte is not adverse to expressing his entire philosophy in terms of short formulae, such as: x = A + F + U + 5. Expressed in words: the schema of life (x) = Anschauung (intuition) + Freiheit (freedom) + Unendlichkeit (infinity) + 5 (Fünffachheit).[102]

Finally, given Fichte's relatively little interest in algebra, one of the most thorough and promising studies of drawing a link between the *Wissenschaftslehre* and the philosophy of mathematics has been surprisingly in the domain of the philosophy of algebra. As we saw in the Introduction, in his 1962 study entitled *La philosophie de l'algèbre*, Jules Vuillemin has argued that Fichte's theory of intuition should be interpreted in the light of the earlier revolution in abstract algebra of employing a priori abstract methods. Looking at this more closely, we see that in sections §12 and §13 of his volume Vuillemin presents his arguments for a number of ›structural parallels‹ between the Fichtean theory of intuition (*Anschauung*) and the method of the French mathematician Joseph-Louis Lagrange (1736–1813), which is chiefly found in his 1770 work: *Réflexions sur la résolution algébrique des équations*. To conclude this section on algebra we will look at Vuillemin's argument in more detail.

Vuillemin starts his analysis by remarking that although it might seem arbitrary to use the method of an algebraist like Lagrange to shed light on the method of a philosopher like Fichte (because there is no historical link between them, Lagrange moved to Paris from Berlin in 1787), he says that nothing prevents us from hypothetically (a priori) trying to see if there is a distant connection between both methods, just as there is a less yet still distant structural connection between the algebraic resolution

opposition. AR would be = –x / but C = A_x, hence x = –A. correctly algebraic.« *Eigne Meditationen* (GA II/3: 64–65).

101 *Ankündigung* (GA I/7: 164).

102 GA II/10: 158–159. See Gaetano Rametta's discussion of this formula in: »La dissolution de la théologie politique chez le dernier Fichte« in: J-C. Goddard & M. Maesschalck eds. *La Philosophie de la Maturité (1804–1814)*, vol. II., (Paris: Vrin, 2003) pp. 31–50.

of equations and the theory of groups. Thus although both domains appear to be entirely independent of one another, there might be, if not an exact relation between the structures, than at least analogies in the problems that they pose.[103]

What was Lagrange's revolution? Lagrange noticed that all earlier, or a posteriori, methods for solving algebraic equations reduced the problem by means of some transformation to the solution of a certain auxiliary equation that had a lower degree. Instead of particular solutions to particular equations, Lagrange found an a priori general method for solving cubic and quartic equations, i.e. of the 3rd and fourth degree. Lagrange's method did not extend to quintic equations (of fifth degree) or higher, because the resolvent had higher degree.[104]

With regard to philosophical methodology, Lagrange's innovativeness was to consider the roots, i.e. the particular numerical values of an equation, in an altogether symbolic or abstract manner. This was in contrast to his predecessors, who had still treated them as numerical values.[105] According to philosophers of mathematics, this was »the first step in an a priori and generalized method that would ultimately lead to the branch of mathematics called group theory – not with particular solutions, but a theory of their structures and laws.«[106]

As Lagrange himself remarks in the introduction to the *Réflexions sur la résolution algébrique des équations*: »Je me propose d'examiner les différentes méthodes que l'on a trouvées jusqu'à présent pour la résolution algébrique des équations, de les réduire à des principes généraux et de faire voir a priori pourquoi ces méthodes réussissent pour le troisième et le quatrième degré, et sont en défaut pour les degrés ultérieurs.«[107]

103 See Jules Vuillemin, *La philosophie de l'algèbre*, p. 102. In a footnote on the same page Vuillemin explains that the sections §§12–13 are not aimed at a demonstration but suggesting historical affinities relating to the Zeitgeist.

104 Cf. Vuillemin, ibid., pp. 99–102; and Jean-Pierre Tignol's account of Lagrange in: *Galois' Theory of Algebraic Equations* (London: World Scientific Publishing Co., 2001), pp. 123–149.

105 See Jules Vuillemin, *La philosophie de l'algèbre*, pp. 81–82; 104–105.

106 Ibid., p. 117. According to Vuillemin here, if Lagrange had furthermore reflected upon the *composition* of these permutations he would have created group theory proper.

It is precisely this abstracting, this generalizing and a priori approach of Lagrange that is of philosophical value to Vuillemin in his comparison with Fichte. Vuillemin outlines three main stages in the methods of Fichte and Lagrange. Starting with Lagrange, Vuillemin analyzes the so-called ›a posteriori' methods of Lagrange's predecessors, and then Lagrange's general solution, before making a comparison with Fichte's method.[108]

For Vuillemin, the first series or stage in Fichte's method is that of derived reflection (*réflexion dérivée*)[109] – it is the first series of philosophical abstraction. It places us at an external point of view by observing how a possible consciousness is constructed. Fichte's method reproduces and adapts the principal stages of the Kantian system, just as the a posteriori method of Lagrange reproduced the principal moments of the resolution in his predecessors. Vuillemin then sees that this systematic deduction leads Fichte to an element that one can view as the analogue to the »resolvent« of Lagrange; that is to say, to an auxiliary equation, an equation of a lower degree, that can be used to solve a more difficult and complex problem. This is precisely Vuillemin's first series of philosophical reflection, since it takes on an external point of view, and there is external data.[110] It concerns possibilities, that is to say, *Denkmöglichkeiten* in Fichte's words (thought possibilities) or possible objects (in the sense of a hypothesis).

In the second series we start with the ›resolvant‹ (or productive imagination), with the fact of representation.[111] In contrast to the first series, the second series therefore proceeds in the opposite direction, and has an internal point of view. The hypothesis of series one is replaced by a ›thesis‹, or the fact of experience replaces the external data. Like the first, it is also synthetic. In Vuillemin's eyes, both the second and third series belong to the genetic method, are a posteriori in the algebraic sense, or concerned with particulars. The third series, is finally the highest level of abstraction, and there is a transition to a new method.[112]

107 J.L. Lagrange, *Réflexions sur la résolution algébrique des équations* (1770/71) p. 206 ; cited in Vuillemin, *La philosophie de l'algèbre*, p. 73.

108 The following analysis is above all based on sections §12 and § 13 of Jules Vuillemin, *La philosophie de l'algèbre*, pp. 102–122.

109 Ibid., p. 105.

110 Ibid., p. 106.

111 Ibid.

112 Ibid., p. 107.

The third series for Vuillemin is a new reflection or a theory of structure (the structural method).[113] This third series is no longer the genetic method, but what can be called the critical or systematic method, hence the progression by both Fichte and Lagrange beyond their predecessors. With regard to Fichte, it is the progression to intellectual intuition, to a mode of intuition higher than the sensible intuition of Kant. The hallmark of this series is this it becomes liberated from the first two contingent and necessary series, and has finally become general and abstract. It does not treat of individual entries, any kind of particulars, but seeks laws and rules.[114] In this sense it is the best illustration of the structural method.

To summarize, in Vuillemin's philosophy of algebra, both Fichte's theory of intellectual intuition and Lagrange's algebraic method should be viewed as related, insofar as both make the transition from sensible, particular cases and constructions, to abstracted and a priori considerations on the activity of constructing itself.[115]

iv. Theory of Infinity

The notion of infinity is an important element in Fichte's philosophy of mathematics in the *Erlanger Logik*. As we saw, in Fichte's theory of the geometry »the extension of space is infinite.«[116] For example, if we picture geometric space in terms of a sphere or even a circle, the geometer can *unconditionally* enlarge, contract or infinitely divide this space. Indeed, it is precisely this geometric process of contracting and expanding that gives rise to the dimensions of space (see Element no. 9 below). However, in geometric operations concerning the real world there are necessarily restrictions on this space so that the determination of a location can be effected.[117] Philosophically, the view of the human I as being both infinite and finite may be said to form the core of the Fichtean theory of striving,

113 Ibid., p. 112.

114 Ibid., p. 114.

115 »L'affirmation de l'intuition intellectuelle chez Fichte n'est autre chose que le passage d'une méthode génétique limitée et naïve à une méthode critique et réfléchie. Il en va de même pour Lagrange quand il superpose les rudiments d'une Théorie des groupes à la recherche indirecte ou directe des solutions des équations algébriques« Jules Vuillemin, *La philosophie de l'algèbre*, p. 116.

116 *Erlanger Logik* (GA II/9: 134).

117 Ibid.

a theory particularly articulated in the *Grundlage der gesammten Wissenschaftslehre* of 1794/95.[118]

With regard to Fichte's theory of infinity in general, I will make a distinction between Fichte's conception of the Infinite as such, and his theory of mathematical infinity.[119] Beginning with the latter, Fichte's concept of mathematical infinity is that of a rule, procedure or task that we can carry out *ad infinitum*. There is no limit to this procedure, and if the process were to terminate or cease, it would for Fichte signify a *finite* or determinate process, and hence one would have therefore ceased dealing with a genuine form of mathematical *infinity*. This is clear from a mathematical dispute that Fichte was involved in from his time in Jena – the dispute over the so-called infinite polygon, more commonly known as the squaring (quadrature) of the circle.[120] It was sparked by Fichte's »Question for the Mathematician« footnote in *Über den Begriff der Wissenschaftslehre* (1794),[121] mentioned above, and concerned the problem of squaring a circle using inscribed or circumscribed polygons with an infinite number of sides. Fichte considered the quadrature of the circle to be impossible. His argument hinges on the heterogeneous nature of the circle and the straight line, in which only straight lines are genuine lines (i.e. as opposed to the line of the circle) because only the concept of a straight line implies the notion of straightness and direction. The element of direction is absent from a circle. Hence the circle, curve or curved line (*krumme Linie*), cannot be classified as lines for Fichte, but only as a continuous stringing together of points that are infinitely close to each other. The original circle is none other than the boundary line of infinite space (*Grenzlinie des*

118 Cf. Fichte, GWL (GA I/2: 394)

119 For a more detailed presentation of this topic, see my article: »Fichte's Conception of Infinity in the *Bestimmung des Menschen*« (forthcoming in the proceedings of the tenth biennial Conference of the North American Fichte Society, held in Lisbon, Portugal, April 2010).

120 For an extensive overview of this dispute, see Paul Ziche, *Mathematische und naturwissenschaftliche Modelle in der Philosophie Schellings und Hegels*. (Stuttgart-Bad Cannstatt: Frommann-Holzboog, 1996), pp. 40–132.

121 Fichte, BWL, GA I/2: 129–139 footnote, p. 135, in section §5. (Cf. Fichte, *EPW*, pp. 120–121).

unendlichen Raums), whereas the finite circle is generated by the synthetic activity of our imagination.[122]

What does this mean for the squaring of the circle by means of a polygon with an infinite number of sides? Fichte is absolutely clear: *squaring the circle is impossible* due to the above heterogeneous natures of a line and a circle, and the unrealizability of a completed infinity. No matter how small the line or side of the polygon becomes in relation to the curved line (circumference) of the circle, the former will always possess something the latter can never have: *direction*. Therefore the complete identification of the straight line with the circumference can never be completed. Fichte elaborates on this in a small text of 1796 entitled: »Was ist von einem UnendlichEk zu halten?» (What are we to make of the infinite polygon?).[123] – Here the task or procedure for dividing the sides of the polygon *ad infinitum* is actually the task of infinitely determining or *measuring*. If you are continually dividing the sides of the polygon, you will never be in the position of actually measuring, since one has to stop to do this. If you stop, however, your polygon will be something determinate or finite; in other words, you will have a polygon with a measurable, finite number of sides. For Fichte, it is therefore a misnomer to call this geometrical object an *infinite* polygon, and he criticized the notion of an infinite polygon for containing »internal contradictions« (those of a completed infinity and directionality in the circumference of the circle). Fichte's own concept of mathematical infinity is that of a procedural rule or task that can never be completed: »What are we to make of the infinite polygon? – It is an internal contradiction. Because infinite simply signifies the progressive rule of proceeding that never, however, terminates. [...] The infinite polygon already contains an inherent *continuity* of direction, and this cannot be done away with. What lies then in the circumference [of the circle]? There is no direction there. [...] However small the quadrature

122 Ibid. See the similar presentation in Fichte's *Practische Philosophie*: »A curved line [circle] and a straight line are wholly *heterogeneous*, and merely the arbitrary procedure of limitation through points gives them something in common. – It appears: however, that every straight line is infinite; and every curved line is finite. [...] The round line [*runde Linie*] arises through a movement at the boundary of infinite space. The I, the *point* moves itself, and never as *it* wants, but always as it does not want. [...] Actually all lines are straight lines.« (GA II/3: 217, 244, 249).

123 See GA II/5: 5.

you use, it still always contains direction: and it belongs to the nature of the circle that it should never have any [direction]. We can never get away from the original synthesis. It is absurd to say: the infinite is. It ›should‹ merely be: it is a task. The whole [problem] boils down to this.«[124]

This small text formed the basis for Fichte's similar comments in the *Grundlage des Naturrechts* (1795/6)[125], where he said that any mathematician who believed he had *actually* arrived at a polygon with an infinite number of sides only possessed an »empty concept of the infinite.«[126] – For such a mathematician had confused the *determinate procedure* for describing the infinite with the infinite itself.[127] Fichte's philosophical inkling or intuition of an inherent contradiction between the straight line and the circumference of a circle, and hence the impossibility of ever completing the measurement of an infinite polygon (of squaring the circle), turned out to be correct. The mathematician Ferdinand Lindemann demonstrated this mathematically in 1882 by showing that the number π is transcendental.[128] π is of course the ratio of the diameter (straight line) of a circle to its circumference. This ratio never terminates; in other words, it is an *infinite* process or procedure, and therefore the classic quadrature of the circle (using a straightedge and compass) is impossible to complete in finite space and time.

In his theory of infinity, Fichte makes a further clear distinction between infinite processes that we find in mathematical progressions or infinite division, and the Infinite (*das Unendlichkeit*) as such. In terms of human knowledge, Fichte explains that each step of our comprehension is

124 Ibid.

125 Cf. Fichte, *Foundations of Natural Right*: »What then is the concept of something that is infinite? Is it the concept *of a task* of dividing the side of the polygon to infinity, and hence the task of *infinite determining*?« (*FNR*, p. 8; GA I/3: 317).

126 As noted above, Fichte's remarks are primarily directed at Jakob Sigismund Beck (cf. Ibid., p. 8). See Günter Buhl, »Über eine Bemerkung Fichtes zur Quadratur des Kreises« *Archiv für Geschichte der Philosophie* 59 (1977): 281–288.

127 »[Y]our polygon is finite and not infinite, as you profess. But because you can comprehend the procedure for describing something that is infinite (i.e. because you can comprehend the empty concept of the infinite) and can label it, for example, with an A, you are no longer concerned about whether you have really acted and can act in this way« (*FNR*, p. 8).

128 Cf. F. Lindemann: »Ueber die Zahl π« in: *Mathematische Annalen* 20 (1882): 213–225.

discursive or conceptual and therefore a *finite* step or a single element of knowledge. We may extend these finite measurable steps as far as we wish, yet we will never reach the end of this infinite process, i.e. never reach the Infinite itself, because the latter cannot be characterized by anything measurable. This distinction is clearly articulated, for example, in the *Bestimmung des Menschen* (1800): »What I comprehend becomes through my mere comprehending something finite; and though infinitely increased and heightened it can never become transformed into the Infinite. You [the Infinite Will] are not different from the finite according to degree [*Grade*], but according to type [*Art*]. Through the above increase they only make you into an ever larger and larger human being; however, never into God, into the Infinite, which is not capable of any measurement. – I only have this discursive progressive consciousness, and cannot conceive of any other.«[129]

It should be clear from this passage that the Hegelian charge of Fichte only holding a conception of ›*schlecht*‹ or ›bad‹ infinity does not stand up to scrutiny.[130] Fichte does not make any sort of ›qualitative‹ distinction about infinity here, or consider the *in*-finite as simply the opposite of the finite. Instead, he makes a fundamental distinction between degree and species or *type*. To the extent that they are unlimited and can never be completed, Fichte's examples of mathematical infinity – such as an infinite approximation or infinite divisibility – are indeed examples of Hegel's bad infinity. Crucially, however, Fichte sees these examples of infinity as having their foundation and ground in *the Infinite as such*. Terms such as measurement, qualitative or quantitative, simply do not apply to the Infinite, for it is beyond all measurement and only has its ground in itself.[131] In

129 (GA I/6: 297).

130 For analyses of Hegel's accusation of Fichte holding a conception of ›schlecht‹ infinity, see Paul Ziche *Mathematische und naturwissenschaftliche Modelle in der Philosophie Schellings und Hegels* (1996), pp. 53–82; and Wayne Martin, »In Defense of Bad Infinity. A Fichtean Response to Hegel's *Differenzschrift*«, *Bulletin of the Hegel Society of Great Britain* 55/56 (2007): 168–187. For an account of Hegel's theory of infinity in general, see Terry Pinkard: »Hegel's Philosophy of Mathematics« *Philosophy and Phenomenological Research* 41 (1981): 452–464.

131 Fichte distinction here recalls Spinoza's different types of infinity in his famous letter to Meyer from April 1663 (Letter 12). Spinoza too argued for an infinite »that must be infinite by its nature or by virtue of its definition« and an infinity

the *Erlanger Logik* the determination of all finite measurable lines can only be understood if we set them against an invariable and fixed line that can be the standard for all measurement – the so-called »Entfernungs-Maas« (distance-measurement) or infinitely long *Ur-perpendikular* (archetypal perpendicular).[132] In the following section we shall see that Fichte's theory of infinity directly applies to certain fundamental elements of geometry, such as the original line, plane and space.

v. Archetypal Elements (Ur-Elemente)

In this excursus on the philosophy of mathematics in the *Erlanger Logik*, Fichte puts forward his most explicit presentation of his theory of an original (*ursprüngliche*) geometry. Fichte gives his model this name because it is based on a number of so-called original (*ursprüngliche*) or archetypal (*Ur*)-elements (the terms are synonymous for Fichte). Fichte's model of geometry forms the focus of the next chapter. However, here we can inquire into the specific nature of these archetypal elements.

What is the difference between these archetypal elements and the other »ordinary« elements of geometry? The elements of »ordinary« (*gewöhnliche*) geometry for Fichte are »finite«, determinate, sensible and derived, insofar as these elements are »external« to the geometer and can be drawn in finite space.[133] These Ur-elements, on the other hand, are the ideal foundation from which the finite elements are derived.

How does Fichte describe these archetypal or original elements in the *Erlanger Logik*? Three archetypal elements are especially singled out: the archetypal point or »absolute« point[134], the archetypal diameter (»der ursprüngliche Diameter«)[135] and the archetypal perpendicular (»ursprünglicher Perpendikel«).[136] All three are »*absolute*«, and because they have to be grasped in intelligible or geometric intuition, they must be es-

that is »unlimited not by virtue of its essence but by virtue of its cause.« See Baruch Spinoza, *The Ethics and Selected Letters*, ed. S. Shirley (Indianapolis, IN: Hackett, 1982), pp. 231–235.

132 Cf. Fichte, *Erlanger Logik* (GA II/9: 130 ff.).

133 See Fichte's description of »ordinary« geometry, ibid., pp. 130f.

134 Ibid. p. 129.

135 Ibid., p. 130.

136 Ibid.

sentially considered as ideal entities.[137] – As Fichte says, his model of original geometry is that of a »geometry posited through pure reason.«[138]

More specifically, the archetypal point has a limiting function within the system of geometry and is therefore termed by Fichte the »absolute limit-point« (*Grenzpunkt*) beyond which we cannot go.[139] Here it is the meeting or centre point of the Ur-diameter and Ur-perpendicular. In this sense it is absolutely invariable, fixed, and every other line or point has a certain distance from this ideal point. Although it is »omnipresent in infinity«, it is grasped in intelligible intuition, and must be situated »within« every person; hence Fichte's assertion: that I myself am this centre or »Mittelpunkt.«[140]

Extending outwards from this archetypal point are two infinite archetypal lines. Firstly, the Ur-Diameter.[141] As the name suggests, it is the diameter of a circle, and in this case – of an infinite circle or infinite sphere. In terms of picturing this diameter, in the *Erlanger Logik* Fichte sets it as a right-angle to the archetypal perpendicular, hence we could picture it as an infinite horizontal line. – This is evident when Fichte says that the philosopher's left and right are the same as the endpoints of the original diameter.[142] It functions as a regulator of all direction[143], that is to say, it is a »normal line« providing a fixed or standard direction (*NormalRichtung*). As such it is »immutably determined«, and in this vein Fichte also calls it the »principle of judgment« for all direction.[144] Here we reach the heart of Fichte's theory of geometrical archetypal elements, since for him the direction or directionality is primary and absolute. The role of the archetypal direction is to provide an absolute standard – or »absolute direc-

137 Ibid., pp. 129, 132–133.

138 Ibid., p. 132. As further evidence of their ideality, Fichte says here that his ideal or »original circle« is one that can only be constructed »through reason.« Ibid., p. 134.

139 This passage on the archetypal point is based on Fichte's description in ibid., pp. 129–130.

140 Ibid.

141 For the archetypal diameter, see Fichte's remarks especially in ibid., pp. 130–134.

142 Ibid., p 135.

143 *das Regulativ aller Richtung*, ibid., p. 130.

144 *Beurtheilungsprincip*, ibid., 130.

tion« – and like the archetypal point it is intuited and everyone brings it with them.

The other archetypal line is the archetypal perpendicular. It may be conceived as an absolute and infinite vertical line extending from the horizontal diameter, and to distinguish it from the former Fichte also calls it the »absolute radius«.[145] If the archetypal diameter is the regulator of direction, the archetypal perpendicular for Fichte is the regulator of location. At this point we see the entry of measurement or the notion of a metrical standard into Fichte's model. In this respect Fichte calls this perpendicular the original line of distance (*Entfernungslinie*), since it functions as the measurement of distance (*Entfernungsmaas*) from the internal ideal diameter, and constitutes a »medium towards the outside.«[146] In other words, it forms the transition to the metrical notion or measurement of space.

Fichte requires these archetypal elements in order to furnish the foundation or normal standards against which all direction and location in ordinary metrical geometry can be measured. According to Fichte, this is the only way to stop circularity in traditional forms of geometry; it provides immutable or invariable fixed points and lines, making geometry into a purely deductive science. If we take these three archetypal elements together as a whole, we arrive at Fichte's original fact or »true principle«[147] of geometry: that is, the archetypal diameter and perpendicular together with the archetypal point at their centre, which can be pictured as the »original right-angle.«[148]

Though most radically and fully expressed in the *Erlanger Logik*, Fichte's theory of original or archetypal elements can already be found in his first writings on the *Wissenschaftslehre*. For example, in *Über den Begriff der Wissenschaftslehre*, we already see Fichte's theory of an archetypal centre-point that can be inwardly intuited – with the I or self taken to be such a centre – from which an infinite number of radii can be drawn, and whose endpoints lie at infinity.[149]

145 Ibid., p. 134.
146 Ibid., p. 130, 134.
147 Ibid., p. 132.
148 Ibid., p. 131.
149 See Fichte, BWL (GA I/2: 129f., 136, in the two footnotes).

vi. Transcendentalism

Fichte's philosophy of mathematics is a *transcendental* science. Because geometry only has »factual evidence«, it needs a philosophy to explain its possibility and conditions. In this sense Fichte's philosophy of mathematics is grounded in his original transcendental enterprise to provide more robust and systematic foundations for Kant's critical philosophy. If the »purpose« of geometry is: »to exhaust the possibility of limiting space within a single universal space,« then the aim of his philosophy of mathematics is to rigorously demonstrate this: »Presupposition: that this possibility has its limits and laws.«[150] As we saw above, the role of the philosopher is to provide the »justifications« for what the geometer does, because the latter works in his profession without paying strict attention to the possibility of his science. In contrast to the philosopher, the geometer works »without justification, and without wondering about it: [...] Question: does the *Wissenschaftslehre* explain this? Certainly.«[151] This view that the philosopher of the *Wissenschaftslehre* is alone capable of explaining the conditions and possibility of geometry is already found in the *Zurich Wissenschaftslehre* of 1794, which states that the »possibility of all [geometric] construction« is grounded in the *Wissenschaftslehre*.[152] Moreover, the *method* of the philosopher of mathematics must also be transcendental with regard to the science of physics, since it supplies the proof of the possibility of the standpoints of both the *physicist* and the geometer: »Only the philosopher can reconcile the two, because he inwardly (not just in appearance) draws the boundaries between both by proving the possibility of the two standpoints.«[153]

Hence, the Fichtean philosophy of mathematics is to *mediate* between geometry and the other sciences, especially between geometry and physics. One of its aims in the *Erlanger Logik* is to establish a link or connection between them, and analyze their common attempts to determine the nature of space: »Does geometry share something in common with physics? – The difference is that there are *forces of nature* in physics. Whereas in geometry, there is one's own *freedom*, which sets down sta-

150 Fichte, *Erlanger Logik* (GA II/9: 125).
151 Ibid.
152 Fichte, *Zurich Wissenschaftslehre* (GA IV/3: 34).
153 Fichte, *Erlanger Logik* (GA II/9: 126).

tionary spaces within universal space. [...] Here, therefore, philosophy provides the connection between these sciences. This is the point where philosophy mediates between the two.«[154] Consequently, the philosophy of mathematics makes a distinction between geometric space and real (physical) space. Geometric space for Fichte is the abstract space that geometers deal with, and has the quality of being »stationary« or »non-extending«, i.e. it is limited and has to hold itself together.[155] *Real* space, on the other hand, is the space of the physical and material world, whose extensions and qualities are investigated by physicists. Hence, the science of physics also requires a philosophy that reflects upon its goals and limits and which Fichte logically calls a »philosophy of physics«. Naturally, the philosophy of mathematics and the philosophy of physics have to be sharply distinguished. The former is concerned with the foundations of geometry and abstract geometric space, whereas the latter is concerned with the foundations of physical space and its natural forces.[156]

vii. The Parallel Problem

The Fichtean philosophy of mathematics also tries to shed light on and provide suggestions for solving one of the most famous problems of geometry – *the so-called parallel problem* – which Fichte castigates as a »dubious blemish on geometry«,[157] that is, the problem of supplying a self-evident foundation for parallel lines. Euclid's fifth postulate or parallel postulate states that there is only one line running parallel to a point outside of it in a two-dimensional plane. Generations of geometers tried to show that Euclid's postulate was not a self-evident indemonstrable postulate at all, and should be treated like a theorem that has to be proved. In trying to prove this »theorem« geometers eventually discovered the existence of other geometries different to Euclid's geometry, so-called non-Euclidean geometries.[158] In the *Erlanger Logik* Fichte rejected certain suggestions for

154 Ibid., p. 128.

155 Ibid., p. 133f.

156 Regarding Fichte's proposal for a philosophy of physics, see the *Erlanger Logik* (GA II/9: 136ff.).

157 Ibid., p. 131.

158 See L. Magnani, *Philosophy and Geometry: Theoretical and Historical Issues* (2001) pp. 67–70.

resolving this issue by other geometers at the time, including Johann Schulz, whose attempt he dismisses as »improbable« and ultimately circular.[159] Fichte put forward his own tentative solution of this problem. Fichte's alternative will be discussed in the next section, suffice to say that he seems to consider a resolution of the parallel problem to be a priority of a genuine philosophy of mathematics, since he believed »we will never arrive at the correct result« if we remain at the current state of the problem.[160]

viii. Dimensions of Space

As mentioned above in our citation from Fichte's letter to Schelling (15th November 1800), one of the key themes that Fichte believed should be in the philosophy of mathematics was the *deduction of the three dimensions* of space.[161] In this letter to Schelling Fichte gave a definition of his theory of original space and argued that it does not possess any dimensions: »Original space, or space as intuition, does not have any dimensions. It is everywhere, small or large, sphere (*Kugel*).«[162] And Fichte concluded from this that the three dimensions must therefore be »universal forms of thought« that are generated by »abstraction«[163], a view he would repeat in the new version of the *Wissenschaftslehre* that he was writing at this time.[164] Fichte also claimed that Schelling's *Naturphilosophie* had to as-

159 Cf. Fichte, *Erlanger Logik* (GA II/9: 131).

160 Ibid.

161 »The deduction of the three dimensions is not at all incumbent upon the pure *Wissenschaftslehre*, but to begin with rather upon the *philosophy of mathematics*.« Fichte, letter to Schelling, November 15, 1800 (GA III/4: 361). Cf. Paul Ziche's article on the debate on the theory of dimensions between Fichte and Schelling in: »Raumkonstruktion, Deduktion der Dimensionen und idealistische Prinzipientheorie. Problemlagen im Fichte-Schelling-Briefwechsel vom November 1800« *Fichte-Studien* 25 (2005), 21–42.

162 GA III/4: 361.

163 »The three dimensions arise through abstract thinking in space: and are nothing more than *the universal forms of thought* itself.« Fichte, letter to Schelling, November 15, 1800 (GA III/4: 361).

164 Towards the very end of the *Neue Bearbeitung der Wissenschaftslehre* (1800), Fichte repeats this view of the generation of the dimensions through abstract thinking: »Space originally does not at all have three dimensions in intuition. They are only in the concept. The three dimensions arise, just like the mathemati-

sume such a deduction in the philosophy of mathematics, to which Schelling disagreed and claimed primacy for his *Naturphilosophie*. Fichte's and Schelling's views on the status and priorities of their respective philosophies soon become diametrically opposed, and in early 1802 lead to a rupture in their friendship and intellectual collaboration.[165] For Schelling the discipline of the philosophy of mathematics is just a pure abstraction of the fundamental geometric forms existing in nature, and hence the task of determining these forms can be best carried out by *Naturphilosophie*: »The line [...], is the schema of arithmetic, whose series also only has this one dimension; the plane is the schema of geometry, and so on. But line, plane and solid only *originally* arise precisely in *Naturphilosophie*, and come first into the philosophy of mathematics by means of abstraction. Hence, *Naturphilosophie* does not presuppose them out of the latter.«[166] In fact, Schelling's brief thoughts on the dimensions (of which Fichte was more curious to know) here in his letter belong to a more sophisticated theory in which there exist a specific set of correspondences between the dimensions of space and the human faculties.[167] Fichte's own ideas on the dimensions remain less developed than

cal point, through thinking. Space, as it actually is, is without doubt incomprehensible, for in it there is something thoroughly infinite.« (GA II/5: 401).

165 The priority dispute of their respective philosophies came to head in 1800–1801 and can be followed in their correspondence and works from this period. Naturally, Fichte believed that in terms of the starting point of philosophy the *Wissenschaftslehre*'s starting point in the intuited activity of the I was preferable to Schelling's *Naturphilosophie* with its starting point in nature. In the *Grundlage der gesammten Wissenschaftslehre* Fichte had already stated the idealistic beginning of his philosophy: »We do not elevate our reason through any kind of natural law or any kind of consequence out of a natural law, but through absolute freedom; i.e. not through a *transition* (*Übergang*), but through a *leap* (*Sprung*). – For this reason in philosophy we must necessarily proceed from the I, because the latter cannot be deduced [from natural processes]; and hence the undertaking of the materialists to define the expressions of reason though natural laws is forever unfeasible.« (GA I/2: 427). In 1800–1801 Fichte was then upset to see Schelling proclaiming a »natural« starting for transcendental philosophy, which he thought should be classified as an »objective idealism« in contrast with Fichte's supposed »subjective idealism« (see especially Schelling's *Darstellung meines System der Philosophie* (Hamburg: Felix Meiner Verlag, 2001), p. 331.

166 F.W.J. Schelling to J.G. Fichte, letter November 19, 1800 (GA III/4: 367).

167 Csech provides a fine outline of Schelling's theory in his *Raumlehre*, pp. 243ff.

Schelling's, but may still be discerned in the *Erlanger Logik*, particularly in the passages on the infinitude of space. This notion of dimension is implicit in the transition from the point to the geometric plane: »From here onwards geometry proceeds in a regular form, from the simple to the composite: point and line. Line to plane, and so on, from the universal laws of the figure to the particular laws.«[168] However, it is unfortunate that Fichte does not elaborate more on the dimensions of space here in 1805, but this is generally the case for the topic of dimensions in his principal lectures and writings.[169] The theory of the dimensions of space clearly held a fascination for Fichte, and his interest in it can be traced back to some of his earliest texts on the Kantian philosophy, where he ponders their relationship to the subject and time.[170] In this regard it is not unusual to see that Fichte had already investigated this subject in 1793/94 in the *Eigne Meditationen* during another examination of time and space. On the one hand, he states his view that time can only have one dimension (like a line), and is »the making objective of a successive subjective ground.«[171] Space, on the other, can be extended in all dimensions. However, both space and time arise for Fichte through a process of limiting and reducing of what is unlimited: »Time is only extended in a *single* dimension; space in all dimensions (thus, the antithesis consists in the unity and totality.«[172] Finally, in 1810 Fichte again touches on the subject of the di-

168 GA II/9: 135. See too the passage: »The extension of space is infinite: this means only geometry can unconditionally enlarge this circle or semi-circle, and must be able to do so (unconditionally enlarge); however, in every genuine operation it necessarily has a restricted space, or all determinations of the location would be at an end.« (GA II/9: 134).

169 See, among others, Fichte's brief discussion of Bardili's view of the three dimensions in *Antwortsschreiben an Herrn Professor Reinhold* (1801) (GA I/7: 299, 305, 312). Interestingly, this discussion is from around the time of Fichte's November 1800 letter to Schelling.

170 For example, see Fichte's very early text from 1792/93: *Bei Lesung der K. Kr.d.r. Vft.*, where he says: »That time has one dimension and space three allows us to conduct some extremely interesting studies: the first refers to the simplicity of the subject and the second to the manifoldness of the 3rd substance according to trigonometry« (GA II/2: 328).

171 GA II/3: 123.

172 Ibid., p. 73. Fichte continues here: »If you limit the unlimitedness on all sides, as on one side, then you still have an unlimited extension of one dimension, time. –

mensions in the *Thatsachen des Bewußtseyns* and links it with the zero-dimensionality of the point and the one-dimensionality of the line. Fichte makes it clear that to fathom the ground of the three dimensions requires passing from the conceptual region of definitions based on concepts to that of *intuition*.[173] We will see in Chapter Three that this is a philosophical move which also holds for his conception of axioms and mathematical cognition in general.

ix. Language of Mathematics

Finally, in the *Erlanger Logik* Fichte was of the opinion that the correct »language of geometry« should be supplied by a philosophy of mathematics. Thus, even with regard to the linguistic formulation of its subject geometry alone cannot understand its assumptions and foundations, but requires help from a higher science. It is Fichte's view that the language of geometry owes its creation and formal perfection to philosophy: »[G]eometry proceeds in a regular form: from the simple to the composite: point and line. Line to plane, and so on, from the universal laws of the figure to the particular: and in every one, e.g. in the triangle again. This formal perfection that exists nowhere else was its triumph. Everyone could do it; there only lacked, among others, a language: but it belongs to philosophy to create its language. – Only philosophy first brings about understanding.«[174]

If you limit the unlimitable as such, then you have space: a defined extension in 3 dimensions. Thus, unity and totality have to be assumed here.« (Ibid.).

173 »People inquire into the ground (*Grund*) of the three dimensions of space. First and foremost we only need the correct conception of dimension, which, once we have provided the ground, will be discovered of itself. You only have to know where to seek this ground; not at all in the region of concepts, but only in the region of intuition: for this is mere intuition, and what is given is a limitation of intuition. To inquire about the ground of the three dimensions does not mean anything else than to say: place me at that standpoint which necessarily gives rise to this intuition.« Fichte, *Thatsachen des Bewußtseyns* (GA II/12: 60).

174 GA II/9: 135. See here Widmann's reflections on the Leibniz-inspired future linguistic system of notation or »symbolic signs« that Fichte mentions at the conclusion of the *Ankündigung* (GA I/7: 164), and was planning for his philosophy but never managed to accomplish: »Exact Concepts: Fichte's Contribution on a Problem of Tomorrow« *Idealistic Studies* 11 (1981), pp. 41–47.

In conclusion, if Fichte's philosophy of mathematics is supposed to furnish a fully-fleshed out theory on the above nine elements or topics, as well as views on number theory and the nature and ontology of infinitesimal calculus for example, then Fichte's *Erlanger Logik* obviously does not provide such a theory. Notwithstanding, it ought to be clear that Fichte still had a philosophical understanding and interpretation of a large number of mathematical issues, which he attempted to articulate at various times throughout his career. The most important of these Fichtean interpretations will be examined in the next section on his model of geometry and in Chapter Four below. There I will show that Fichte not only employed the conventional geometric explanations for the point, line and plane, but also had original and interesting philosophical interpretations of them, which in the truest sense constitute an integral part of his philosophy of mathematics.

Although this excursus on the philosophy of mathematics in the *Erlanger Logik* is only brief and comprises no more than 14 pages, it remains unique in Fichte's oeuvre, since it not only contains sustained reflections on the theoretical underpinnings of mathematics, but also puts forward his model of an original or archetypal geometry. We will now turn to this Fichtean model of original geometry and compare it with both the Kantian and Euclidean models.

Chapter Two: Fichte's Theory of »Original« Geometry

1. »Ordinary« and »Original« Geometry

a. Synthetic and Analytic in Geometry

In the *Critique of Pure Reason* Kant applies the term synthetic to the propositions of geometry to explain that we must »go beyond« the concepts of the propositions and »seek assistance« in their corresponding intuitions.[1] One is so accustomed in transcendental philosophy to the Kantian definition of the term »synthetic« that it is easy to forget that it already had a long and distinguished tradition in the history of geometry well prior to Kant. Thus, before discussing the Kantian and Fichtean conceptions of geometry it is worth briefly summarizing the earlier, traditional meaning of the terms analytic and synthetic in the history of geometry and mathematics.

Historically speaking, the distinction between the two terms can be traced back to Greek philosophy and science as far back as the school of Alexandria with Euclid at its head. In the history of geometry the primary meaning of synthetic and analytic did not refer so much to the content as to the method and manner of reasoning. Analysis is based on a prior speculative assumption and the reasoning is continued until the truth is obtained; synthesis starts with a given and infers until it is proved. But both methods seek to arrive at the same result and proof, and it is frequently the case that their meanings are understood antithetically, i.e. by placing one in opposition with the other: »The terms synthesis and analysis are used in mathematics in a more special sense than in logic. In ancient mathematics they had a different meaning from what they now have. The oldest definition of mathematical analysis as opposed to syn-

1 Kant, *Critique of Pure Reason* (B15), p. 144. See too Kant's comments in A46/B64-A49/B66 and A716/B744-A727/B755, pp. 170f., 631–637.

thesis is that given in Euclid XIII.5, which in all probability was framed by Eudoxus: ›Analysis is the obtaining of the thing sought by assuming it and so reasoning up to an admitted truth; synthesis is the obtaining of the thing sought by reasoning up to the inference and proof of it.' The analytic is not conclusive unless all operations involved in it are known to be reversible. To remove all doubt, the Greeks, as a rule, added to the analytic process a synthetic one, consisting of a reversion of all operations in the analysis. Thus the aim of analysis was to aid in the discovery of synthetic proofs or solutions.«[2]

Around 1600 the meanings of the terms changed slightly with the popularization and integration of algebraic methods in geometry. Any geometry departing from the traditional Euclidean model of proof, or which *especially employed algebra, co-ordinates and calculus* to solve geometric problems, became designated as »analytic« geometry. Attempts were made by certain prominent geometers, including Girard Desargues and Blaise Pascal, to forgo the analytic trend, and to only employ »pure« synthetic methods. – However, René Descartes exposition of geometry that includes co-ordinates and algebraic equations soon gained in popularity and became the prime exemplar of analytic geometry at the time. It aimed to solve geometrical problems by first translating them into algebraic equations to see if they could not be resolved better that way.[3]

2 Florian Cajori. *A History of Mathematics* (New York: Macmillan, 1919; reprint: Muschamp Press, 2007), pp. 30f. Cajori devotes a chapter to synthetic geometry (pp. 293–307), and defines synthetic method as: »the continual direct viewing of figures as existing in space.« (Ibid., p. 307). According to Carl Boyer, Euclid's *Elements* is the Greek culmination of the synthetic method: »The work of Euclid represents the final synthetic form of all mathematical thought – the elaboration of deductive reasoning of the logical implications of a set of premises.« *The History of Calculus and Its Conceptual Development* (New York: Dover Publications, 1959) p. 48.

3 The French geometer Girard Desargues (1591–1661) was a friend of René Descartes (1596–1650). The former developed his own brand of pure synthetic geometry, but the latter's co-ordinate geometry eventually gained more followers, and Desargues' methods fell into oblivion, before being rediscovered in the nineteenth century. Desargues's geometry had the support of Blaise Pascal who managed to produce some exciting work at a very young age including his famous »Mystic Hexagon« or, Pascal's Theorem, which has fascinated generations of ge-

Hence, when one traditionally spoke of *synthetic geometry* and synthetic methods in geometry it above all referred to the method of reasoning and proof. Such reasoning and proof champions the systematic *exclusion* of algebra, co-ordinates and algebraic equations. Synthetic geometry, therefore, aims to restrict itself to the use of so-called »pure« methods, that is to say, the traditional Euclidean method of the construction of figures and the deduction and solving of problems entirely without recourse to algebra and calculus.[4] Any theorem that can be demonstrated by means of the two traditional instruments (the straightedge and compass) is said to have been »constructed« in a valid synthetic manner; whereas if the construction or problem requires measurement or algebraic intervention, the demonstration is deemed to have been carried out in an analytic manner. And this is the general meaning that these two terms have retained in the field of geometry today: »The difference between synthetic and analytic geometry resides in the language, in the models and in the constructions. Analytic geometry is a geometry whose language is based on the algebraic constructions – vector spaces over vector fields, addition, multiplication etc. With this language go algebraic models and morphisms that preserve this algebra [...] Synthetic geometry is a geometry whose language is based on geometric constructions – intersecting lines, points on planes, equal distances etc. With the geometric objects go morphisms which preserve these constructions – collineations, congruences etc.«[5]

Another significant aspect, especially for Fichte's theory on account of its imaginative aspect, is the retention and primacy in synthetic and Euclidean geometry of *actual figures and geometric shapes*. The latter tend only to play a secondary role in analytic geometry, due to its focus on the corresponding algebraic equations of these figures. Despite the clear dominance of analytic methods in geometry and the obvious power of differential calculus, the pure synthetic approach to geometry continued to be developed in the history of geometry and reached its height in Germany

ometers due to its simplicity and beauty. See Florian Cajori. *A History of Mathematics*, pp. 174–178.

4 In fact, *synthetic* Euclidean geometry restricts itself to construction with a straightedge (a ruler without markings) and compass. Shabel gives a good overview of these constructions in *Mathematics in Kant's Critical Philosophy*, pp. 9–20.

5 D. Kueker and Carl Smith, *Learning and Geometry* (Boston: Birkhäuser, 1995), pp. 122f.

and France in the early 19th century, with what is perhaps the purest version of synthetic geometry – *projective* geometry.[6] Having its distant origins in the use of perspective in the arts of painting and drawing, projective geometry is defined as: »that branch of geometrical science which deals with those properties of figures which are unaltered by radical projection from plane to plane or space to space no matter what the number of dimensions involved.«[7] One of the main features of this geometry is its reliance on geometrical intuition and the relations of incidence, i.e. the intersection of entities, such as lines and planes, and even the total absence of measurement and distance, including that of angles and areas. – The role of geometrical intuition comes especially to the fore in the works of the Swiss Jacob Steiner (1796–1863) and the German Karl Georg Christian von Staudt (1798–1867).[8] The search for a metric-free system of geometry, i.e. one not dependent on measurement, can be found in von Staudt's works *Geometrie der Lage* (1847) and *Beiträge zur Geometrie der Lage* (1856–60). Von Staudt wrote his masterpieces while a professor at the University of Erlangen, where his chief aim was to provide a geometry not based on metrical elements but solely on the principle of incidence.[9] In von Staudt's own words concerning the metric-free nature of the *Geometrie der Lage* or the ›geometry of position‹: »I have attempted in this writing to make the geometry of position into an independent science, which does not require measuring.«[10] Not long after, another geometer in Erlangen, Felix Klein, in his celebrated *Erlanger Programm* of 1872, demonstrated that projective geometry is the most elementary geometry of all the geometries. It was more elementary because it contained *fewer* axioms, and

6 See J.L. Coolidge, »The Rise and Fall of Projective Geometry«, *American Mathematical Monthly* 41 (1934): 217–228.

7 Ibid., p. 217.

8 Cf. von Staudt introduction: »Every geometrical instruction has to proceed from general considerations that acquaint the student with different kinds of geometrical configurations and exercise his faculty of intuition.« K.C. von Staudt, *Geometrie der Lage* (Nuremberg: Bauer & Raspe, 1847), p. III. See too Jacob Steiner, *Systematische Entwickelung der Abhängigkeit geometrischer Gestalten von einander* (Berlin: G. Fincke, 1832).

9 As Coolidge summarizes it: »Here we have at last projectivity defined with no relation to distance«, J.L. Coolidge, »The Rise and Fall of Projective Geometry«, p. 223.

10 K.C. von Staudt, *Geometrie der Lage* (Nuremberg: Bauer & Raspe, 1847), p. III.

every other geometry, including Euclidean geometry, could be developed from its foundations.[11]

To summarize: analytic geometry employs algebra and algebraic methods in order to solve its problems and constructions; synthetic geometry, on the other hand, eschews the deployment of related algebraic methods, and retains the primacy of *figures* and visual constructions and only permits the employment of the straightedge and compass. As we shall see in the next section, it is precisely geometry's focus on constructions and actual figures, as opposed to the dominance of algebraic equations, which appealed to Fichte. In conclusion, it is an interesting fact of history that Fichte gave his most detailed presentation on geometry in 1805 precisely at the University of Erlangen, which became in 1847 and 1872 with von Staudt and Felix Klein respectively, the location of some of the most important advances in pure synthetic geometry.

b. Kant's Conception of Geometry

In a certain respect one of the goals of Kant's *Critique of Pure Reason* was to provide a satisfactory answer to the question: »How is pure mathematics possible?»[12] This is because Kant equates this question with the central inquiry of the 1st Critique: »How are synthetic judgments a priori possible?»[13] These kinds of judgments, Kant argues, must exist and be real because of the simple »fact« (*Factum*) that we have such sciences as pure mathematics. (B 128). We are therefore justified in asking how these judgements, and consequently pure mathematics itself, are at all possible. (B 20–21).

Kant puts the question concerning the possibility of pure mathematics on a par with his question concerning the possibility of synthetic a priori judgments, and draws upon two concrete examples from the science of mathematics to illustrate and prove his argument that such judgements exist. One is an example from arithmetic: 7 + 5 = 12; and the other is from geometry: the proposition that a straight line is the shortest distance be-

11 See L. Magnani, *Philosophy and Geometry: Theoretical and Historical Issues (2001)*, pp. 122, 219.

12 Kant, *Critique of Pure Reason*, B20, p. 147.

13 Ibid., B 19, p. 146.

tween two points.[14] This second example of course is a classic principle from Euclidean geometry and Kant declares it to be the paradigm of a synthetic proposition. This is because with this proposition we need to add the predicate »shortest« to the concept, that is to say, an additional element that cannot be acquired from any analysis or decomposition of it: »That the straightest line between two points is the shortest is a synthetic proposition. For my concept of straight contains nothing of quantity, but only of quality. The concept of the shortest is therefore entirely additional to it, and cannot be extracted out of the straight line by any kind of analysis. Help must here be gotten from intuition, by means of which alone the synthesis is possible.«[15]

In fact, Kant will go on to claim that all the propositions of geometry are synthetic. Kant's meaning of the terms synthetic and analytic are obviously not the same as the sense discussed above in analytic and synthetic geometry. Thus, we will examine this difference between his meanings and the application of them in his theory of geometry to help us with our main goal in the next section of presenting Fichte's model of geometry. In order to do this let us turn to Kant's *Prolegomena to Any Future Metaphysics* (1783).

In section §2 »Of the Kind of Cognition that Alone Can be Metaphysical«, Kant explains what he means by the *content* of judgments, stressing that this content may either be analytic or synthetic. An analytic judgment, for Kant, is one that may be termed »explicative«. This is because it merely *explains* and adds nothing to the actual content of the cognition, the content remains the same. A synthetic judgment, on the other hand, may be termed ›ampliative‹; that is, it adds to or *enlarges* the cognition. Kant further explains his theory by using an example from the domain of natural science – that of an extended or heavy body. To say a body is extended does not amplify my concept of body because extension is a quality already thought before the judgment, and is therefore analytic.[16] For

14 Ibid., B 16, pp. 144f.

15 Ibid., B16, p. 145.

16 »Analytic judgments say nothing in the predicate that was not already actually thought in the concept of the subject, though not so clearly and with the same consciousness. If I say: all bodies are extended, I have not amplified my concept of body in the least, but only analysed it. Extension, though not explicitly said of that concept, was already actually thought by it before the judgment. The judg-

Kant, all analytic judgments rest on the principle of contradiction, whereas synthetic judgments require a different principle that must still be in conformity with the principle of contradiction.[17] For instance, judgments of experience are always synthetic, as well as all mathematical judgements and pure metaphysical judgments.[18]

Although most scholarly attention on Kant's conception of geometry is devoted to the *Critique of Pure Reason* or the *Prolegomena*, a number of Kant's ideas on the synthetic nature of geometry are clearly evident in some of his early essays. In contrast, these early writings do not focus on the synthetic nature of judgments but on the synthetic *method*; or more precisely, on the differences in the *methods* of philosophy and geometry (mathematics). According to Kant, the method of mathematics is synthetic and that of philosophy analytic. It was only later in the Introduction to the *Critique of Pure Reason* that Kant turned his attention to the nature of analytic and synthetic *judgements* in the two disciplines. Why does Kant consider the *method* of geometry to be synthetic?

Kant on the Synthetic Method of Mathematics

In the *Inquiry Concerning the Distinctness of the Principles of Natural Theology and Morality, being an Answer to the Question proposed for Consideration by the Berlin Royal Academy of Sciences for the Year 1763 (1764)* Kant compared the methods of cognition in metaphysics and geometry.[19] In section §1 he claimed that the definitions of mathematics are arrived at synthetically because they are not given prior to the definition, whereas in

ment is thus analytic. On the other hand the proposition: some bodies are heavy, contains something in the predicate that is not actually thought in the universal concept of body. It thus enlarges my cognition in that it adds something to my concept, and hence must be called a synthetic judgment.« Kant, *Prolegomena to Any Future Metaphysics*, ed. G. Zöller (Oxford: Oxford University Press, 2004), p. 72 (AA 4: 266f.).

17 Ibid., p. 73. (AA 4: 267ff.).

18 Ibid. pp. 73–76 (AA 4:267–274).

19 *Untersuchung über die Deutlichkeit der Grundsätze der natürlichen Theologie und der Moral. Zur Beantwortung der Frage, welche die Königl. Akademie der Wissenschaften zu Berlin auf das Jahr 1763 aufgegeben hat (1764)* (AA 2: 276–301). An English translation is present in *Kant: Theoretical Philosophy 1755–1770*, pp. 243–275.

philosophy definitions are arrived at analytically. In short, the synthetic process for Kant is: »the arbitrary combination (*willkürliche Verbindung*) of concepts.« – He selects an example from geometry, that of the trapezium, to illustrate his definition of synthesis. The definition of a trapezium is: four arbitrary non-parallel lines bounding a plane surface. According to Kant when we imagine the figure of a trapezium its concept is not given to us prior to the definition of the trapezium, but in fact requires this definition to come into existence, and thus the definition is synthetic.[20] The method of synthesis is proper to geometry and entails that the concept of the definition not be given before the definition. To render the process of synthesis even clearer Kant adopts the common strategy of contrasting the method of synthesis with its opposite method: *analysis*. He asserts that analysis is the method in philosophy of arriving at a general concept by »separating out« what is present in the given cognition.[21] In this essay he considers the concept of the things in philosophy to still be given, »albeit confusedly or in an insufficiently determinate fashion.«[22] Kant explanation hinges on the notion of separating out the »characteristic marks« of the concept, in contrast to mathematics which combines and compares concepts of »magnitude.«[23] Accordingly, the method of synthesis is an advancing progression from a given part to the whole; while analysis is a regression from the whole to the parts, or a »subdivision of a given compound.«[24] Kant forcefully concludes: »What they do not notice at all is the fact that geometers acquire their concepts by means of synthesis, whereas philosophers can only acquire their concepts by means of analysis – and that completely changes the method of thought.«[25]

In the Introduction to the *Critique of Pure Reason* (1787) Kant introduced another aspect to his distinction between synthesis and analysis, for as mentioned above it is now no longer a question of the difference in method but of a distinction between synthetic and analytic *judgements*,

20 Kant, *Inquiry*, in: *Theoretical Philosophy 1755–1770*, p. 248.

21 Ibid.

22 Ibid.

23 See the closing words in section 2 (Ibid., pp. 248ff.).

24 Kant, *Inaugural Dissertation*, in: *Theoretical Philosophy 1755–1770*, p. 378 (footnote).

25 Ibid., p. 262.

with those of geometry considered to be synthetic and those of philosophy analytic. In the section entitled »On the difference between analytic and synthetic judgements« Kant calls a judgment analytic if the predicate belongs to the subject as something *covertly contained in* the concept. This is not the case in a synthetic judgement where the predicate is connected with the concept but belongs *outside* of it.[26] Moreover, in the text from 1790, *Über eine Entdeckung nach der alle neue Kritik der reinen Vernunft durch einen ältere entbehrlich gemacht werden soll*, Kant adds a further nuance to this view and places more emphasis on the respective *functions* of the two judgments. The function of analytic judgments is simply to explicate a given concept or to render clear what is already *implicit* in this concept, whereas the function of synthetic judgments is to *extend* our knowledge beyond the concept in question.[27]

Yet why does Kant consider that all mathematical and geometrical judgments are without exception synthetic? First of all, this is because mathematical propositions are *a priori* on the one hand, that is to say, they bear a necessity with them and cannot be taken from experience and are therefore not empirical.[28] As we saw, Kant argues that the geometrical proposition of the straight line between two points being the shortest is synthetic because the concept of straight is not a magnitude but simply a quality. It therefore requires something to be *added* to it – the concept of the shortest – which cannot be drawn from any analysis of the concept of the straight line. Here we see the importance of Kant's theory of *intuition* (*Anschauung*) in geometry, which alone makes the synthesis possible. An intuition for Kant is an immediate and singular representation of a particular object supplied by the senses.[29] A concept, is contrast, is an abstract and general representation formed by the intellect; a concept is merely a »mark of intuition«, since it is the intuition that relates to the object.[30] As mentioned, mathematical cognition cannot remain with the concept as it

26 Cf. Kant, *Critique of Pure Reason* (A6–7/B10–11), p. 130.

27 Kant, *Über eine Entdeckung* (AA VIII: 228ff.). See Henry Allison's illuminating analysis of these nuances, in: *The Kant-Eberhard Controversy*, pp. 48f.

28 See Kant, *Prolegomena to Any Future Metaphysics*, p. 74 (AA 4: 268).

29 Cf. Kant, *Critique of Pure Reason*, p. 155 (A19-A20/B33-B34).

30 See Kant's notes inserted into his own copy of the 1st Critique translated by Paul Guyer and Allen W. Wood in the Cambridge edition (*Critique of Pure Reason*, p. 155). For a discussion of Kant's theory of intuition see Chapter Three below.

is but requires something additional. – With a proposition of mathematics we have to go beyond the concept to its corresponding intuition, and this is why all the propositions of mathematics for Kant »can and should never originate from analysis of concepts, i.e. analytically, and are therefore without exception synthetic.«[31]

In the Transcendental Aesthetic of the *Critique of Pure Reason* Kant again argues that mathematical propositions involve conceptual synthesis, and this highlights another crucial factor in his synthetic theory of geometry.[32] Kant's conception of mathematical cognition is indivisibly linked to construction, and consequently, mathematical concepts and propositions are *constructible*. Using the example of a triangle, Kant says that in mathematical construction the geometer puts together through pure intuition a manifold that belongs to the schema and concept of the triangle. It is in vain to reflect upon a triangle philosophically (discursively). We need an intuitive use of reason »through the construction of concepts.«[33]

In summary: the propositions of geometry are without fail synthetic for Kant, since on the one hand we have to pass beyond the concept to the corresponding *intuition*, and on the other, these propositions are dependent on construction. Because of these two factors, the principles of geometry and mathematics cannot be analytic. How does this relate to the Fichtean model of geometry?

c. Fichte's Theory of »Original« Geometry

In this section I will examine the Fichtean model of geometry and demonstrate how it departs from both classical Euclidean geometry and the Kantian conception of geometry. I will begin by outlining how the Fichtean model differs from the Euclidean in three significant ways: 1) First of all Fichte suggests an entirely *new starting point* for geometry. Instead of the »ordinary« Euclidean beginning of a line joining two points (Postulate 1 of Book One of the *Elements*), Fichte proposes starting with the »original« (or archetypal) diameter and perpendicular joined together with an archetypal point at their centre. This gives rise to what he calls

31 Ibid., pp. 74f. (AA 4: 269–272).

32 Cf. L. Shabel, »Kant's Philosophy of Mathematics«, in: ed. Paul Guyer, *The Cambridge Companion to Kant and Modern Philosophy*, p. 98.

33 Kant, *Critique of Pure Reason* (A718/B746-A719-B747), p. 633.

»true, original« geometry. The infinite ideal or archetypal elements mentioned in Chapter One constitute the foundations of this geometry. 2) Fichte finds fault with the assumption underlying Euclid's fifth or so-called *parallel* postulate, and additionally posits a normal line with a normalizing direction to remedy the parallel problem. It will be shown that this normal line is the same as the archetypal diameter. 3) Fichte's conception of the *straight line* is at odds with the Euclidean straight line. Fichte possesses a synthetic and *continuous* conception of the geometrical line as opposed to the Euclidean version of a line as a set of *discrete* points that can be indefinitely extended. All three of these differences are discussed in Fichte's excursus on geometry in the *Erlanger Logik* 1805. And to conclude this section, I will try to situate Fichte's theory of original geometry in the history of geometry.

(1) The True Starting Principle of an Original Geometry

Before discussing Fichte's starting point for geometry, we should preface this with an even more fundamental question: How does geometry actually come into existence for Fichte? In the *Zurich Wissenschaftslehre* Fichte asserts that space and a point are not sufficient for geometry to arise in human culture. We require one further ingredient that for him is tacitly assumed by the Euclidean system – *freedom*. We have to presuppose the free acting of the human being in order to move the point around in space and into a line: »Geometry arises through the free acting of my I, by moving the point into a line in space. Geometry could never arise solely from space and the point.«[34]

If we now turn to Fichte's starting point for geometry, we see that it is completely different to that of traditional Euclidean geometry. Euclid famously starts with the drawing of a line between two points. That is to say, the first postulate of Euclid in Book One of the *Elements* is: »To draw a straight line from any point to any point.«[35] Fichte's starting point in contrast begins with what he calls an *Ur-diameter*, an infinite (horizontal) line, which has an archetypal point at its centre. Extending vertically from

34 GA IV/3: 24.

35 Euclid, *The Elements*, translated by Sir Thomas Heath (Illinois: University of Chicago, 1952), p. 2.

this archetypal point and forming a right-angle with the diameter is the archetypal perpendicular. (We shall see that Fichte's archetypal point and lines resemble the ideal elements of projective geometry). In consequence, Fichte's philosophy of mathematics makes a distinction between two different types of geometry – »ordinary« geometry and »true, original« geometry.

What is »ordinary« geometry for Fichte? Exactly like in Euclidean geometry, Fichte's version of a so-called ordinary geometry starts with the drawing of a line. In Fichte's formulation: »The first thing in ordinary geometry is that a line is drawn with absolute freedom external to the subject«; or again: »The fact with which ordinary geometry commences is the freely constructed line.«[36] It should be clear that Fichte's ordinary geometry is none other than standard *Euclidean* geometry. Why is Fichte dissatisfied with ordinary Euclidean geometry? The short answer is because he thought its beginning point was mechanical, external and circular. He argued as follows: if we commence geometry with a line that is externally drawn to the subject as proposed by Euclid, say on a surface or piece of paper, and then we draw a second line to the first line, it can create one of three angles: a right-angle, an obtuse angle or an acute angle. Yet this leads to a circular foundation – for the slope of the line (the angle the line makes with a horizontal axis) determines the angle and the angle in turn determines the slope.[37] Euclidean geometry lacks an absolute standard line against which all the other lines may be measured. To avoid this Euclidean circularity Fichte proposed another beginning for geometry, which he called the »true principle« of geometry. What is this true principle of the Fichtean model of geometry? – It is the Fichtean *Ur*-diameter with the *Ur*-perpendicular attached to its centre, and he also calls it, the »original fact« of geometry.[38] Moreover, Fichte's model does not assume or employ any kind of »mechanical acts« like ordinary geometry, i.e. the

36 *Erlanger Logik* 1805 (GA II/9: 134).

37 »This angle is determined by the slope of the lines to one another. The slope of the lines is in turn determined by this angle, and this [argument] is circular.« Ibid., p. 132.

38 Cf. Ibid., 132, and 133f. The importance of the perpendicular for Fichte is already apparent in the 1794 text *Über den Begriff der Wissenschaftslehre*, where Fichte uses the example of the right-angle made by a perpendicular as the paradigmatic example of a scientific proposition from geometry. (GA I/2: 113f.).

tools of a straightedge and compass, but solely presupposes the use of *pure reason*. Hence, in contrast to the external line of ordinary geometry the Fichtean archetypal line is *internal* to the subject and can only be grasped by geometrical intuition. As Fichte says in the *Erlanger Logik*: »[T]he fact with which *ordinary geometry* commences is the freely constructible line. [...] However, I say: the fact with which *true original geometry* commences, i.e. geometry posited through pure reason, is an entirely different [fact]. Ordinary geometry cannot dispense with this fact, for ordinary geometry is only possible through this fact and within this fact. It only does not penetrate it with knowledge, but carries it out mechanically, like a law of nature. We have just penetrated this fact and thereby furnished *geometry with its true principle. It is the original diameter, with the perpendicular at its centre, which everyone brings with them to geometry.*«[39]

What are these original or archetypal elements? In his theory of geometry, we saw that Fichte assumes a number of archetypal elements, including the *Ur*-diameter, *Ur*-perpendicular and the archetypal point. They are ideal, infinite and absolute entities that the Fichtean geometer has to assume or »bring with him«. In other words, these elements have to be grasped in intelligible or geometric intuition, and Fichte's model of original geometry therefore is none other than a rational geometry, a »geometry posited through pure reason.« [40]

Summarizing briefly our discussion from Chapter One: the archetypal point is the meeting or centre point of the Ur-diameter and Ur-perpendicular; it has a limiting function within the system of geometry, and is called by Fichte the »absolute limit-point« (*Grenzpunkt*). All the lines of geometry proceed from this absolute point, and as such it forms an absolute location or fixed point of reference. Everyone has such an ideal point within themselves, and which is grasped by means of geometrical intuition: »Look into yourself. You have a point in yourself [...] Everyone has an absolute limit-point, of all distance, out of which infinitely many lines may be drawn, and the originally drawn lines. Without this point, which is absolutely invariable in itself, and co-borne and omnipresent in infinity,

39 Fichte, *Erlanger Logik* (GA II/9: 132) (my emphasis).
40 Ibid., pp. 129, 134.

nothing has a location and we would not find anything again in infinite space.«[41]

The *Ur-diameter* is an archetypal line, and it is to be considered as an infinite horizontal line. Using the human being, ourselves, as the centre of this infinite straight line, we can say that it extends into infinity on the left and into infinity on the right. This can be seen from Fichte's reflections on the original nature of right and left in the *Erlanger Logik*, in contrast to ordinary left and right.[42]

The function of the archetypal diameter is that of a regulator of direction. It is an absolute standard or »normal line« for furnishing a fixed direction (*NormalRichtung*). The archetypal perpendicular, on the other hand, may be viewed as an »absolute radius«, i.e. as a vertical line extending upwards from the horizontal diameter.[43] In the Fichtean model of geometry, if the archetypal diameter is the standard for regulating all direction, the archetypal perpendicular regulates the location of geometric objects. This second archetypal line constitutes the transition from a non-metric line to metrical elements, and is also termed by Fichte as the *Entfernungsmaas* or *Entfernungslinie*, that is, as a measurement line of distance serving as a »medium towards the outside.«[44] Like with the archetypal point, both lines are intuited in geometric intuition: »Just like you did with the earlier point, so you also bring a line through itself with you, of which the first point is the centre: i.e. [this line is] a *diameter*: and standing on this is the first mentioned distance-measurement [*EntfernungsMaas*], as directional-measurement [*RichtungsMaas*], and bearer, the *perpendicular line* [*perpendikular*] [...] And determined direction is only possible through this absolute diameter and this perpendicular, absolute orientation concerning the location and direction. The original diameter is the *regulator* of all direction, the perpendicular on it is the medium toward the outside.«[45]

41 Ibid., p. 129.

42 »A philosopher: *right*, and *left*. 1). The endpoints of the original diameter. *Right*, that quadrant of the semi-circle in which everyone who is left to their own devices directly measures. Conventional *right* and *left* is something different. For the conventional left-handed person, left is his right.« Ibid., p. 135).

43 Ibid., p. 134.

44 Ibid., p. 130, 134.

45 Ibid., p. 130.

These are the three principal elements on which Fichte builds his conception of geometry. He believes they supply a non-metrical foundation and absolute standard for all metrical geometry, and hence put an end to the circularity that he believes plagues ordinary geometry. Moreover, further archetypal elements arise from these basic three elements above, such as the »original right-angle«[46], formed by the intersection of the two archetypal lines. This original right-angle is again the absolute standard for every finite and measureable right-angle. Hence we see once again the intimate relation between Fichte's system of geometry and Euclidean geometry, with the Fichtean starting point not only providing the »original direction« (*UrRichtung*) but the »absolute orientation«[47] for the system as a whole, i.e. a regulating structure for all direction and all location.

In answer to the question, what does Fichtean original geometry look like? It is a system of geometry taking into account all the above archetypal entities. This species of geometry contains elements that are infinite and purely ideal, i.e. purely rational entities – and is therefore a rigorously deductive and synthetic system of geometry. But why does Fichte need to assume elements of this kind at the start of his geometry? Fichte chose this new ideal starting point to avoid what he perceived as the circularity of the angle and slope, believing that his new principle would give directional stability to all geometry. This is because his new starting principle is *absolute*. According to Fichte, ordinary Euclidean geometry unconsciously presupposes this fact but does not articulate it, and Fichte sees the task of his true geometry to articulate and clarify precisely this notion of direction. Here we see the ultimate *Grundsatz* or grounding principle of Fichte's model of geometry – it is not based on angle, measurement or distance, but on the absolute and primary concept of *direction*: »The direction is *primary* and absolute [...] The original diameter with the perpendicular at its centre [...] is creatively included in everything in geometry: especially the first freely drawn line, which for true geometry is not the first element: it generates the possibility of drawing and positing with which geometry is solely concerned – *its location and its direction.*«[48]

46 Ibid., p. 131.

47 Ibid., p. 130, 132.

48 Ibid., p. 131f. Fichte also cites the importance of the semi-circle as an original absolute constructed through reason (ibid., p. 135).

This new foundation and starting point is of course related to Fichte's conception of science outlined at the beginning of his excursus on geometry in the *Erlanger Logik*. For Fichte believes his new point of departure will allow a non-circular science: »In this manner science becomes *deductive*, purely derived, not turning in circles.«[49] A second important reason why Fichte felt the need to assume archetypal elements was to solve one of the most intractable problems of geometry – the parallel problem.

(2) Fichte's Proposal for Resolving the Parallel Problem

As mentioned above in element no. vii of our exposition of Fichte's philosophy of mathematics, Fichte was unhappy about the contemporary state of geometrical knowledge concerning the famous problem of parallel lines. His interest in the problem can already be dated back to *Über den Begriff der Wissenschaftslehre* (1794), where his example for an indisputable scientific fact is the directional argument that a perpendicular column forms a right-angle with a horizontal plane.[50] For many geometers, Euclid's fifth or parallel postulate did not seem to be a postulate at all and therefore had to be proved, like any other theorem. Despite the efforts of countless able geometers for over two millennia the problem did not seem to be closer to being resolved in 1805.[51] Somewhat astonishingly perhaps, in the *Erlanger Logik* Fichte put forward his own proposal for resolving the impasse, and in this respect his model of geometry again diverges from the traditional Euclidean model of geometry. Fichte also mentions other attempts to solve the parallel problem, including that of the Königsberg mathematician Johann Friedrich Schultz. In the preface to his *Entdekte Theorie der Parallelen nebst einer Untersuchung über den Ursprung ihrer bisherigen Schwierigkeit* (Königsberg 1784) Schultz's theory states that magnitude or »size (*Größe*) of a plane angle is determined by the size of the plane lying between its infinitely extended side.«[52] Fichte says of

49 *Erlanger Logik* (GA II/9: 133).

50 GA I/2: 112.

51 L. Magnani provides a good overview of the history. Cf. *Philosophy and Geometry: Theoretical and Historical Issues*, pp. 67–70.

52 Quoted in: *Erlanger Logik* (GA II/9: 131, footnote).

Schultz's attempt that it »does not accomplish anything, and his entire theory is improbable«, because it too is based on circular reasoning.[53]

Fichte's suggestion was to attack the issue at its root, and like Euclid's starting point, he also viewed Euclid's famous fifth postulate itself as being circular.[54] In essence, Fichte argued that the parallel problem must be some kind of *proton pseudos*; that it to say, it must contain some kind of erroneous proposition or starting premise[55] (a claim he had made four years earlier to dismiss the starting point of Schelling's *Darstellung meines Systems der Philosophie.*)[56] Accordingly, any conclusion drawn from such a false premise would in turn have to be false. This is why the parallel problem had never been solved satisfactorily, because geometers (or philosophers) had not properly investigated the foundations and assumptions of Euclid's original postulate. Fichte's suggestion for avoiding the circularity of the original postulate was an initial investigation into the *directional* aspect of the angle. In ordinary geometry, the direction of the line depends on the angle, yet the angle in turn is defined by the direction of the line. This is circular. For Fichte, in contrast, the angle depends on the direction, and the direction is primary in his system, as we saw above. The angle and direction do not reciprocally depend on one another, and this therefore allows Fichte to put a halt to the circularity of angle and direction. Such a move of course requires a fixed and absolute element in a system, and this is where Fichte deploys the above-mentioned *Ur*-diameter and *Ur*-perpendicular as central normalizing concepts. Combined with

53 Ibid., p. 131.

54 Cf. GA II/9: 131.

55 The term ›*proton pseudos*‹ is an original error in a premise and relates to Aristotle's theory of the syllogism. It states if the premises of an argument are false, or if there is an original error, then the conclusion must necessarily be false despite the soundness of the intermediary reasoning.

56 Ibid., p. 130. See Fichte's letter to Schelling (Oct. 8th 1801) where he says: »I can tell you our differences in a few words. – You say »the absolute« (concerning *which* and whose *determination* I completely agree with you, and whose intuition I have also possessed for a long time) »as I claimed in my *Presentation, exists under the form of quantitative difference.*« This is clearly what you assert; and it is *precisely because of this* that I found your system to be in error and rejected the *Presentation* of your system – because no conclusion and exposition can be correct which is not in line with this principle. Spinoza does exactly the same, and all dogmatism in general, and this is the *proton pseudos.*« (GA III/5: 91).

this new view of angle, Fichte therefore believed that the key to the parallel problem lay in turn in a deeper comprehension of the nature of *direction*. Parallelism, for him, likewise implicitly implied direction. He thought that what was missing in the solution was some original fixed or »normalizing« line that all other parallel lines could be measured against. This is essentially why geometry requires such *Ur*-diameter and *Ur*-perpendicular lines, since they would furnish the absolute fixed directions for all parallel lines and their corresponding angles.

With Fichte's employment of an archetypal diameter and archetypal perpendicular we clearly see how his theory of geometry is based on a number of *a priori assumptions* and *strategies*. Like his conception of the straight line, the *Ur*-diameter also has to be implicitly assumed as a continuous whole and brought to geometry by the geometer himself or herself. In order to acquire the further elements of his theory, the geometer has to then divide up space into two equal parts to generate the secondary directions, for the primary direction is already given by the *Ur*-diameter and Ur-perpendicular. The establishment of this original perpendicular yields an archetypal right-angle, and all other possible angles (obtuse, acute etc.) are derived in turn from this archetypal right-angle. Indeed, the right-angle itself arises from the synthesis of a »non-original line« with the *Ur*-diameter. *Thus, what is assumed as absolute and primary in the Fichtean model of geometry is the direction of the Ur-diameter*. This strategy allows Fichte to posit a »primary, normalizing line« with which to compare the direction of all other lines and determine their parallel nature. Due to its importance, I will quote this passage at length: »On parallelity. – Direction through the angle: angle through the direction of the lines: this runs through the whole of geometry, and you cannot arrive at the correct view with a *proton pseudos*. – What is it [the correct view] then? I can tell you in a few words: the direction is primary and it is absolute: and it is the direction of the original diameter that everyone brings with him: this is the straight line and straight direction *per se*. In order to measure all the other directions against it, you divide the space above it into two equal parts, and this yields the original right angle. This in turn is not determined by the direction of the perpendicular, but determines the direction itself, as [something] secondary, and it is not the primary and immediate direction; but it is determined by the equal parts of space: this perpendicular with its original right-angle now determines all other right angles or non-right angles, i.e. obtuse or acute angles, and measures them through itself. The first parallel runs as a second parallel to the archetypal diameter [*Urdiameter*], [which is] the primary or normal direction; and this parallelity is equal to the two right angles on it: All [lines] drawn with this parallel line are parallel among themselves, not because they are as such, but because they are parallel to the normal line [*Normal Linie*]. Now if an oblique line intersects this whole through the centre of the archetypal diameter, then the known relation will result, everywhere in the same manner, out of the quality of its line and the lines intersecting it, in any case not only relatively, but absolutely, and therefore qualitatively oblique lines are the result: and you won't be tempted, in alternating proofs, to make the result again into the principle of its own principle. – In the former series the absolute direction of the archetypal diameter is what is primary. The angle, and indeed, the right-angle first arises in the synthesis of *other* non-original lines with this one, and its direction is estimated according to the first one, as the first and absolute on the centre of the archetypal diameter; its direction has to be judged according to the angle on the external line and by means of the same.«[57]

The essence of Fichte's solution therefore is to posit a normal or normalizing line as an absolute standard for determining the angle and establishing the correct parallelism of any given line. As Fichte himself admits in this excursus on geometry in the *Erlanger Logik*, the whole of his geometry and the solution to this problem rests upon the positing of archetypal and ideal elements – in this instance, above all the diameter that can provide us with the original direction.[58] The relative and specific directions, as well as all angles, are then derived in turn from this archetypal or absolute direction. This and only this provides a fixed and absolute base and allows us to avoid the circularity of the Euclidean parallel problem: »This angle, therefore, is determined through the *Ur*-direction itself; however, it is only determinate for derived and relative directions. This of course is how the above circularity and arbitrariness can be avoided.«[59] Thus, just as Archimedes posited his famed lever that was capable of moving the entire earth if it were only possible to find a suitable location on which to stand, so the meeting point of the *Ur*-diameter and *Ur*-perpendicular becomes

57 Ibid., pp. 131f. See the Appendix below for a complete translation of this passage.

58 Ibid., p. 131.

59 Ibid., p. 132.

Fichte's *absolute* location to determine and ground all relative and specific directionality. With the discovery of non-Euclidean geometry Fichte was in one sense later proved right. The fifth postulate of Euclid did have some implicit assumptions, insofar as it assumed that only *one single line* could be drawn through a point in a plane parallel to another line, an assumption spectacularly overturned by non-Euclidean geometry, which declares that there does not have to be any line at all running through the point external to the parallel line, or that there may be an infinite number of lines. So Fichte's suggestion to investigate the underlying assumptions of the fifth postulate and his assertion that parallelism first presupposes a better understanding of the *nature of direction*, not only indicates that he (like many geometers) was dissatisfied with Euclid's formulation or wording, but that his proposal to re-examine its assumptions and to posit an archetypal and normalizing direction did not diverge too far from the path of the future development of the philosophy of geometry.

(3) The Fichtean View of the Straight Line

Fichte formulated a conception of the *straight line* that is distinctly at odds with the Euclidean straight line. Hence this is a third departure from the classic Euclidean model of geometry. Fichte possesses a synthetic and *unified* conception of the geometrical line, as opposed to the Euclidean version of a straight line as a collection of points.[60] In Chapter Four below I will supply a more exhaustive account of Fichte's view of the geometrical line, as well as his accompanying philosophical interpretation of it. For the present I will simply highlight Fichte's differences from the traditional Euclidean model.

As mentioned, Euclidean geometry assumes the discrete nature of the straight line, i.e., it can be broken down and is composed of an infinite number of points. In this model the straight line is a one-dimensional figure that has no thickness and extends infinitely in both directions.[61] Fichte disputes this Euclidean conception and considered its underlying atomistic model to be an absurdity, because it would signify a transformation

60 »A *straight line* is a line which lies evenly with the points on itself«, *The Elements of Euclid*, Book One, Definition 4, p. 1.

61 Cf. Ibid., Book One, Definitions 2 and 4, and Postulates 2 and 5, pp. 1f.

in our *qualitative* notion of the line. Such a conception, he maintained, would entail a revision in our very idea of the meaning of *straightness*. As a result, the smallest part of the line for him is always still a line and never a point: »The infinitely smallest part of the line, *is a line*: not at all a point. Wanting to dissolve the line into points through an infinite division does not imply that the quantity is extended, in which geometry alone is unlimited, but it involves *transforming the quality*. [...] However, the line can only be constructed as a line, never, though, out of points. [...] The geometer brings it with him *directly and absolutely* as *a line*.«[62]

Why does Fichte object to the straight line being composed of points? As mentioned, the notion of a point for Fichte contradicts and conflicts with the idea of straightness that is the essence of the notion of the line: the two concepts cannot be reduced to one another, and therefore the predicate of straightness must already be inherently *contained in* or belong to the concept of the line. Fichte's reflections here in the *Erlanger Logik* hark back to the infamous question he had already posed in a footnote to the 1794 *Über den Begriff der Wissenschaftslehre*: »A question for the mathematician: Is not the concept of straightness already included in the concept of a line? Is there any other sort of line except a straight one?»[63] Fichte's fundamental objection to the Euclidean view of the line is that one would have to pass through an infinite number of points in order to complete the line, any line at all. That is, in a small segment of line there are infinitely many points. Hence, passing from one to another and then through them all is an infinite process and would require an infinite amount of time. Such an infinite feat, however, could never be finished or terminated in finite time. Naturally, this is not to say that the line cannot contain points or be broken down into points. We only have to take an empirical line and cut it into small pieces. But in Fichte's conception these points only enter into the line in a *subsequent* or later movement, i.e. after the cutting or intellectual abstracting activity of the geometer, and do not therefore form part of the *original* nature of the line. This gives rise to his conception in the *Erlanger Logik 1805* of the line as a *continuous whole*:

62 Fichte, *Erlanger Logik* (GA II/9: 127; my emphasis).

63 BWL (GA I/2: 135); English translation by Daniel Breazeale in: *Fichte: Early Philosophical Writings*, p. 120. See too Fichte's reflections in the text: *Was ist von einem UnendlichEk zu halten?* (GA II/5: 5).

»The geometer brings it [the line] with him directly and absolutely as *a line*. If it were otherwise we would have to traverse and *complete* an infinite number of points in order to pass from one point to another, which would be absurd. I can never infinitely approach a point by means of points, but have to assume a piece of line, however small it may be: and for me, given over to myself, it would always be *straight* [...] [The geometer] brings the line as a line purely holding it together and the points flow into it.«[64]

d. Where to Place Fichte's Theory in the History of Geometry?

With his conception of an »original geometry«, the question now to answer is where does Fichte's model of geometry fit into the history of geometry? For all its emphasis on distinguishing itself from ordinary Euclidean geometry, is it still to be classified as ultimately Euclidean in nature? Or is it perhaps some kind of proto-form of non-Euclidean geometry?

Let us begin by re-examining the notion of synthesis, but this time with regard to Fichte's model of geometry. What does Fichte mean by the term »synthetic«? Is it the same as Kant's meaning? In one sense the answer is clearly affirmative, because following Kant's lead Fichte too makes a distinction between the synthetic *method* in philosophy and geometry on the one hand, and synthetic *judgment* on the other. Fichte likewise contrasted the synthetic method with the analytic method and ultimately adheres to the Kantian synthetic conception of geometry. For instance, in the early *Eigne Meditationen*, these two methods are elucidated at the very start of the text and at this period Fichte was still not sure if it might be possible to combine the two methods in such a manner that each would act as a test of the other like in ancient Greek geometry.[65] Compared with

64 Fichte, *Erlanger Logik* (GA II/9: 127). See the similar argument in the *Wissenschaftslehre 1801/1802*: »If the line is the traversing of infinitely many points, then the line is impossible. The points disintegrate. Rather, they hang together through it.« (GA II/6: 231).

65 Regarding a possible union of the two methods, in the *Eigne Meditationen* Fichte also hits upon an idea for understanding Descartes' cogito: »I have just come up with the idea to test it with analytic propositions until I am driven to synthetic propositions.« (GA II/3: 91)

an analytic method solely based on the principle of contradiction, in the end Fichte considered his philosophical method to be synthetic, since he preferred to start with the unity of apperception and descend to the practical legislation of reason.[66] In the *Eigne Meditationen* Fichte also believed that he had discovered that the I's *activity* in bringing forth the elements of geometry, such as the point and line, is *also synthetic*: »Just as the I bears the point to the Not-I: so through synthesis the Not-I bears the line over to the I. Brilliant hypothesis. The I not only synthetically brings forth the point, but *all the forms of intuition*. [...] And philosophy also has to demonstrate this.«[67] In this connection it must be noted that not only with regard to geometry but the *Wissenschaftslehre* as a whole (and especially the intellectual intuition upon which it is based) has synthesis as its methodology. Fichte's philosophy is founded on the intuiting of one's own I, and this process can never occur analytically, since it involves intuitions and not concepts. The intuiting of the laws and acts of the I requires a synthetic methodology, as Fichte explains in the *Wissenschaftslehre nova methodo*: »The *Wissenschaftslehre* first of all is erected upon an I, but it does not wish to analyse it; that would be an empty philosophy, instead, it allows the I to act according to its own laws, and to thereby construct a world. This is not an analysis, but always a progressive synthesis [...] we have to inwardly act, and then observe this acting.«[68]

But of course, once this process of synthesis – the establishment of the acts of the mind – is completed, they can in turn be subject to analysis, i.e. to philosophical reflection. In Fichte's brief formulation from the beginning of Part Two of the *Grundlage der gesammten Wissenschaftslehre*:

66 »Sometimes one proceeds synthetically and sometimes analytically. [...] If one is able to think of a single path that leads upwards and ascends from the unity of apperception to the practical legislation of reason and then descends in turn from the latter to the former, then the former would be the path followed by the synthetic method and the latter that of the analytic method. Could these not serve as a test of each other? [...] My inner feeling tells me that the proposition that has to be proven is indeed true, but how in the world can this proposition be proven? – The principle of contradiction is purely analytic. – My procedure is synthetic.« *Eigne Meditationen* (GA II/3: 22, 26, 29).

67 Ibid., p. 217.

68 *Wissenschaftslehre nova methodo*, GA IV/2: 28f. (Cf. GA IV/3: 343f.).

»The acts that are established are *synthetic*; however, the reflection that they established, is *analytic*.«[69]

Furthermore, in both Fichte's Jena lectures on logic and metaphysics using Platner's *Aphorisms* (in: GA II/4) and in the *Kollegnachschrift* that are entitled: *Vorlesungen über Logik und Metaphysik SS 1797* (in: GA IV/1), Fichte talks about the different procedures of science and philosophy and compares it to mathematics. Like Kant, he mostly confines himself to a discussion of analytic and synthetic *judgements*. But in the section with the title »On Judgments« (*Vom Urtheilen*), Fichte states that the difference between a synthetic and analytic judgment is *relative*, and draws an analogy with pedagogy: when decomposing a proposition in order to instruct someone else I judge analytically; however, if I am compelled to first learn it myself and determine it in consciousness then I judge synthetically.[70] In this pedagogical respect Fichte believed the Kantian exposition of these judgements to be too complicated and obscure and proposed instead a simpler definition: synthesis is what is *real*, analysis is what is *abstract*.[71]

Moreover, another crucial related aspect is that Fichte's theory of the line is based on a different conception of *infinity* to that of the Euclidean model. If Euclid considers that the straight line can be produced »continuously« (Postulate 2), Fichte considers the original straight line to be not only continuous but infinite. This is the case for the archetypal perpendic-

69 Fichte, GWL (GA I/2: 284).

70 »In order to acquire a correct insight into this occupation of the human mind we have to once again place ourselves at the standpoint of the original synthetic judgments, not however at that of the analytic. – If I already now possess something and have determined it (and have already posited it within consciousness) and I now decompose it again for whatever purpose, say to instruct someone else, to find something new and so on, then I judge analytically. If, however, I first *learn* it and first determine it in consciousness, then I judge *synthetically*. It follows that the same judgment can be synthetic for one person and analytic for another person./ Relationship between the teacher and listener.« Fichte, *Zu Platners ›Philosophischen Aphorismen‹* (GA II/4: 184f.).

71 »The matter has been treated by Kant in a highly obscure manner, and the dispute is extremely intricate. – *Synthetic* is whatever is grounded in a synthesis: they have reality. Analytic only to the extent they are also grounded on it, otherwise they are not – if an *arbitrary* abstraction is examined.« Ibid., p. 220.

ular, for example, which is the »absolute radius« of the circle at infinity.[72] This is also the implication of Fichte's mathematical footnote to *Über den Begriff der Wissenschaftslehre* (1794) where he imagines the I as a centre with an infinite number of radii. – Due to our limited or finite imagination we can assign a terminus to these radii, yet can never carry out or complete the infinite approximation of the resultant infinite or original curve if we take these points together as a whole.[73]

Where can we find similar conceptions in the history of geometry? Fichte's designation of these terminal points of the radii corresponds to what in projective geometry is called the »line at infinity«. This »ideal« line has exactly the same characteristics as Fichte's original curved line. It is a composite of all the ideal points, or the points at infinity, which when strung together form a new line, the so-called »line at infinity«. Traditional Euclidean geometry does not possess such a line at infinity and we must turn instead to synthetic or projective geometry for a model of geometry that posits precisely these infinite elements. In fact, in projective geometry it is a strange fact that parallel lines meet – they meet in an ideal point at infinity.[74] Hence, with his usage of the ideal line, or line at infinity (or original line), Fichte is following in the footsteps of the great synthetic ge-

72 »(The space of geometric determinations is necessarily a limited space). This limit (*Grenze*) is *one and the same*, and it has a certain distance from the centre-point: i.e. it is the line of a circle described around the centre-point by the original distancing line that is a perpendicular (and here the absolute radius), and which everyone brings with them.« *Erlanger Logik* (GA II/9: 134).

73 Cf. GA I/2: 135 (Fichte, *Early Philosophical Writings*, p. 121).

74 And this is true for all parallel lines, if we attribute to them: »a *point of intersection*. Such a point is not, of course, a point in the ordinary sense; it is to be regarded as an *ideal point*, which we suppose two parallel points to have in common. Its introduction [into geometry] merely amounts to a change in the ordinary terminology. Such an ideal point we call a *point at infinity*; and we suppose one such point to exist on every line. [It is necessary] to attach a meaning to the notion of two points at infinity. Such a line we call a *line at infinity*, and think of it as consisting of all the points at infinity in a plane.« From the essay, »A Mathematical Science«, by Oswald Veblen and John Wesley Young; reprinted in: *The World of Mathematics*, ed. J. Newman (New York: Dover Publications, 2000, Volume 3, pp. 1696–1707), p. 1702.

ometer Girard Desargues, who along with Kepler, was the first to introduce these notions into geometry two hundred years prior to Fichte.[75]

However, as mentioned at the start of the chapter, it is perhaps most fruitful to view Fichte's model of geometry in the context of other 19th century attempts in pure geometry. For example, insofar as Fichte's model of geometry is based on geometrical intuition, it shares a deep affinity with the work of the Swiss-German mathematician Jacob Steiner. Like Fichte, Jacob Steiner was a friend (and even pupil) of the Swiss pedagogue Pestalozzi. Recalling Fichte's attempts in philosophy, in his work from 1832: *Systematische Entwickelung der Abhängigkeit geometrischer Gestalten von einander*, Jacob Steiner did not seek simply an abstract geometrical system but attempted to proceed in an »organic« manner in order to find the basic foundational principles of geometry and their relation to the real world of nature. He created his model of geometry synthetically starting from four fundamental geometric entities: »Space, the plane, the line (straight line) and the point.«[76] Unlike Fichte, however, Jacob Steiner conceives of the straight line as a collection of points, though his line also extends to infinity in both directions.[77] Jacob Steiner placed enormous weight on the cultivating of geometrical intuition in comparison with analysis, and at times even sought to impress upon the student the value of imagining the geometric figures without the aim of external drawings.[78]

Philosophically speaking, on the one hand Fichte interpreted geometry along Kantian lines – and viewed its propositions, method and cognition as *synthetic*. That is to say, its cognition cannot be based on the analysis of mere concepts alone but requires intuition in addition. The role of

75 Desargues posited a point at infinity where parallel lines meet: »Desargues believed that the straight line may be regarded as a circle whose centre is at infinity; hence the two extremities of a straight line may be considered as meeting at infinity. Parallels differ from other pairs of lines only in having their points of intersection at infinity.« Florian Cajori, *A History of Mathematics*, p. 252.

76 Cf. Jacob Steiner, *Systematische Entwickelung der Abhängigkeit geometrischer Gestalten von einander* (Berlin: G. Fincke, 1832), pp. V-VI.

77 »I. *The straight line.* In the straight line an innumerable number of points immediately succeeding one another are conceivable, which, starting from any of them, extends on the two opposite sides out into infinity.« Ibid., p. XIII.

78 See Felix Klein, *Vorlesungen über die Entwicklung der Mathematik im 19. Jahrhundert* (New York: Chelsea, 1967), pp. 127–128.

the philosopher of mathematics is to grasp and inwardly penetrate the conditions and possibilities of this science. On the other hand, the first *propositions* of geometry are also synthetic in the sense of pure geometry, as Fichte says in the *Erlanger Logik*: »To express it: The beginning of geometry is *synthetic*; neither point, or line, but both, and both comprehensible through one another. 2) [...] Once again this is a fact that geometry does not need to grasp in its possibility.«[79] And so, similar to Jacob Steiner, Fichte's geometrical *method* is also synthetic. Like in his conception of a rigorous science, the geometric method synthetically and systematically passes from the point, to the line, to the plane: »[Geometry] ascends synthetically: and as a strict science it does not shirk this for the sake of the external goal of popularity, but it is necessitated by *its own* nature. This ascension has to occur rigorously and systematically: without leaps, without circles, without standing still, but with rigorous determinacy.«[80]

Even if he was not conscious of this fact, with his emphasis on synthetic methods to the exclusion of algebra and analysis, as well as his stress on the importance of geometric intuition and absolute and ideal entities at infinity, Fichte is signalling a debt to traditional synthetic geometry in the historical sense delineated above. Consequently, Fichte is acknowledging, like Schelling explicitly does, a relationship between the synthetic methods in transcendental philosophy and the sphere of geometry.[81] And again, this view endorsed by Fichte that the beginning of geometry starts with two fundamental entities, the point and the line, which cannot be reduced to one other, also belongs to synthetic geometry, and was employed most fruitfully by a number of German geometers who lived after Fichte, particularly Karl Georg Christian von Staudt.[82] Von Staudt's 1847 masterpiece: *Die Geometrie der Lage* is considered to be the

79 Fichte, *Erlanger Logik* (GA II/9: 127f.).

80 Ibid. p. 135f.

81 See Schelling's *Further Presentations from the System of Philosophy*, where he thinks that philosophy has borrowed its methods from mathematics: »At this point I think it is useful to discuss the distinction between analytic and synthetic method that has been clumsily imported into philosophy from mathematics« *Philosophical Forum* 32 (2001), pp. 389f., translated by Michael Vater.

82 Cf. Coolidge, »The Rise and Fall of Projective Geometry«, *American Mathematical Monthly* 41 (1934), p. 223.

first rigorous attempt at non-metrical geometry.[83] Fichte's system is therefore especially similar to the projective geometric intentions of von Staudt, insofar as its basis is also supposed to supply a non-metrical foundation for all metric geometry. Like von Staudt's project, it could be furthermore argued that the Fichtean model of geometry is based on the principle of incidence, as its »true« principle is the incidence or meeting point of the two original lines (the original diameter and original perpendicular). And because Fichte's archetypal elements such as the »normal line« are absolute and invariable and the standard for measuring, it seems that he is well aware of the necessity of a form of non-metrical geometry, or rather a geometry that is the foundation of all metric geometry.

Again, Fichte's suggestions for solving the parallel problem place him among those seeking an alternative to Euclidean geometry. And as we have seen, by positing different models of geometry, from Euclidean geometry up to a universal non-metrical geometry, one would have to view his system as a forerunner to the geometric ideas and intentions of Jacob Steiner, Karl Christian von Staudt, and Felix Klein. The latter ranks different systems of geometry according to their universality in his 1872 »Erlanger Programm«, seeing metric geometry as only one special part of projective geometry.[84]

There are a number of problems with this hypothesis of course. Fichte's model of geometry is obviously not rigorous or worked out in any detail, but is fragmentary, and it could be argued that it is only confined to the sphere of pure philosophical speculation. Again, in terms of the epoch, Fichte's ideas were conceived in Erlangen in the year 1805, four decades *before* the work of von Staudt in Erlangen. How can we scientifically explain this? In his *La philosophie de l'algèbre* Jules Vuillemin states that the revolutions in mathematics are often only explained by philosophers centuries later, and uses the example of Fichte's theory of intuition to elucidate the earlier discovery of abstract algebra.[85] However, with regard to Fichte's model of geometry, the opposite appears to be the case: Fichte's

83 See von Staudt, *Die Geometrie der Lage* ((Nuremberg: Bauer & Raspe, 1847).

84 See Felix Klein, *Vorlesungen über die Entwicklung der Mathematik im 19. Jahrhundert* (New York: Chelsea, 1967), pp. 133 ff.

85 Cf. Jules Vuillemin, *La philosophie de l'algèbre*, p. 4.

philosophical speculations on a consistent, ideal and absolute model of original geometry seem to be in advance of the geometers themselves.

2. Fichte's Mathematical Platonism

To conclude Chapter Two, I will raise again a question first posed at the outset: If we had to classify the foundations of Fichte's theory of mathematics and model of geometry according to earlier and current classifications in the philosophy of mathematics, how could we classify Fichte's theory? Though there exist many schools of thought and interpretations of the foundations of mathematics, I will examine four main classifications in the current philosophy of mathematics. These are logicism, formalism, intuitionism, and Platonism.[86] It is in fact possible to find textual support for all these four traditions and streams in Fichte's writings, but I will argue that Fichte's philosophy of mathematics in general, and model of geometry in particular, should be classified as mainly Platonistic in character. This is especially the case if our primary concern is with the *essence*, existence and ontology of the entities that constitute Fichte's philosophy of mathematics.

a. Formalism, Logicism and Intuitionism

Let us examine these three classifications in turn. Firstly, what is logicism in the philosophy of mathematics? Logicism essentially refers to the theory that the content of mathematics can be reduced to logic. Representatives of the logicist view of mathematics include Gottlieb Frege (especially for arithmetic), Rudolf Carnap and Bertrand Russell. In an influential article from 1931, for example, Carnap neatly summarized the standpoint of logicism by arguing two main points. 1. that »the *concepts* of mathematics can be derived from logical concepts« by means of explicit definitions, and 2. that »the *theorems* of mathematics can be derived from logical axioms« by means of logical deductions.[87]

86 For an overview of these different schools and traditions in the philosophy of mathematics, see *Philosophy of Mathematics: Selected Readings* (1983), eds. Paul Benacerraf and Hilary Putnam, pp. 5–24.

87 Rudolf Carnap, »The Logicist Foundations of Mathematics« in: *Philosophy of Mathematics: Selected Readings*, eds. P. Benacerraf and H. Putnam, (Englewood Cliffs, NJ: Prentice-Hall, 1964), p. 31.

However, in line with our exposition above, we saw that Fichte's theory of mathematics *is opposed to mere logicism*, because unlike mathematics which has content, logic for Fichte is merely a speculative tool devoid of any content.[88] This is especially clear in the *Zurich Wissenschaftslehre* and *Über den Begriff der Wissenschaftslehre* where Fichte distinguishes his philosophy from both logic and geometry, since logic is without content and geometry is concerned with limiting the forms of space.[89] As a mere tool, Fichte »would even banish pure logic entirely from the sphere of philosophy.«[90] Hence this lack of content makes it impossible for logic to form the basis of mathematics for Fichte. Mathematics not only has »material content« for him, it is furthermore »the only completely scientific undertaking that exists.«[91]

Secondly, what about formalism? As we saw above in Chapter One, Fichte acknowledges the formal power and usefulness of mathematics, especially the symbolism to be found in algebra and arithmetic. However, on account of its creative, imaginative and ideal aspects, mathematics for Fichte *cannot be reduced to formalism* or the formulas of a strictly symbolic science. For in the formalist view there are no mathematical objects but only formalized axioms and rules *that do not refer to anything*. For example, one of Hilbert's aims was to show the consistency of an axiomatic system, and hence the consistency of mathematics in general, by expressing it formally in a specific language.[92] In the formalist philosophy of mathematics, the content therefore is no longer the most important aspect, but rather, the operations and procedures for manipulating the symbols and rules. In other words, *the method of proof* becomes salient, and traditional mathematics is philosophically viewed by the formalist as nothing more than a »game« played out using the »symbols« of mathematics. Johann von Neumann summarizes mathematical formalism in exactly this sense: »We must regard classical mathematics as a combinatorial game played with the primitive symbols, and we must determine in a

88 Cf. Fichte, *Wissenschaftslehre nova methodo* (GA IV/2: 265).

89 See especially Fichte, *Zurich Wissenschaftslehre* (GA IV/3: 32–37) and *Über den Begriff der Wissenschaftslehre* (GA I/2: 137–140).

90 Fichte, *Ankündigung* (GA I/7: 159).

91 Ibid., p. 160.

92 Regarding Hilbert and the history and details of formalism as a tradition, see P. Davis and R. Hersh, *The Mathematical Experience* (1999), pp. 318 ff.

finitary combinatorial way to which combinations of primitive symbols the construction methods or ›proofs‹ lead.«[93]

Considerations of Fichte's theories within a formalistic framework have found some support in the reception of his philosophy. Although attracted by Fichte's philosophical theory of intuition, Hermann Weyl and Fritz Medicus viewed the Fichtean foundations as ›constructivistic‹, and ultimately in a formalistic light, on account of the role of *active* construction in Fichte's theory of knowledge.[94] In Fichte research, this view has also held sway somewhat. Wolfgang Schüler, for example, uses Fichte's transcendental philosophical views on the nature of consciousness to support a formalist and logical reading of the foundations of mathematics in the sense of Frege and Hilbert.[95] While Joachim Widmann too, as mentioned in the Introduction, has considered Fichte's *Grundsatz* in terms of its logical and formal structure, and compared it with the axiomatic principles of David Hilbert.[96]

Thirdly, Fichte's model of cognition (like the Kantian) does indeed require what he calls »intuition«, yet Fichte's theory of intuition is to be *sharply distinguished* from L.E.J. Brouwer's species of intuitionism. Brouwer only sees the objects of mathematics as subjective mental constructs and does not grant them any independent ideal existence. Mathematical entities are only »in the human intellect«[97], i.e. they only exist insofar as they are thought. The human mind naturally creates the products of mathematics, but we cannot ascribe any kind of objective (external) existence to them. As the intuitionist Arend Heyting puts it: »The intuition-

93 Johann von Neumann, »The Formalist Foundations of Mathematics«, in: *Philosophy of Mathematics: Selected Readings*, eds. P. Benacerraf and H. Putnam, (Englewood Cliffs, NJ: Prentice-Hall, 1964), p. 51.

94 I refer the reader here to Norman Sieroka's detailed analysis of the ›holistic‹ formalism of Weyl and the influence of Fichte. See *Symbolischer Konstruktivismus im Anschluss an Hermann Weyl und Fritz Medicus* (Zurich: Chronos, 2010), esp. pp. 97–111.

95 Wolfgang Schüler, *Grundlegungen der Mathematik in transzendentaler Kritik. Frege und Hilbert* (Hamburg: Felix Meiner Verlag: 1983), esp. pp. 87–117.

96 Joachim Widmann, *Die Grundstruktur des Transzendentalen Wissens nach Joh. Gottl. Fichtes Wissenschaftslehre 1804*[2] (Hamburg: Felix Meiner Verlag, 1977), pp. 21ff.

97 L. E. J. Brouwer, »Intuitionism and Formalism«, in: *Philosophy of Mathematics: Selected Readings* (1983), eds. Paul Benacerraf and Hilary Putnam, p. 67.

ist mathematician proposes to do mathematics as a natural function of his intellect, as a free, vital activity of thought. For him, mathematics is a production of the human mind. [...] mathematical objects are by their very nature dependent on human thought. Their existence is guaranteed only insofar as they can be determined by thought.«[98] Fichte, on the other hand, admits objective existence, universal validity and non-temporal laws to the sphere of mathematics.[99]

b. Fichte's Mathematical Platonism

Finally, however, in my opinion Fichte's theory of mathematics and geometry is closest in spirit to the Platonistic conception of mathematics. This is above all on account of Fichte's conception of the *existence* of absolute and invariable mathematical objects such as the archetypal point and archetypal lines. However, what exactly is Platonism in the philosophy of mathematics? Without going into all the possible objections to this view, a Platonistic conception of mathematics asserts that the objects or elements of mathematics reside in an ideal, non-temporal and non-causal realm, and ontologically speaking, exist as eternal and unchangeable archetypes or ideas.[100]

Of course, the modern label of mathematical Platonism can be traced back to views in the works of Plato himself. For example, in the *Republic* we find that knowledge of mathematics forms an important part of the education of the philosopher and the future ruler, and can serve as the foundation for every other kind of knowledge. Although geometry has a practical usefulness, e.g. in mapping or military plans, Plato here above all emphasizes the intellectual benefits of geometry. A study of geometry increases the facility with our reason, drawing the mind upwards, and aids us in grasping the ideal forms. We come to the realization: »that the ob-

98 Arend Heyting, »The Intuitionist Foundations of Mathematics«, in: Ibid., p. 42.

99 Cf. for example, Fichte, *Einleitung in die Wissenschaftslehre 1813* (forthcoming in: GA II/17: 257-258), and the *Ankündigung*, (GA I/7: 160).

100 For detailed definitions and the arguments pro and contra the Platonistic view of mathematics, see Mark Balaguer, *Platonism and Anti-Platonism in Mathematics* (Oxford: Oxford University Press, 1998); and Paul Bernays, »On Platonism in Mathematics« in *Philosophy of Mathematics: Selected Readings*, eds. P. Benacerraf and H. Putnam, pp. 258–272.

jects of that knowledge are eternal and not liable to change and decay«; and thus in terms of their existence or ontology, for Plato: »the objects of geometrical knowledge are eternal.«[101] Again, in the Seventh Letter, we have an illustration of the five levels of Platonic knowledge using the example of a geometric circle. Firstly, we have the name of the object (Kreis, circle, annular etc.); secondly, the definition (circle = an object whose distance from its circumference to its centre is everywhere equal); thirdly, the image (the object represented or depicted on a blackboard for example); fourthly, our human knowledge of this object; fifthly, the object in itself. The latter is never wholly the same as the earlier stages, for at this fifth stage the circle is the perfect circle, it is »permanent« and »unchanging«.[102]

According to the Platonistic conception of mathematics, particularly as classified by modern philosophies of mathematics, our knowledge of mathematics is therefore universally applicable and unchangeable, i.e. mathematical cognition is not subject to the fluctuations of history and convention. – A correctly established mathematical truth will be just as valid in another two thousand years as it is today. In line with the above characterizations, I would argue that Fichte's model of geometry most closely approaches mathematical Platonism. This is for at least four specific reasons:

(1) Like in mathematical Platonism, Fichte asserts that *our mathematical knowledge is invariable and universal.* That is to say, genuinely knowledge in this domain is not subject to change. As Fichte says in 1803 on the nature of the triangle: »For what is invariably identical is not the triangle itself, but my knowledge of the triangle.«[103] Fichte's view might seem to differ from standard Platonism insofar as Fichte seems to be claiming here that the geometric objects themselves are not eternal and invariable. Yet as we shall see, this reference is more pointing more to the variability of the empirical triangle and all knowledge associated with it. What is clear

101 Plato, *The Republic*, trans. Desmond Lee (New York: Penguin Books, 1980), pp. 334–335. See too the *Meno*, where Plato uses the example of geometrical knowledge in a slave boy to illustrate that all learning is a recovery of the soul's pre-existing knowledge. (Cf. Plato, *Protagoras and Meno*, pp. 129–138).

102 Cf. Plato, Epistle VII, in: *Timaeus. Critias. Cleitophon. Menexenus. Epistles* (Cambridge MA: Harvard University Press, 1999), p. 477–573.

103 *Privatissimum für G.D* (1803) (GA II/6: 334).

is that *our knowledge and the laws* of these mathematical objects and their attendant postulates are universal and invariable, and it is precisely in this sense that his theory of mathematics is to be classified as Platonistic. Fichte repeats this thought in 1813 when contrasting the lack of stability in the knowledge of empiricists. In contrast, the invariable law of the triangle can forms the epistemological basis for a mode of philosophical knowledge that is also invariable: »They [the philosophers of empirical induction] do not notice that in the infinite determinability of the act of knowing there might be a solid and invariable law; just like in the infinite determinability of the Δ [triangle].«[104]

(2) Again following Plato, Fichte grants a higher existence or reality to ideal objects than to sense or natural objects: »We have just said that the absolute concept is real and genuine being, and hence the doctrine of ethics is the doctrine of being as pure theory. Plato too had already done this, when he said things are the reflections, appearances, of the Ideas; that the Ideas are the models of the real world.«[105] And thus, as we just saw with his theory of the *Ur*-perpendicular and *Ur*-diameter, the fundamental elements of mathematics for Fichte are posited as objective *Urbilder* or *geistige Bilder*. That is to say, these elements are archetypes or spiritual images wholly in the tradition of the Platonic forms, and are therefore to be distinguished from all their empirical or material instances or images in time and space. In the Fichtean theory the latter are merely empirical ›bodies‹ for the higher ideal or ›spiritual‹ images that can only be grasped in geometric intuition: »Do not believe that what is sketched on the blackboard or impressed in copper is the object of the intuition that is demanded of you; the former is merely a body (*Körper*) to guide your imagination in producing for yourself the spiritual image (*geistiges Bild*). A mathematical line or a triangle that arises out of lines of this kind, or a triangle as such (*ein Dreyeck überhaupt*) that is neither acute, right-angled or

104 *Einleitung in die Wissenschaftslehre 1813* (forthcoming in GA II/17: 258): cf. Fichte, *Ultima Inquirenda*, p. 37.

105 The original German: »Wir haben gesagt, der absolute Begriff sei das reale und wahrhaftige Sein, darum sei die S.-L. Seinslehre, als reine Theorie. So auch schon Plato, indem er sagt: die Dinge seien Abspiegelungen der Ideen, der Gesichte; die Ideen seien die Vorbilder der wirklichen Welt.« *Das System der Sittenlehre* (1812) (SW XI: 42). (Cf. GA II/13: 338)

obtuse, or again, the schema of a geometrical demonstration that is universally valid and subsumes all possible cases, cannot understandably be marked out with a straightedge and compass (*Zirkel und Lineal*) but can only be inwardly intuited. Whoever does not possess the skill to elevate himself to this inner intuition has therefore uselessly studied this science merely in accordance with its letter.«[106]

What is the most explicit example of Fichte's mathematical *Urbild* or spiritual image? – None other than the original or archetypal elements that constitute the foundations of his theory of original geometry in the *Erlanger Logik*.[107] It is precisely these geometrical elements that possess the above characteristics of mathematical Platonism. They are »absolute«, »invariable« and »infinite« geometric elements that serve as the ideal models for all empirical and measurable geometric objects. They cannot be depicted or drawn in the physical world, but are only accessible through pure reason. They cannot initially be grasped conceptually, but require the faculty of intelligible intuition. Not only the archetypal elements of Fichte's system belong to this class of ideal geometric entities, but also the original configurations and figures of geometry. That is to say, the perfect triangle and circle, or the triangle or circle in themselves. As Fichte says of this triangle in the *Einleitung in die Wissenschaftslehre* (1813) – the »triangle »as such« that can »never be empirically drawn« but must be intuited and constructed in intuition. Moreover, it contains within it the »possibility« for every single finite and empirical triangle.[108]

(3) Like Plato himself (and later recalling the theories of Schelling and Maimon), Fichte also proposes a divine origin for mathematics.[109] Here it

106 *Die am Ende des Winters [Sommers] 1794 von Fichten Gehaltene Schlussvorlesung als Beylage der Verantwortung* (1797) (GA I/4: 413). Among others, also see *Die Thatsachen des Bewußtseins* (1813) (SW IX: 496).

107 Regarding Fichte's theory of archetypal elements, see Chapter One above which is based on Fichte's exposition in the *Erlanger Logik* (GA II/9: 124–137).

108 (*Kollegnachschrift* Halle: GA IV/6: 381); cf. the parallel passage: *Ultima Inquirenda*, p. 128.

109 Regarding Schelling's Platonism: »If there were no God, there would be no geometry« (*Abhandlung über die Quelle der ewigen Wahrheiten*) (*Sämmtliche Werke* II, 1, 528). Maimon also expresses our likeness to God, since in the construction of the objects of mathematics we also are involved in an activity where creating and

is God who has »generated« mathematics from out of his inner nature: »All fundamental moments of human history through which the human race actually advances (not through the hesitations of individual freedom) reside in the law of appearance. Just as with Christ, so God has also eternally generated, for example, mathematics and philosophy, from out of his inner being.«[110] And this view is also implicit in Fichte's extraordinary mathematical characterization of the human being in a letter to Schelling from 1801. Here the human being is conceived as a divine mathematical entity: »[man is a] rational square of an irrational root that is present is the whole spirit world«, and where the latter in turn is a »rational square of [...] an irrational square = to the immanent light or God.«[111]

(4) Just as Plato demanded students of his philosophy to practise geometry as an intellectual preparation before entering the gates of the Platonic Academy, Fichte likewise held an identical view about mathematics for the *Wissenschaftslehre* (cf. Chapter Five below). Mathematics provides a training ground for understanding the nature of self-evident propositions, which for Fichte forms the starting point of his philosophy: »I wish people had some knowledge of mathematics before embarking on a study of the *Wissenschaftslehre.* [...] In other words, I would hope that no one begins a study of the *Wissenschaftslehre* without first obtaining a clear insight into the foundation of the *immediate self-evidence* and *universality* of mathematical postulates and theorems.«[112]

thinking are one: »All concepts of mathematics are thought and simultaneously presented by us as real objects by means of a priori construction. In this we are similar to God.« Quoted in G. Freudenthal: *Maimon's* in *Philosophy of Geometry*, p. 13. Freudenthal then makes the following remark about Maimon theory of mathematics: »It stands in Maimon's philosophy for the possibility of basing human knowledge on pure understanding, of dispensing with sensuality and imagination which are dependent on corporeality and elevating Man to the company of the angels and God rather than associating him with the beasts« (ibid., pp. 13f.).

110 *Staatslehre* (1813) (SW IV: 550). (GA II/16: 147) See too Fichte's early view in the ›Closing Remark‹ to the *Versuch einer Kritik aller Offenbarung* (1792) (GA I/1: 54).

111 See Fichte's letter to Schelling, May 31st – August 7th 1801 (GA III/5: 49).

112 *Ankündigung* (GA I/7: 160).

c. Platonism, Neo-Platonism and the Wissenschaftslehre

These are the four principal reasons why I would classify Fichte's philosophy of mathematics as Platonistic in spirit. At this point we could ask: what about Fichte himself and his philosophical system in general – should he and the *Wissenschaftslehre* also be classified as Platonistic? I think that would be far too simplistic, and would neglect the entire Kantian heritage and transcendental component to Fichte's system. Nevertheless, as I have attempted to show for Fichte's theory of geometry in particular, there are other Platonic or neo-Platonic elements and aspects of Fichte's thought. Moreover, the connection with Platonism has been clearly admitted by Fichte himself. For example, in the *Die Anweisung zum seligen Leben* (1806) Fichte calls Plato both a forerunner to his system and an »exemplar« of higher morality, and furthermore places the *Wissenschaftslehre* in the spiritual tradition of the writings of the author of the Gospel of Saint John: »Among the Greeks, Plato was on this path [to the *Wissenschaftslehre*]. The Johannine Christ says entirely the same as what we teach and prove.«[113]

It is not simply the morality of the Platonic system that appealed to Fichte, but its lofty *idealism*. The life and world of the highest ideas and ideals, a living and weaving in ideality reveals the spiritual affinity between the *Wissenschaftslehre* and Plato's philosophy. As we saw, this concern with eternal ideals and absolute ideality forms the basis for Fichte's philosophy of mathematics and model of geometry. Indeed, in the *Erlanger Logik*, it is exactly this connection with the idealistic heritage of Platonism that is underscored by Fichte in the lectures preceding his exposition on original geometry. »With your permission I might add, that it is a gross untruth to state that what is called idealism etc. has only been in the world since Kant's Critique. To the extent that the human race has been enthused, it has been on account of religious men like Plato and Jesus: and they have used almost the same words as we speak.«[114]

Is it then fair to call Fichte a Platonist? Not exactly, and in this respect the last point is crucial. – Fichte did not see himself as simply a Platonist,

113 Fichte, *Die Anweisung zum seligen Leben*, lecture two (GA I/9: 73); and in lecture five: »among the ancient philosophers Plato may have had an inkling [of our view]« Ibid., p. 110.

114 Fichte, *Erlanger Logik* (GA II/9: 96).

because philosophy itself had progressed since the days of Plato. As mentioned above, it is important to bear in mind that the *Wissenschaftslehre* is above all a form of *transcendental philosophy*, and in this respect radically differs from the ancient form of Platonistic philosophy. Fichte's makes mention of this fact when he himself points out what Plato was lacking. Both Fichte and Plato ultimately distinguished between objective and pure knowledge, or knowledge of the natural sense world and knowledge of the ideal world. But unlike Plato, Fichte furthermore made a clear distinction between the two forms of the objective sense world: the world as a product of freedom in practical moral striving, and the given empirical world that is devoid of any connection with freedom. As a consequence, in the end Fichte decided that he could not say he was a Platonist, but that he was *more than* a Platonist, since there is a transcendental basis to his philosophy of knowledge: »In this antithesis [between external things and the Ideas] it is entirely obvious that he [Plato] made a distinction between objective and pure knowledge (*Wissen*) and in this respect he puts the *Naturphilosophen* to shame. However, it is hard to ascertain if he clearly distinguished between the two forms of the objective world, of the world-form as a product of freedom created in a practical manner, and the given empirical world form that has absolutely no relation to freedom. [...] Am I a *Platonist*? I believe I am more than one.«[115]

The Platonistic or neo-Platonistic heritage in Fichte's philosophy was already recognized by Fichte's contemporaries, and also has not gone unnoticed by some of his commentators, including a number of leading mathematical minds. As Werner Beierwaltes has pointed out, Novalis was one of the first of the German idealists and romantics to renew an interest in the neo-Platonic philosophy of Plotinus.[116] Here Novalis underscored the »*idealistic* similarity« between Fichte and Plotinus[117], as thinkers who sought to combine a »critical idealist and realist path« in their methodology.[118] As already noted, in this neo-Platonic vein Novalis even proposed

115 *Das System der Sittenlehre* (1812) (SW XI: 42f.). (Cf. GA II/13:338).

116 Werner Beierwaltes, *Platonismus und Idealismus* (Frankfurt am Main: Klostermann, 2nd ed., 2004), pp. 84f.

117 Cf. Novalis' letter to Friedrich Schlegel, December 10, 1798 (HKA IV: 269).

118 See Novalis, *Allgemeines Brouillon*, nos. 924, 1096 (HKA III: 445, 469). For a study of the Platonic background to Fichte's mathematical thinking in connection with Novalis, see: W. Csech »Von der Reichweite mathematischen Denkens bei Fichte

in 1798 a reading of the structure of Fichte's system as a »mathematical *Wissenschaftslehre*«.[119] Many modern commentators also situate certain aspects of Fichte's philosophy in the stream of neo-Platonism, especially his attempt to reformulate his philosophy in the *Wissenschaftslehre 1804/II*.[120] Precisely in this presentation Helmut Girndt views Fichte as a logical successor to the neo-Platonism of Nicholas Cusanus[121], whereas the Fichtean presentation of the One and Many in this same text and the doctrine of the logos in the *Anweisung zum seligen Leben* (1806) is also a renewal of neo-Platonic thought for Emmanuel Cattin.[122]

A mathematical mind of the calibre of Edmund Husserl has also situated Fichte's thought in the Platonic current. In a series of lectures from 1917 entitled: »Fichtes Menschheitsideal« Husserl above all called attention to the »spiritual dignity and power« of Fichte's idealism which for him recalled that of Plato.[123] For Husserl, Fichte's lofty view of God and the idea of the Good in his later writings are none other than »the resurrection of the glorious, if not yet mature, type of platonic interpretation of the world«[124] This is chiefly because Fichte also privileges the primacy of the practical (ethical/moral) life over the theoretical: »Fichte resembles Plato also in that for him, who, as we earlier stated, is a thoroughly idealist practical person, the theoretical interpretation of the world as foundation-

und Novalis« *Theologie und Philosophie. Vierteljahrschrift* 70 (1995), esp. pp. 41–45.

119 For a more detailed presentation of Novalis's reading, see my article: »›*Mathematische Wissenschaftslehre*‹: On a Late Fichtean Reflection of Novalis«).

120 For a recent overview of Fichte's relation to Plato and Platonism, see Christoph Asmuth: *Interpretation – Transformation. Das Platonbild bei Fichte, Schelling, Hegel, Schleiermacher und Schopenhauer und das Legitimationsproblem der Philosophiegeschichte* (2006), pp. 23–46. Here Asmuth argues that Fichte above all saw Plato as a »*vorchristlicher Wissenschaftslehrer*«.

121 Cf. Helmut Girndt, »Strukturprobleme der Wissenschaftslehre 1804«, *Fichte-Studien*, 17 (2000): 159.

122 Cf. Emmanuel Cattin, »Le néoplatonisme de Fichte« in: J-C. Goddard and A. Schnell (eds.), *L'être et le phénomène. Sein und Erscheinung. J. G. Fichte. Die Wissenschaftslehre (1804)* (Paris: Vrin, 2009), pp. 439–451.

123 E. Husserl, »Fichtes Menschheitsideal«, in: *Husserliana*, XXV (1986): 267–293. English translation: »Fichte's Ideal of Humanity« in: *Husserl Studies* 12 (1995): 111–133.

124 E. Husserl, »Fichte's Ideal of Humanity« in: *Husserl Studies* 12 (1995): 120.

al is supposed to be for a practical elevation and salvation of humanity.«[125] Husserl also considers the different elaborations of the Fichtean doctrine of religion as a modern illustration of the historical transition from Platonism to neo-Platonism, and it is here we find »the kinship the Platonic One has with the Fichtean I.«[126]

Drawing similar conclusions to Husserl, the Swiss mathematician Andreas Speiser also places the later Fichtean philosophy in the stream of neo-Platonic thought. In fact, Speiser views Fichte's »mathematical mode of thought« in the *Wissenschaftslehre 1804/II* as the perfect introduction to understanding the role of the One in Plato's Parmenides dialogue.[127] Speiser sees Fichte's Platonistic mathematical thinking to consist in his presentation of the *Fünffachheit* of modes of knowledge, which like Plato's attempt in the *Republic*, ultimately makes a sharp distinction between empirical historical knowledge and the knowledge resulting from the unified »inner connection« between all the sciences.[128] However, Fichte's presentation of the knowledge of the One and Many (or difference) in the *Wissenschaftslehre 1804/II* is inverted with regard to Plato's Parmenides dialogue, and for this reason the Fichtean theory of knowledge constitutes a perfect introduction to Plato for Speiser: »[Numbers 15–17 of Plato's *Parmenides*] contain a descent down from the being One [*vom Seienden Eins*], from reason or the light, to the intellect, which is determined through the concept of difference. Here Plato is actually obscure, because he hardly provides more than the briefest of indications. However, Fichte treats precisely this part in the greatest detail, and indeed, in the opposite ascending direction, insofar as he begins with ordinary knowledge and then ascends to transcendental knowledge.«[129]

These studies on the *Wissenschaftslehre*'s general relationship to Platonism and neo-Platonism help pave the way for an understanding of the

125 Ibid.

126 Ibid., p. 123. For an overview of Husserl's interpretation of Fichte's Platonism, see Olivier Lahbib: »Husserl, lecteur de Fichte«, *Archives de Philosophie* 67 (2004): 421–443; and *De Husserl à Fichte: Liberté et réflexivité dans le phénomène* (Paris: L'Harmattan, 2009).

127 See Andreas Speiser, *Ein Parmenideskommentar – Studien zur platonischen Dialektik*, 2nd ed. (Leipzig: K.F. Koehler, 1959), esp. pp. 73–106.

128 Ibid., pp. 74f.

129 Ibid., pp. 6f.

Platonism of Fichte's theory of mathematics. I believe it is above all Fichte's conception of archetypal geometric elements that most explicitly reveals his Platonistic heritage. If the type of philosophy one chooses depends on the type of person one is, then the same holds for the conception of mathematics one chooses. And Fichte's philosophy of mathematics and model of geometry are ultimately a supreme form of absolute and neo-Platonistic idealism.

To summarize Fichte's mathematical Platonism: Although Fichte's model of geometry differs from the Euclidean model in a number of fundamental respects, his model is not strong enough to be classified as a non-Euclidean geometry. Like many other thinkers and mathematicians, Fichte harboured doubts about the fifth postulate's formulation and he questioned its underlying assumptions, but he did not replace it or find a substitute for this postulate, which would be a strict minimum for a non-Euclidean geometry. On the one hand, Fichte's model of geometry is indebted to the Kantian conception insofar as he follows Kant in distinguishing between synthetic and analytic methods and propositions in geometry. On the other, Fichte's model of geometry also follows the synthetic method in the classic sense; that is, it privileges the exclusion of algebraic proofs and relies instead on the method of pure construction. Moreover, his starting point is not the Kantian and Euclidean starting point of the line drawn between two points, but commences with an ideal original diameter and perpendicular which would supply a normalizing and absolute direction for all derived directions. For this reason Fichte employs *other definitions* and ideal elements not to be found in Euclid, such as the original line at infinity, which corresponds more with synthetic geometry's definition of a line. And finally, if we had to classify Fichte's theory of mathematics and geometry, it would have to be interpreted as a species of mathematical Platonism, above all due to his theory of archetypal elements. Like Plato's forms or the perfect circle, Fichte also posited a number of geometric entities that are ideal, absolute and infinite, providing the basis and standard for all finite and empirical instances, and which can only be grasped in pure geometric intuition.

On account of these significant departures from Euclid and Kant, and because of Fichte's integration of original, infinite and Platonistic elements into his mathematical theories, we can no longer speak of Fichte's model of geometry as identical with the traditional Euclidean model of geometry or with a Kantian version of the same. Instead, one is justified in speaking of an original *Fichtean* model of geometry, or of a model of »true original geometry« as Fichte calls it in the *Erlanger Logik* 1805.[130]

130 »*wahre ursprüngliche Geometrie*« (GA II/9: 132).

Chapter Three: Axioms, Intuition and Construction

1. Is Fichte's *Grundsatz* equivalent to a Geometric Postulate?

In this first section of Chapter Three I will investigate one of the more striking similarities between Fichte's *Wissenschaftslehre* and geometry – its employment of first principles (*Grundsätze*) in a manner that seems to recall the postulates and axioms of mathematics and geometry. Here we shall see that although Fichte forcefully insisted upon the employment of *Grundsätze* early on in his career – which reached a highpoint in 1800 with the *Neue Bearbeitung der Wissenschaftslehre* and its attempt at a presentation *more geometrico* – Fichte ultimately abandoned this geometric presentation and structure and did not return to it in the later versions of his philosophy.

The main questions to be discussed in this section are: what is the relationship between the *Grundsätze* or first principles of Fichte's philosophy and the axioms and postulates of geometry? Are the former based on the latter? If so, to what extent? And might the opposite be the case? – That is to say, might the axioms of geometry be in Fichte's eyes more limited versions of the axioms of philosophy? Answering these questions will require beginning with an analysis of the axioms and postulates of Euclidean geometry. Unlike in Chapters One and Two, where there was a relative paucity of literature on Fichte's philosophy of mathematics, here we shall see that the relationship between Fichte's *Grundsätze* and the postulates of geometry has attracted greater attention from Fichte scholars, most notably Joachim Widmann, Reinhard Koch and Daniel Breazeale.[1]

1 For example, see Widmann's *Die Grundstruktur des Transzendentalen Wissens* (1977), pp. 15–30; Reinhard Koch's *Fichtes Theorie des Selbstbewusstseins* (1989), pp. 82–100, 115–126; and Daniel Breazeale's 2008 essay, »Toward a *Wissenschaftslehre more geometrico* (1800–1801)«, pp. 3–40.

a. Definitions, Axioms and Postulates: The First Principles of Geometry

Traditional Euclidean geometry begins with a number of first principles. What is an axiom for Euclid? It is conventional to say that geometry is a deductive science, and that Euclidean geometry in particular starts with axioms and is an axiomatic science. In fact, Euclid did not employ the term »axiom« (αξίωμα/*axioma*), but chose instead »αἰτήματα« (*aitemata*), which is commonly translated as »postulate« in English. The root word for axiom in Greek is *axios*, which means worthy or deserving.[2] In this sense an axiom or postulate is a principle that is to satisfy certain conditions: it has to be self-evident; that is, its truth does not need to be deduced or formally proved but is so transparent and clear to the mind that we can accept it without any preconditions. We are convinced of the correctness of such a principle through our own experience and therefore accept it as true. Unlike theorems, which need to be proved, axioms and postulates cannot be proved – they are *indemonstrable*. First principles of this kind are vital and indispensable to geometry and it cannot do without them. Otherwise the system itself would involve an infinite regress, with its first principles requiring other principles to justify or support them, which in turn would require further principles, and so on *ad infinitum*. To avoid this infinite regress one has to start with a few basic principles that do not rest on or require any other principles to justify them. Their truth rests on nothing but their own content and self-evidence. This furthermore means that a philosophical inquiry beginning with axioms or axiomatic first principles explicitly acknowledges a limit to its knowledge, or to be more precise: the limits of logical proof. Such an inquiry provisionally ignores these problems of the undefinability and the justification of its axioms.[3] Moreover, an axiom is a summons: It expresses in a formal manner (by means of signs or words or symbols) what we are to do, to carry out, what and how we are to think in a particular situation. Precisely this aspect was of importance to Fichte.

2 See Darius Koriako, *Kants Philosophie der Mathematik*, pp. 46.ff.

3 Cf. Jules Vuillemin's remarks »The axiomatic method precisely determines a field by listing a set of premises from which theorems follow with certainty. But it ignores the nature of its undefinable concepts and the justification of its indemonstrable principles.« *What are Philosophical Systems?* (Cambridge: Cambridge University Press, 1986), p. 104).

The Notion of a Definition

Before presenting Euclid's postulates, however, it should not be overlooked that Euclid does not start with a list of postulates in the *Elements* but first of all states 23 *definitions* of basic mathematical terms. Euclid's definitions are formulated in such a manner that they are also to be accepted as self-evident and are likewise indemonstrable. His definitions range from the first that defines the point: »A *point* is that which has no part«; to number twenty-three which states the definition of parallel lines: »*Parallel* straight lines are straight lines which, being in the same plane and being produced indefinitely in both directions, do not meet one another in either direction.«[4] Euclid's definitions have the role of introducing us to the fundamental entities of geometry by means of words. He characterizes the objects of geometry using simple but precise mathematical vocabulary; and again, these definitions are to be simply accepted as true.

With Euclid's definitions in the *Elements*, we should be crystal clear about what exactly is being defined, as well as about what is *not* being defined or is undefinable. For example, in Euclid's first definition of the point quoted above, the key and often overlooked element in this definition is the notion of a part, which is here not defined. *Prima facie*, the definition of a point seems to function negatively, insofar as one could interpret a point as simply absence and therefore equivalent to nothing. Yet what we have to bear in mind is that the definition is actually based on an intuitive notion of something else that can be defined, and this is a »part«. – Thus, even though it naturally could be defined, the notion of a part is *not* explicitly defined in Euclid's first definition of the point. As Lisa Shabel says: »Euclid uses the undefined notion of part to define a geometric object which we would take to be undefinable: point.«[5] The Euclidean axiomatic system therefore clearly demonstrates the practical limits of definability in a system of knowledge. This is because in any group of definitions some terms must remain undefined or otherwise we embark on an infinite regress – a notion we will encounter again with the axioms.

4 Euclid, *The Elements*, Book One, pp. 1f.

5 Shabel, *Mathematics in Kant's Critical Philosophy*, p. 13.

In the *Critique of Pure Reason Kant* also sets out his theory of definition and shows that philosophical definitions are different to the definitions of mathematics. This is because for Kant mathematics (geometry) begins with transparent definitions that it constructs, whereas philosophy begins with concepts. Philosophy employs definitions that usually occur at the end of a philosophical inquiry, and hence for Kant philosophers should not try to »imitate« the definitions of mathematics.[6] In the 1764 *Inquiry Concerning the Distinctness of the Principles of Natural Theology and Morality* Kant had earlier argued that mathematical definitions come about by »synthesis«, i.e. by the *combining* of concepts, whereas those of philosophy occur through analysis, or the »separating out« process carried out by cognition.[7] In Kant's theory of mathematics the definition depends on its concept being first *constructed* in cognition and only comes into being in this manner. Here »construction« of a concept means »to exhibit a priori the intuition corresponding to it.«[8] Thus, a mathematical definition for Kant only comes into being after it is constructed by the mathematician.[9] This has a consequence for the certainty of the mathematical definition, as Kant stresses in the section entitled »The discipline of pure reason in dogmatic use« in the *Critique of Pure Reason*. This is because the certainty of the concept itself is given through the definition and can therefore only contain what is cognized by means of it, whereas the analytic definitions of philosophy can be erroneous by not being exhaustive enough or by incorporating extraneous elements into the concepts.[10]

6 Kant, *Critique of Pure Reason* A727/B755, p. 637, and A730/B758, pp. 638f.

7 Kant, AA 2: 276–278. English translation in: Kant, *Theoretical Philosophy 1755–1770*, pp. 248ff.

8 Kant, *Critique of Pure Reason*, A713/B741, p. 630.

9 »The concept which I am defining (*erkläre*) is not given prior to the definition itself; on the contrary, it only comes into existence as a result of that definition.« Kant, *Inquiry Concerning the Distinctness of the Principles of Natural Theology and Morality* (1964) (AA 2: 276). English translation in: Kant, *Theoretical Philosophy 1755–1770*, p. 248.

10 »Mathematical definitions can never err. For since the concept is first given through the definition, it contains just what the definition would think through it. [...] Analytic definitions, on the contrary, can err in many ways, either by bringing in marks that really do not lie in the concepts or by lacking the exhaustiveness that constitutes what is essential in definitions, since one cannot be so entirely

In Widmann's interpretation, Fichte also began various versions of the *Wissenschaftslehre* with *definitions*, and hence shares this starting procedure in common with Euclid's *Elements*.[11] Of course, in contrast to Euclid, Fichte's principal task is not to define the objects of mathematics and geometry, but to precisely indicate the conditions, the nature of the possibility and activity of cognition, so that his students would be properly guided into his system.

Fichte's employment of the definition in the sense of the Euclidean system is particularly apparent in the *Zurich Wissenschaftslehre*, which begins the first lecture with a number of definitions and precise reflections on the role of a definition. According to Fichte, there are certain things that the science of philosophy cannot define, such as consciousness; philosophy has to be content with merely »describing« them.[12] Yet if we do define something the definition should indicate what Fichte calls the »*genus*« (*Gattung*) of what is to be defined as well as the identifying marker or »mark« (*Merkmahl*) that distinguishes it.[13] Similarly, in the *Erlanger Logik* of 1805 Fichte states that we can only define something if we know its so-called »genetic law«, i.e. how it *comes into being*, and he quotes the definition of a triangle to support his claim.[14] With a triangle we can precisely state what it is because we can penetrate into it and »grasp the law of its construction«, it does not contain any arbitrariness or chance (*Zufälligkeit*). And if we cannot do this, such as with the definition of »human being« (*Mensch*) or the »pure I«, then we »have to be content with

certain of the completeness of their analysis. For this reason the mathematical method of definition cannot be imitated in philosophy.« Kant, *Critique of Pure Reason* (A 731/B759-A732–760), pp. 639f.

11 Joachim Widmann, *Die Grundstruktur des Transzendentalen Wissens* (1977), pp. 22f. However, I am not in agreement with Widmann's contention that Fichte's text *Über den Begriff der Wissenschaftslehre* (1794) had the function of acting like a »definition« for the 1794 *Grundlage* (cf. ibid. footnote 52, p. 316). Widmann's view of a definition is far too broad. I would suggest that the text *Über den Begriff der Wissenschaftslehre* is much more elaborate than a simple definition and is more a propedeutic piece of writing. However, I will argue below that in a number of texts, such as the *Neue Bearbeitung der Wissenschaftslehre 1800*, Fichte also employed definitions partly in the Euclidean manner.

12 *Zurich Wissenschaftslehre* (GA IV/3: 21).

13 Ibid.

14 GA II/9: 81.

descriptions.«[15] We will come across Fichte's genetic theory again in his conception of construction and in his discussion of self-evidence.

As already discussed, it is highly probable that Fichte's knowledge and view of the subject of geometry, especially its technical aspects, was influenced in part by the mathematician Johann Schultz. The latter's most influential textbook on mathematics, the *Anfangsgründe der reinen Mathesis* (Königsberg: Hartung, 1790), is not solely a dry mathematical textbook, but is interspersed with some interesting philosophical reflections on the nature of mathematics in general and on definitions in particular. Here Schultz draws attention to their chief characteristics. Definitions are: 1) detailed with the marks (*Merkmale*) of the object; 2) precise and original; and 3) not derived. His conception of a definition shares a number of technical points in common with Fichte's view, especially with regard to the marks and »genetic« method. ›Genetic‹ signifies for Schultz the »possibility« of a thing: »A *Definition* (*Erklärung*) means the original and precise determination of the detailed concept of a thing. The definition should determine the object in such a way that everyone thinks the same concept under it. Consequently, the concept (1) must be detailed, i.e. it has to completely contain the marks (*Merkmale*) of the object. (2) It has to be exact or precise, [...] (3) It has to be original (*ursprünglich*), i.e. it must not first require a derivation or a deduction, but be absolutely and thoroughly posited. If the definition also exhibits the possibility of the thing (*Sache*), e.g. a circle is a planar figure is described when a straight line moves around a fixed point, then it is called a »genetic« definition.«[16]

As noted above, the version of the *Wissenschaftslehre* that most approaches the external form and method of Euclid's system is the *Neue Bearbeitung der Wissenschaftslehre* 1800 (GA II/5: 331–401). This text only partly holds for the Fichtean notion of a definition, because although the

15 Ibid., p. 81f. Fichte talks about the undefinability of the I in *Einige Vorlesungen über die Bestimmung des Gelehrten* (1794/95), saying that it can only be intuited and not defined; or that it can only be represented in a negative manner: »What would actually the spiritual element in man be, the pure I – absolutely in itself – isolated – and apart from any relation to something external to the same? – This question is unanswerable – and taken in itself is self-contradictory. [...] The pure I can only be negatively represented; as the opposite of the Not-I.« (GA I/3: 28f.).

16 Johann Schultz, *Anfangsgründe der reinen Mathesis* (Königsberg: Hartung, 1790), p. 16.

idea is evoked in the text, it does not possess as much weight as the Euclidean definitions (or Schultz's presentation of the same). Fichte is here displaying his habitual distrust of definitions and considers them to be not much better than descriptions. For unlike Euclid and Schultz, Fichte does *not* start with a list of definitions but proceeds to directly enumerate his postulates. In the *Neue Bearbeitung* a definition only has the role of *clarifying* and *directing* certain activities of the mind so that the student acquires a definite intuition. As we shall see with the axioms and postulates, all definitions and demonstrations are *explanations*, and not to be blindly accepted as fixed, but simply function as »devices« or »signposts« (*Hinleitungen*) leading us to intuitions and providing the description with meaning.[17] In this respect the *Wissenschaftslehre* is similar to the method of mathematics: »The manner of proceeding is exactly the same as in *mathematics*. It is demonstrated in intuition. I can do nothing else. Mathematical demonstrations are only signposts for leading us to the point of intuition.«[18]

The Axioms and Postulates of Geometry

The following is a list of the first four postulates in Book One of Euclid's *Elements*. Euclid presents his postulates in an entirely direct manner without any long introduction:

»*Postulates*

Let the following be postulated:

(1) To draw a straight line from any point to any point.

(2) To produce a finite straight line continuously in a straight line.

17 Cf. Fichte's *Neue Bearbeitung* (GA II/5: 335). In this text Fichte uses the example of the definition of »determined« (*bestimmt*): »Definition (*Erklärung*). We call »determined« whatever is thought as a limited portion of a certain sphere, and which in a certain respect is opposed to everything else that is external to this sphere. – This is a mere explanation of the words. The kind of thinking just described of the thought of the determined has to be found by everyone in his own intuition. And the given description first procures its meaning and lucidity from this intuition« (GA II/5: 352f.). I am interpreting and translating the German term *Erklärung* as *Definition*, which is also the case for the editors of the Cambridge Edition of Kant's works in English. See the translation indications of the editors in the footnotes to the *Critique of Pure Reason*, p. 639.

18 GA II/5: 335.

(3) To describe a circle with any centre and distance.
(4) That all right angles are equal to one another.«[19]

It is apparent that the aim of these first four postulates is simply to state the most basic and uncontroversial principles for a system of geometry. These postulates are expressed in short sentences that cannot be proved, but must be experienced by every thinking person. Everything in Euclid's system of geometry is to be deduced from these postulates, as well as making use of the definitions. And just like the definitions, all the subsequent theorems to be proved in Euclid's system are to follow from these un-provable and indemonstrable first principles. These first principles genuinely ›postulate‹, i.e. they say or tell us that certain actions and acts can or may be carried out, that they are *possible*.

However, Euclid's first book of the *Elements* has of course five postulates and not four. The fifth postulate, the famous parallel postulate or postulate about parallel lines, is as follows:
»(5) That, if a straight line falling on two straight lines makes the interior angles on the same side less than two right angles, the two straight lines, if produced infinitely, meet on that side on which the angles are less than the two right angles.«[20]

One immediately notices that Euclid's fifth postulate is unlike the other four postulates. It is not expressed in a short simple sentence but is long and appears to be unnecessarily complicated, and somewhat astonishingly perhaps, it does not include the term »parallel.« Upon first reading it is certainly arguable that its truth is *not* immediately transparent to the mind and therefore does not satisfy one of the criteria for a postulate – a transparent, self-evident truth. As we saw in Chapter One, generations of geometers believed that Euclid's fifth postulate was not on a par with the simplicity and self-evident nature of the other four postulates and hence thought that the fifth postulate must be a theorem. Consequently, they either sought to reformulate it into a more straightforward statement incorporating the term »parallel«, such as »parallel lines never meet.«[21] Or they

19 Euclid, *The Elements*, p. 2.
20 Ibid.
21 Davis and Hersch have a list of four other formulations of Euclid's fifth postulate. It includes the famous Playfair postulate, named after the British mathematician John Playfair (1748–1819): »Given, in a plane, a line L and a point P not on L, then

thought it ought to be capable of proof and set about trying to prove it, as was the case with Johann Schultz in his *Anfangsgründe der reinen Mathesis*.[22] This ultimately led to the different formulations of the fifth postulate which resulted in non-Euclidean geometries.

Philosophically speaking, that such indemonstrable principles do in fact exist and that certain basic and fundamental principles cannot be proved, was a notion also admitted by ancient philosophers from around the same period as Euclid. Aristotle, for example, claimed that the law of non-contradiction was precisely such an indemonstrable principle. Its truth was so self-evident and obvious that we had to simply accept it as true and forgo any attempt at proving it if we are to avoid an infinite regress. [23] Even though a number of modern philosophers and mathematicians claim that some of Euclid's first principles or postulates are not explicit enough and actually contain a number of logical »gaps« and »deficiencies«, it is still almost universally granted that there exist certain principles beyond proof and which do not rest on further principles, and these principles are our so-called postulates or axioms. [24]

through P there exists one and only one line parallel to L.« See *The Mathematical Experience*, p. 243.

22 In this work Schultz's definition of a parallel line is first preceded by »definition 25«, and then his so-called parallel »proof« (*Beweis*) follows from theorem (*Lehrsatz*) 21. See Johann Schultz *Anfangsgründe der reinen Mathesis* (Königsberg: Hartung, 1790), pp. 284ff. As already noted, Fichte was critical of Schultz's theory and remarked that it »does not accomplish anything and his entire theory is improbable« (*Erlanger Logik*, GA II/9: 131).

23 In Aristotle's eyes the law of non-contradiction is »the most certain of all principles«, and because »a demonstration of everything is impossible« *The Metaphysics* (1006a). In book gamma of the *Metaphysics* Aristotle formulated the law of non-contradiction in the following manner: »It is impossible that the same thing can at the same time both belong and not belong to the same object and in the same respect, and all other specifications that might be made, let them be added to meet local objections.« (1005b19–23). *The Metaphysics*, trans. Hugh Lawson-Tancred (London: Penguin, 1999), pp. 87ff. See too Widmann's study of the principle of non-contradiction in a Fichtean context: »Dialektik more geometrico. Fichtes Prinzip des Systemgenesis« (1983), pp. 188ff.

24 For example, Shabel cites the complaints of Morris Kline and Marvin Greenberg against Euclid's first principles, which involve the standard charge that in terms of modern logic Euclid's postulates were lacking in total transparency and explicitness (see Lisa Shabel, *Mathematics in Kant's Critical Philosophy*, pp. 9ff.). Howev-

We might now ask: what did Fichte have to say about the axioms of geometry in the sense of purely geometric principles? That is to say, not specifically with regard to *Grundsätze* and the starting point of the *Wissenschaftslehre* (which we will discuss in a moment), but with regard to the notion of a geometric axiom as such? It appears that Fichte did not use the word »axiom« much at all, and when speaking of geometry he preferred other terms like *Satz* (proposition) or *Theorem*, and *Postulat* (postulate). Interestingly, Fichte uses the term axiom in one of his most public writings, the *Verantwortungsschriften gegen die Anklage des Atheismus* (1799). – However, here the term does not refer to geometry but is employed in connection with logic where two of the headings are entitled »First Logical Axiom« and »Second Logical Axiom.«[25] Why did Fichte choose the term axiom here? In my opinion he did so because it relates directly to his notion of the role of propositions in geometry and the schematic nature of our thought, which for Fichte depends on a process of proof and construction, and the deducing of unknown truths from known ones. In contrast with the employment of the term in geometry (which Fichte mentions in this context), Fichte here seems to use the term axiom in a *pejorative* sense, as an example of a rigid and inflexible belief.[26] This is also how the term is used in *Bericht über den Begriff der Wissenschaftslehre und die bisherigen Schicksale derselben* (1806/07), where Fichte speaks of the prejudices of his opponents and the: »absolute axiom and premises of all their presupposed judgements.«[27] Again in the *Anweisung zum seligen Leben* (1806), Fichte criticizes those who dogmatically assert the primacy of empirical truths and reject any investigation of in-

er, as Shabel herself points out in reply, Euclid's system of postulates should not be interpreted in our modern sense as a strictly logical deductive system, and »since Euclid did not use the term ›axiom‹ there is no reason for us to suppose that his postulates and common notions should function as the axioms of a modern mathematical system do.« (ibid. p. 10).

25 For example, the first axiom runs: »Whoever denies certain determinations of a thing (in a concept) does not thereby necessarily remove the thing (the concept) itself.« – Fichte, *Verantwortungsschriften gegen die Anklage des Atheismus* (1799). (GA I/6: 44).

26 Cf. Ibid., pp. 43ff.

27 Fichte, *Bericht über den Begriff der Wissenschaftslehre und die bisherigen Schicksale derselben* (GA II/10: 41).

tellectual or non-empirical principles.[28] There are of course some exceptions to this judgment, for instance in the 1807 unpublished text with the name: *Nebenbemerkungen zu 1.*, where Fichte establishes an axiom for representations (*Vorstellungen*).[29] However, more often than not the term »axiom« for Fichte seems to be used in the sense of an infallible truth to which people blindly cling.[30]

Kant, in contrast, preferred the term »axiom« for the first principles of geometry. In the »Doctrine of Method« in the *Critique of Pure Reason* Kant states what he considers the nature of an axiom to be, and he appears to deploy it in a manner similar to the axioms of Euclidean geometry, insofar as he lays stress on their absolute certitude and transparency. But this is misleading with regard to the efforts of Fichte to establish philosophy based on a first principle or axiom, because Kant in fact rejects precisely this view of philosophy based on axioms in the same manner as geometry. Kant sees transcendental philosophy as rational cognition based on concepts, and since concepts require »additional« information and »deduction« any synthetic principles based on them are »discursive« in contrast with the intuitive and immediately certain axioms of geometry. According to Kant, it is therefore futile to attempt to establish mathematical-like axioms or axiomatic first principles within the domain of philosophy: »On axioms. These are synthetic a priori principles, insofar as they are immediately certain. [...] Now since philosophy is merely rational cognition in accordance with concepts, no principle is to be encountered in it that deserves the name of an axiom. [...] Mathematics, on the contrary, is capable of axioms, e.g. that three points always lie in a plane, because of the construction of concepts in intuition of the object it can

28 See GA I/9: 83.

29 »We can establish as an axiom: in the representation (in the concepts of the understanding) there is nothing that is not something (according to the analogy of the Kantian proposition). Being (*das Sein*) is above all an absurdity (or non-thing: *Unding*). This is why the understanding (*Verstand*) is absolute.« (GA II/10: 226) According to the editors of Fichte's *Gesamtausgabe* this text is mostly likely a collection of notes relating to the *Wissenschaftslehre* given in Königsberg in 1807 (ibid., p. 221).

30 See too Fichte's remarks in the *Annalen des philosophischen Tons* (1797) (GA I/4: 294). In the *Erlanger Logik* 1805 Fichte also uses the term axiom in a negative sense as something equivalent to a dogmatic truth (cf. GA II/9: 95).

connect the predicates of the latter a priori and immediately. [...] Philosophy thus has no axioms and can never simply offer its *a priori* principles as such, but must content itself with justifying their authority through a thorough deduction.«[31]

Turning to the mathematician Johann Schultz, he preferred to label the first principles of geometry »axioms« in his *Anfangsgründe der reinen Mathesis*, and they are entirely in the spirit of Euclid.[32] However, Schultz also mentions his Königsberg colleague Kant in this text, and precisely with regard to the status of »*Sätze*« and »propositions«, which he likewise defines as the determination of the relations between two items of knowledge.[33] Distinguishing between analytic and synthetic propositions in strict Kantian fashion, he concludes that »pure mathematics rests entirely on synthetic principles a priori, as I have shown in detail in my *Prüfung der Kantischen Critik der reinen Vernunft.*«[34]

To confuse matters slightly, while Euclid had of course preferred the term »postulate« to axiom, Kant famously uses the term postulate in an altogether different fashion. For example, Kant enumerates the »postulates of empirical thinking« in the *Critique of Pure Reason*, as well as the postulates of practical reason, such as immortality, freedom and the existence of God, in the *Critique of Practical Reason*. A postulate for Kant is not a theoretical dogma but a »presupposition« or a *Grundsatz* from the practical standpoint.[35] It is an idea that must be accepted as true despite the fact that it cannot be proved in a theoretical manner. Practical postulates of this kind are required to live a moral life, and as Kant asserts in the

31 Kant, *Critique of Pure Reason*, A 732/B760-A734-B762, pp. 640f..

32 For example, his first two axioms of geometry formulate his view of space as 1) »infinitely extended on all possible sides«; 2) space is »continuous« (*stetig*), i.e. there are no breaks and the limits and boundaries of one body or solid is the boundary of the next.« Johann Schultz, *Anfangsgründe der reinen Mathesis*, p. 245.

33 Ibid., p. 17.

34 See Schultz *Prüfung der Kantischen Critik der reinen Vernunft* (Königsberg: Nicolovius, 1792). Cf. Henry Allison, *The Kant-Eberhard Controversy*, p. 9; and Darius Koriako, *Kants Philosophie der Mathematik*, pp. 279f.

35 In this sense Kant defines a postulate in his *Logik* as: »a practical, immediately certain proposition or a *Grundsatz* (fundamental proposition) which determines a possible action of which it is presupposed that the manner of executing it is immediately certain« (AA IX: 112).

Critique of Practical Reason, they are *necessary hypotheses*.[36] Kant was conscious of the fact that he was giving a new meaning to the term postulate that was not identical to the one in geometry: »Pure geometry has postulates which are practical propositions, but contain nothing further than the assumption that we can do something if it is required that we should do it, and these are the only geometrical propositions that concern actual existence. They are, then, practical rules under a problematical condition of the will; but here the rule says: We absolutely must proceed in a certain manner. The practical rule is, therefore, unconditional, and hence it is conceived a priori as a categorically practical proposition by which the will is objectively determined absolutely and immediately (by the practical rule itself, which thus is in this case a law); for pure reason practical of itself is here directly legislative.«[37]

Finally, Kant also employs the term »Satz« (proposition or theorem) in the *Critique of Practical Reason* in a different (but related) sense to traditional Euclidean geometry. Kant says a theorem does not have the same meaning as the above use of »postulate«, yet may be compared with the postulates of Euclid. This is because it is still a practical proposition and can be therefore used to derive further theorems, just as one does in geometry.[38] As mentioned, in contrast to Kant, Fichte does not favour the term axiom but the term postulate or proposition (*Satz*) when discussing

36 »The expression of a postulate of pure practical reason might give most occasion to misapprehension in case the reader confounded it with the signification of the postulates in pure mathematics, which carry apodictic certainty with them. These, however, postulate the possibility of an action, the object of which has been previously recognized a priori in theory as possible, and that with perfect certainty. But the former postulates the possibility of an object itself (God and the immortality of the soul) from apodictic practical laws, and therefore only for the purposes of a practical reason. This certainty of the postulated possibility then is not at all theoretic, and consequently not apodictic; that is to say, it is not a known necessity as regards the object, but a necessary supposition as regards the subject, necessary for the obedience to its objective but practical laws. It is, therefore, merely a necessary hypothesis. I could find no better expression for this rational necessity, which is subjective, but yet true and unconditional« (AA V: 11). (Translated by Thomas Kingsmill Abbott).

37 Kant, *Critique of Practical Reason* (AA V: 31) (Translated by Thomas Kingsmill Abbott).

38 Kant, ibid. See too Kant's remarks in the *Prolegomena* (AA IV: 301).

geometry. Following Reinhold, and against the letter of the Kantian critical philosophy, it was Fichte who most popularized and championed the notion of a *Grundsatz* or first principle in philosophy. However, is Fichte's *Grundsatz* identical to an axiom, *Satz*, or postulate in geometry? It is to this question that we will now turn.

b. Fichte's Grundsatz and the Postulates of Geometry

> *Grundsatz*: The I originally posits absolutely its own being.
>
> J.G. Fichte.[39]

> First Postulate: Think of yourself, construct the concept of your self and attend to how you do this.
>
> J.G. Fichte.[40]

This second section of Chapter Three will examine Fichte's idea of a *Grundsatz* – a first or grounding principle – and try to determine in more detail its relationship with the axioms and postulates of geometry. Due to his insistence on first principles Fichte's system has often been labelled as a type of »foundationalism« and a large body of literature has sprung up investigating the early reactions, rejections or anti-foundationalism of Fichte's students and opponents during the Jena period.[41] This foundational debate will not concern us here, except insofar as it is relates to the status and role of geometrical first principles and axioms in philosophy. Precisely this question as to the exact nature of Fichte's *Grundsatz* has generated an entire library in itself. But as I said, I will examine it in a specific sense: to what extent is Fichte's *Grundsatz* equivalent to a geometrical postulate, axiom, or theorem? The question is not as straightforward as it may perhaps appear, and there is no consensus even within Fichte research. As Erich Fuchs has recently remarked: »In my view, even among

39 »Das Ich sezt ursprünglich schlechthin sein eigenes Seyn«, in: §1 of the *Grundlage der gesammten Wissenschaftslehre* 1794/95 (GA I/2: 261; SW I: 98).

40 Fichte, *Zweite Einleitung in die Wissenschaftslehre* (GA I/4: 213).

41 Among others, see especially Manfred Frank, *Unendliche Annäherung* (1998); Frederick Beiser, *German Idealism* (2002), pp. 235–272; and Günter Zöller, *Fichte's Transcendental Philosophy: The Original Duplicity of Intelligence and Will* (Cambridge: Cambridge University Press, 1998), pp. 45–48.

Fichte scholars it is still debatable whether the fundamental act presented in § 1 of the *Foundations of the Entire Wissenschaftslehre* is a hypothesis, a postulate, a belief, or an insight accomplished in immediate intellectual intuition.«[42]

A *Grundsatz* or first principle is something that is basic, primary and indemonstrable, and cannot be derived from and is not based on any further principle. Insofar as Fichte uses the term »postulate« for a first principle, it tells us, or it ›postulates‹, that something may be carried out.[43] As we saw, the idea of indemonstrable first principles in philosophy goes back at least to Aristotle. Fichte asserted that philosophy needed a *Grundsatz* in particular to render the discipline of metaphysics more scientific and systematic: »Philosophy has lacked up to now a supreme and universally accepted *Grundsatz*, and only after the establishment of one can philosophy raise itself up to the rank of a science.«[44] In his Zurich letters and in the *Zurich Wissenschaftslehre*, and above all in his early Jena writings, Fichte championed a foundational philosophy based on such a first principle. In this respect he believed himself to be following in the footsteps (and spirit) of both Kant and Reinhold, and happily acknowledged his debt to them. Writing to Reinhold in early 1795 Fichte said: »You have, just like Kant, brought something to humanity that will eternally remain in it. He proceeded from the investigation of the subject, while you examined how it had to be carried out from a single *Grundsatz*.«[45] While in the 1794 *Über den Begriff der Wissenschaftslehre*, Fichte recounts that this single *Grundsatz* of a science has to be absolutely certain (*gewiss*) and equally linked to every other principle in the science: »A principle of this kind that is certain prior to and independently of its linking with the other principles is called a *Grundsatz*. Every science must have a *Grundsatz*.«[46] In the *Zurich Wissenschaftslehre* of February 1794 Fichte pointed out that a *Grundsatz* should have certain characteristics or »identifying marks«: (1)

42 Erich Fuchs, »Fichte: A System of Freedom? Biographical-Philosophical Reflections« *Idealistic Studies* 41 (2011): 116.

43 For instance, see Fichte, *Zweite Einleitung in die Wissenschaftslehre* (GA I/4: 213).

44 Fichte, *Recension des Aenesidemus* (1794) (GA I/2: 42).

45 J. G. Fichte to K. L. Reinhold, end of March/ beginning of April 1795 (GA III/2: 282). In the *Grundlage* Fichte again hinted at his debt to Kant, yet also criticized him for failing to provide a *Grundsatz* (cf. GA I/2: 262).

46 GA I/2: 115.

It should be immediate and absolutely certain. (2) Consequently, it should not require any explanation or proof, and is in fact so clear and transparent that it is beyond proof.[47] Following Kant, Fichte here took a proof or demonstration to mean a showing or establishing (*vorweisen*) of something in a sensible representation (*sinnliche Vorstellung*) by means of decomposition.[48] As many other scholars have already pointed out, Fichte adopted a number of these conditions of his *Grundsatz* directly from Reinhold, especially the *Grundsatz's* characteristics of being self-evident and transparent and the foundation and sufficient reason for every other principle.[49]

What is Fichte's Grundsatz?

It is clear that a first principle was of prime importance to Fichte at this time. However, what exactly is Fichte's first principle or *Grundsatz*? In his early Zurich and Jena writings we find a number of different formulations, but they are all similar in form and content. Fichte's aim was for each student to become aware of the activity of own »I« (*Ich*) or conscious ego, and this was the *starting point* of his whole philosophy. This *freely* undertaken intuiting of the I and the raising of it into consciousness constitutes the very first principle of the Fichtean system. Already in the Zurich *Eigne Meditationen* we see Fichte testing out a number of different formulations of his first principle. Alternating between the language of logic and the language of geometry (propositions, postulates, corollaries etc.), they all revolve around the student's conscious intuiting of his own I: »The first proposition is that of the ›I‹. [...] 1st proposition./ ›The I can be intuited.‹ ›Intuit your I.‹ ›You are conscious of your I.‹ (1st postulate. – ›We can become conscious of our I.‹) [...] Corollary. Everyone is only immediately certain of *his own* I; for he can only intuit this.«[50] Almost iden-

47 See lecture three of the *Zurich Wissenschaftslehre*: »Such a grounding principle should possess the following identifying marks: it should be certain in itself, *immediately* and without any proof; it should be clear without any explanation.« (GA IV/3: 30).

48 Cf. Ibid., p. 33.

49 Concerning Fichte's intellectual debt to Reinhold, see Frederick Beiser, *The Fate of Reason*, pp. 244ff.

50 Fichte, *Eigne Meditationen* (GA II/3: 26ff.).

tical formulations are found in Fichte's major Jena writings, including the *Grundlage der gesammten Wissenschaftslehre* of 1794/95[51] and the *First* and *Second Introductions to the Wissenschaftslehre.*[52] It is significant to note that Fichte did *not* think it paramount to provide a definition or explanation of exactly what is meant by the term »I« (*Ich*), but that he left it up to the individual to experience and discover this for himself or herself. In fact, Fichte stated that whoever is unable to discover the existence of such an I within himself is not fit for any kind of intellectual analysis or the business of philosophizing in general.[53]

Due to its reliance on the conscious intuiting of the I, many commentators have pointed out the parallels between Descartes *cogito* and Fichte's foundational *Grundsatz.*[54] Fichte himself noted this resemblance in the *Grundlage*, but claimed both the system of Descartes and Spinoza were mere propedeutics to the *Wissenschaftslehre*, because they lacked a genuinely supreme first principle.[55] Fichte views his attempt as a rigorously systematic and *transcendental* system of philosophy, insofar as it not only invokes a *cogito*, but posits a self-evident first principle from which one is to derive the necessary conditions of its possibility, and from these conditions one should derive in turn further conditions, and so on.[56]

51 »The I originally posits absolutely its own being« (GA I/2: 261). Here Fichte also discusses it in relation to the so-called »*Grundsatz* of logic«. I.e. »A = A« (GA I/2: 261).

52 For instance, in the *Zweite Einleitung in die Wissenschaftslehre*, 1797: »First Postulate: Think of yourself, construct the concept of your self and attend to how you do this.« (GA I/4: 213).

53 See, for example, Fichte's words from the *Eigne Meditationen*: »Remark. – What the I is, that is, how can it be thought, is not the question here. It cannot be defined for anyone who doesn't already know it, and it will not be defined. Whoever cannot become conscious of his I will doubtlessly have no pretensions to philosophizing.« (GA II/3: 27).

54 See for example Reinhard Lauth's *Transzendentale Entwicklungslinien von Descartes bis zu Marx und Dostojewski* (Hamburg: Meiner, 1989); and Frederick Beiser's *German Idealism*, p. 235.

55 In this passage of the *Grundlage*, not only Descartes and Spinoza, but also Leibniz, Kant, Reinhold and Maimon are all considered as historical precursors to Fichte's system (cf. GA I/2: 261ff.).

56 Cf. Fichte, BWL (GA I/2: 120f.).

Yet why did Fichte see the *need* for a *Grundsatz* at all in philosophy? On the one hand, Fichte believed that such a *Grundsatz* would be the only way for philosophy to become *scientific*, to raise metaphysical speculation to the level of a science. On the other, such a scientific philosophy would be able to refute the sceptic philosophers, with whom Fichte had explicitly set himself in opposition right at the outset of his career. This point is particularly obvious in the *Zurich Wissenschaftslehre*, where Fichte first discussed the discipline of geometry. Here Fichte asserted that every genuine science had to have a *Grundsatz* or grounding principle that was modelled on geometrical postulates or axioms insofar as they were self-evident, indisputable, and did not require any proof. All further principles of the science were then to be derived in a deductive manner from this first principle, which in turn would give rise to further principles. And like in geometry, all these principle are »linked« together into a concatenation and form a single whole which Fichte considers to be the essence of a science. It is scientific not only with respect to its form but also its content.[57] Yet even the *Grundsatz* of mathematics, or of natural science, are not the supreme first principles of science, but each in turn is to be derived from the first or grounding principle of the *Wissenschaftslehre*. According to Fichte, this is because all the other sciences are grounded upon »facts« (*Thatsachen*), whereas the *Wissenschaftslehre* is grounded upon a fact/act or a *Tathandlung*, i.e. the first principle stated above that is the actively free intuiting of one's own I.[58]

57 »Every science has a *grounding principle* (*Grundsatz*), on the basis of which all the propositions are proven, and which itself stands in need of no proof. Everything that presents itself within a science must be precisely determined and must be derived from the grounding principle either directly or indirectly. The concatenation of all the propositions into a single whole makes a science into a system. This interlinking and concatenation of various propositions into a *single whole* has been considered to be the *essence* of science. / One determines the *how* of the presentation; its *what* remains undecided. / *Form*, however, can only be a *means*, not the *end*.« Fichte, *Zurich Wissenschaftslehre* (GA IV/3: 19).

58 »The *Grundsatz* of mathematics and that of pure natural science, for example, must be contained in the *Wissenschaftslehre*. But if these two types of inferences were not sharply distinguished, then everything would be *Wissenschaftslehre*. There would be no special sciences. It will become evident that all sciences are grounded upon *facts* (*Thatsachen*), which in the case of the *Wissenschaftslehre* are fact/acts (*Thathandlungen*).« Fichte, *Zurich Wissenschaftslehre* (GA IV/3: 23).

Finally, why did Fichte choose this particular idealistic *Grundsatz* of the self-intuiting I as the basis of his philosophy, and not some other different axiom or first principle? Well, the best way to answer this is to refer to what is probably Fichte's most famous expression: that the type of philosophy one chooses for oneself, depends on what kind of person one is.[59] – As Heinrich Scholz remarked in 1934 about this particular expression: »Fichte once said in a frequently cited passage: *What kind of philosophy one chooses depends on what sort of person one is*. I will refrain from commenting on this passage here. [...] In any event, we may say of the final axioms of a person, what Fichte once declared to hold for philosophy in general. – The choice of a final axiom is indeed decisive for what sort of person one is.«[60]

Grundsätze compared with Geometric Axioms and Postulates

We have now ascertained the nature of Fichte's *Grundsatz*, but what is its relationship to geometric axioms or postulates? – To begin with, Fichte's strategy was to try and present his *Grundsatz* in such a manner that would convince his contemporaries that it shared a number of characteristics in common with the axioms of geometry, especially the characteristic of *self-evidence*. Recalling his early letter to Flatt (in November of December 1793): »I am convinced that philosophy can only become a science through the development out of a single *Grundsatz*, but that it must receive the *self-evidence* of geometry, and that there exists a *Grundsatz* of this kind. However, it has only not yet been established *as such*: I believe I have found such a *Grundsatz*.«[61] By January 1794 Fichte was clearly buoyed by the success of his project, and the text of the *Eigne Meditationen* is the first attempt at building a philosophy upon a geometric-like *Grundsatz*: »I have really already built my outline on my *Grundsatz*, and found the transition to the practical part.«[62] With regard to the term »postulate« during this Zurich period, Fichte uses it both in the sense of Euclidean geometry as well as in the Kantian sense of a practical postulate.

59 Cf. Fichte, *Erste Einleitung* 1797, GA I,4: 195.

60 Heinrich Scholz, »Warum Mathematik«, in: *Mathesis Universalis: Abhandlungen zur Philosophie als Strenger Wissenschaft*, p. 316.

61 GA III/2: 18.

62 Letter to Reinhard, January 1794 (GA III/2: 38).

For example, in the text entitled *Practische Philosophie* that is devoted to an exploration of moral philosophy, Fichte likewise uses the term »postulate« in the Kantian sense of the postulates of practical reason.[63] On the other hand, in the *Eigne Mediationen* which dates from the same period, Fichte uses »*Satz*« (or »Saz«), »*Postulat*« and *Grundsatz* interchangeably, but primarily in the sense of a first and fundamental principle of philosophy. Hence, in differing but overlapping senses both these long notebook-like texts are already imprinted with the technical language of geometry, including the terms proposition (*Satz*), corollary (*Corollar*), axiomatic requirement (*Heischesatz*) and »Q.E.D.« – the classic formulation of a completed geometrical proof.[64]

Later works, such as the *Neue Bearbeitung der WL 1800* and the *Erlanger Logik 1805*, shed even more light on the geometric quality of Fichte's *Grundsatz*. Indeed, in the *Erlanger Logik* Fichte explicitly compares his *Grundsatz* of the *Thathandlung*, the intellectual intuition of the I, with the first postulate of Euclidean geometry, i.e. that of drawing a line between two points. There are also some crucial and important differences, according to Fichte. – In geometry our freedom is only theoretical or *formal*, and does »not condition the content«, whereas in the *Wissenschaftslehre* the content is »living«, and based on the primacy of practical reason.[65] As we shall see below this is ultimately the key difference: the

63 »However, how is an experience *postulated*? *Quo jure*? Perhaps this opens up to us the path to all possible practical postulates.« (GA II/3: 191).

64 »1st Postulate (*Postulat*) – one can become conscious of one's I. [...] Corollary. Everyone is only immediately certain of *his own* I, for he can intuit only this. It is precisely the aim of an *Elementary Philosophy* to show how this immediate certainty is compatible with the conviction of the existence of external things and, by means of the latter, also compatible with conviction concerning the existence of rational beings outside of oneself. – That is why this proposition is a requirement (*Heischesaz*): If anyone wants to philosophize with us, then he is expected to become conscious of his I by means of intuition.« (GA II/3: 27f.). Fichte employs the Latin geometrical term Q.E.D. (*quod erat demonstrandum*) meaning »that which was to be demonstrated« for the conclusion of a mathematical proof several times in the *Eigne Meditationen* (see GA II/3: 47, 65, 81).

65 Fichte, *Erlanger Logik* (GA II/9: 45). This difference between formal and absolute (living) principles is prominent in the *Eigne Meditationen*: »The first principle remains the first principle. [...] A spark of light: I cannot make any progress with the purely *formal* grounding principle (*Grundsatz*); thus, it drives me endlessly

first postulates of geometry are theoretical, whereas the first principle of the *Wissenschaftslehre* is practical.

Thus, to summarize their similarities or equivalence: Fichte's *Grundsatz* is equivalent to a geometric axiom to the extent that they are both the first principles of their own respective sciences, and cannot be reduced to any other principle. Both are self-evident and cannot be proved, but have to be accepted, assumed and grasped in inner intuition by every student. Moreover, both exhibit the characteristic of universality – that is to say, their content can be grasped by every rational being.[66] And finally, similar to Euclid's postulates, Fichte's *Grundsatz* is a formal expression or summons that tells us it is possible to carry out a mental specific act or activity. Both do not make any factual statements or assertions about something, but are only *invitations* to an act; here we are asked to freely carry out an inner activity. This latter aspect is especially underscored in the *Wissenschaftslehre nova methodo*, as I will explain.

Beck's Postulate and the Wissenschaftslehre nova methodo

In his 2005 book *All or Nothing: Systematicity, Transcendental Arguments, and Skepticism in German Idealism*, Paul Franks has likewise discussed the notion of a postulate in Fichte (and Schelling).[67] However, Franks argues that Fichte's choice of the term »postulate« stems from the influence of a discussion of the term in a 1796 volume of philosophy by Jakob Sigismund Beck (1761–1840) rather than from Euclidean geometry or mathematics as such.[68] I have to disagree with this statement, because as we have just seen, Fichte employs the term both in its specific geometric sense in his philosophy as early as 1793/94 (in his private notes entitled *Eigne Med-*

round in a circle and does not make possible the fact (*das Faktum*). – Therefore, I require an unconditioned, absolute (*Grundsatz*), a supreme unity« (GA II/3: 48).

66 Cf. Fichte, *Ankündigung* (GA I/7: 160f.)

67 Paul Franks, *All or Nothing: Systematicity, Transcendental Arguments, and Skepticism in German Idealism*. (Cambridge Mass: Harvard University Press, 2005), pp. 350 ff.

68 See Jakob Sigismund Beck, *Einzig-möglicher Standpunkt, aus welchem die kritische Philosophie beurteilt werden muß* (Riga: 1796): »Above all one has to note that the highest *Grundsatz* of philosophy cannot have any other form than a postulate.« (p. 27)

itationen[69]), as well as equating his first principle with a postulate and the act of postulating in the published 1794 *Grundlage der gesammten Wissenschaftslehre.*[70] Franks also points out the two different senses of the term »postulate« used in German idealism as we have done above – one in its moral sense (in practical philosophy) and the other theoretical with its geometric undertones: »On the first interpretation, the term postulate is understood with reference to Kant's moral philosophy. On the second, it is understood with reference to its use in geometry. Fichte and Schelling started using the term postulate in a geometrical sense in response to Jakob Sigismund Beck, who, in the third volume of his semiofficial exposition of Kant's critical philosophy, *The Standpoint from which the Critical Philosophy is to be Judged* [Riga, 1796], which Fichte recommends as the best preparation for his own writings, attacks the idea of a systematic derivation from an absolute first principle, declaring that an adequate system must begin with a postulate. In response to Beck's attacks, [...] Fichte and Schelling protest that they have been misunderstood and that their first principle was always intended as a postulate.«[71]

Schelling and Fichte did indeed protest and both replied to Beck's charge. For his part, Schelling will soon argue that freedom is the first unprovable principle or postulate of philosophy along the lines of Fichte's argument in the *Erste Einleitung in die Wissenschaftslehre* (1797)[72], where Fichte had dismissed the »materialistic dogmatists« who reject the absolute independence of the I upon which all idealism is founded. For Schelling, first principles or postulates are equivalent in philosophy and

69 See Fichte, *Eigne Meditationen* (GA II/3: 27f.).

70 On Fichte's remarks on postulating and the *Grundsatz* as a postulate in the *Grundlage*, see for example: »Whoever cannot become conscious of the postulated fact – one can certainly know that this will not be the case for any finite rational being – for such a being the whole science does not have any content, it would be empty: yet, he has to admit its formal correctness.« (GA I/2: 390). See too ibid., pp. 323, 361f, 390; 398f.

71 Paul Franks, *All or Nothing*, p. 350. Here Paul Franks is referring to Fichte's comment about Beck in the *Wissenschaftslehre nova methodo* (cf. GA IV/2: 28f.).

72 Fichte, *Erste Einleitung*: »[The dogmatist] entirely denies the independence of the I on which the idealist builds, and turns it simply into a product of things, into an accident of the world; the consistent dogmatist is necessarily also a materialist. He can only be refuted through the postulate of freedom and the independence of the I; but it is precisely this that he denies.« (GA I/4: 192).

geometry in the conventional sense of both being indemonstrable and incapable of proof. They must be assumed, just as the first postulate of geometry must be assumed as a given.[73]

This is the interpretation we can likewise draw from the *Wissenschaftslehre nova methodo* (1798/99) which contains Fichte's reply to Beck. As the title suggests, the *Wissenschaftslehre nova methodo* is a fresh presentation of the Fichtean system. Fichte reformulates the first principles and method of his philosophy and begins with what he calls a »postulate.« According to scholars such as Ives Radrizzani and Günter Zöller, what is new about Fichte's method is not so much that he begins with a postulate but that the art of postulating itself is now described as an *idealistic* and *experimental* operation – it is an »internal experiment« to be actively and scientifically undertaken by his students, and those who realise this postulate have succeeded in elevating themselves to the »philosophical standpoint.«[74] Thus it is a formally expressed summons, an invitation for the reader to carry out the indicated intellectual act. Thus, in contrast to Beck, who claimed that the postulate of philosophy had to have it basis in »facts« (*Thatsache*) and whose postulate was »original representing«,[75] Fichte argued that the first postulate of philosophy should be based on a »*Thathandlung*« or a fact/act, the latter signifying that »one should inwardly act and pay attention to this acting.«[76] Thus, in the *Wissenschaft-*

73 Schelling, »The first postulate of all philosophy, to act freely, seemed to him as necessary as the first principle of geometry, to draw a straight line. And just as little as the master of geometry should try to prove the existence of a line, so just as little should the philosopher try to prove freedom« (quoted in Paul Franks, *All or Nothing*, p. 350).

74 See Ives Radrizzani, *Vers la fondation de l'intersubjectivité chez Fichte. Des Principes à la Nova Methodo* (Paris: Vrin, 1993), pp. 82–90, 123–126; and Günter Zöller: *Fichte's Transcendental Philosophy: The Original Duplicity of Intelligence and Will*, pp. 32–38.

75 »Hence, even if we hold the postulate: ›to originally represent an object‹, for the supreme *Grundsatz* of all philosophy, we imply by this an assertion, which upon closer inspection is the first principle itself. We namely assert that all philosophy must be grounded on facts.« Jakob Sigismund Beck, *Einzig-möglicher Standpunkt, aus welchem die kritische Philosophie beurteilt werden muß* (1796), p. 125f.

76 »In any event it is correct that philosophy has to start with a postulate; the *Wissenschaftslehre* also does this, and expresses it by *Thathandlung*. This word has not been understood; it means, and should not mean anything else than that one

slehre nova methodo Fichte explicitly replies to Beck's charge about a lack of a postulate in his philosophical system by pointing out that his indemonstrable first principle known as a »*Grundsatz*« is in fact identical to a postulate: »Professor Beck complains about the attempt to discover a first principle (*Grundsatz*) and contends instead that philosophy must begin with a *postulate*. But a postulate is also a starting point, which moreover is also not proven, and therefore it is also a first principle. A first principle is any cognition that cannot be further proven in turn. Hence, anyone who states a postulate also states a first principle.«[77] Fichte continues his argument about the equivalence of the *Grundsatz* and a postulate, and in unambiguous words asks that it be interpreted *in the manner of a geometric postulate*. That is to say, both the first postulates of geometry and the first postulates of the *Wissenschaftslehre* are a free summons to engage in an intellectual act or activity. The former with respect to the description of space, the latter with respect to the conceptual activity of thinking the I: »The first *Grundsatz* is a postulate. Just as the teaching of geometry proceeds from the postulate of describing space, so in philosophy the reader or listener must also do something similar. Whoever understands the first principle is placed into the philosophical mood. Postulate: One thinks the concept of the I and thereby thinks of oneself.«[78]

In conclusion we might add that it is precisely in this sense that Schelling agreed with the Fichtean appropriation into philosophy of geometrical axioms or postulates. For around 1796–1797 Schelling reminded his audience that the origin of the term »postulate« is to be sought in geometry and not philosophy. Moreover, Schelling explicitly did this in a text that was to provide an analysis of Fichte's system: the *Erläuterung der Idealismus der Wissenschaftslehre* (1796/97). Here Schelling underscored Fichte's debt to pure geometry with regard to first principles, arguing that in order to correctly understand the usage and meaning of the term postulate we need to turn to the science of mathematics: »The expression ›postulate‹ is borrowed from mathematics. In geometry the most primor-

should inwardly act and pay attention to this acting.« *Wissenschaftslehre nova methodo* (GA IV/2: 28).

77 *Wissenschaftslehre nova methodo* (GA IV/2: 28f.). (cf. GA IV/3: 343–350, 483).

78 Fichte, *Wissenschaftslehre nova methodo* (GA IV 2: 28f.) In this text Fichte also refers to mathematics with regard to the divisibility of space and the indivisibility of the mathematical point (ibid., p. 105).

dial construction is not demonstrated but postulated. This most primordial (simplest) construction in space is the extended point or the straight line. [...] Mathematics thus offers philosophy the example of a primordial intuition from which every science must proceed that wants to make a claim for evidence. It does not proceed from a demonstrable principle but from what is indemonstrable and primordially intuitable.«[79]

Geometric Postulates presuppose Fichte's first Postulate

Here we see not only the similarities between Fichte's postulate and the postulates of geometry, but of course one of their greatest differences.[80] Their *forms* are similar: both are self-evident, universal or accessible to every rational being and transparent. However, their *contents* are different. The postulates of geometry receive their content from the outside and are concerning with the nature and limiting of space. In Fichte's postulate the content is self-generated – or to be more precise – content and form are one and the same: the activity of the I is the content of the I. The geometric postulate is essentially concerned with theory, i.e. with the theoretical constitution of space, the Fichtean postulate is inherently practical, and grounded in the practical (moral) nature and activity of the human being. These differences should not be underestimated, nor the ultimately practical foundation to Fichte's *philosophia prima* be forgotten.

The activity of our I can be expressed and represented in seemingly innumerable ways. Indeed, it could be argued that the content of our entire science and arts are none other than the outward expression and representation of our innermost activity. As we saw, according to Fichte, this also holds for the science of geometry and mathematics, which rely on the activity of the I to even come into being.[81] Moreover, the very activity of our I can also be *represented geometrically*. For example, in the *Grundlage der gesammten Wissenschaftslehre* of 1794/95 the pure activity or striving

79 In: *Idealism and the Endgame of Theory: Three Essays by F.W.J. Schelling*, trans. Thomas Pfau (Albany, N.Y.: SUNY Press, 1994), p. 133.

80 In this regard see Widmann's short overview of the parallels between Euclid's *Elements* and Fichte's philosophy, where he emphasizes the different roles of propositions (*Sätze*) and concepts (*Begriffe*), in: *Grundstruktur*, pp. 21–24

81 See, Fichte, *Zurich Wissenschaftslehre*: (Lecture, February 27, 1794; GA IV/3: 24).

of the I is represented using the image of an infinite straight line.[82] While in the *Bestimmung des Menschen* (1800), the act of *drawing* the geometric line (not the fixed or finished line) provides Fichte with a simple and directly intuitable image of our living and dynamic intellectual activity: »My spiritual faculty appears to inwardly move itself to and fro, to quickly travel from one [element] to another; in short: it appears to me as a *drawing of a line* [*Linienziehen*]. – Determined thinking makes a point on this line.«[83] Here we see another intimate relationship between not only the form of the first postulate of Euclid (to draw a straight line from one point to another one) and Fichte's first postulate of positing the I, but also their content. In this respect any argument for an ultimate rupture between the beginning of the *Grundlage* of 1794/95 or the *Wissenschaftslehre nova methodo* and the starting of the *Wissenschaftslehre 1801/02*[84] (which commences with the activity of drawing a line), loses some of its force. With respect to its form, the activity of constructing a line for Fichte is none other than a preparatory activity for understanding the constructing of the self. With regard to its content, the activity of drawing a line is an external representation of the internal spiritual activity of the I. The former is an accessible sensible intuition of an intellectual intuition. And in fact, it is I who freely draws the line in geometry, and in this sense every sensible intuition of geometry presupposes an intellectual intuition.[85] This is the transition from the activity of geometry to the activity of philosophy. The postulates of geometry for Fichte are therefore more limited sensible instances of the more universal postulate that (explicitly) constitutes the starting point of the Jena versions of the *Wissenschaftslehre*. Or as Fichte says in the *Zurich Wissenschaftslehre*: »all the intuitions of geometry are grounded in the intuitions of the *Wissenschaftslehre*.«[86]

How is this possible? How could the intuitions or postulates of geometry be grounded in the *Wissenschaftslehre*? Should it not be the opposite,

82 See Fichte, GWL: »We can picture this activity [of the I] extending into infinity using the image of a straight line, passing from A through B toward C, and so on.« (GA I/2: 379).

83 Fichte, *Die Bestimmung des Menschen*, Book Two (GA I/6: 237f.).

84 Regarding the example of drawing a line, see section §1 of Fichte's introduction to the *Wissenschaftslehre 1801/02* (GA II/6: 135 ff.).

85 Cf. Fichte, *Zur Ausarbeitung der Wissenschaftslehre von 1801/02* (GA II/6: 51).

86 Fichte, *Zurich Wissenschaftslehre* (GA IV/3: 34).

as Schelling has argued above – that the scientific status of the axioms of geometry are the basis for the first *Grundsatz* or postulate of philosophy? The answer is no for Fichte. In the *Grundlage* of 1794 we see him explicitly arguing this same point for the axioms of logic, that is to say, that they presuppose and are dependent on his first *Grundsatz.*

What is Fichte's argument for the axioms of logic?[87] Fichte begins with A = A, the identity proposition, which he believes that everyone »must admit« as self-evident and certain.[88] He equates it with another identity proposition: »I = I«. Because this is a justified a priori proposition, Fichte argues that to immediately grasp its truth requires rationality and that we are ultimately rational beings. Yet the validity of the identity proposition might be contested. Fichte therefore searches for something even more fundamental than this logical proposition and arrives at the solution of the act of judging: »Through the proposition A = A is judged [*wird geurtheilt*]. All judging, however, according to empirical consciousness, is an acting of the human spirit; [...] Now this acting does not lie grounded in anything higher than X = I am.« The ground of this acting is therefore Fichte's own first proposition which is not a theoretical proposition like A = A or the first axiom of geometry, but a thoroughly practical one. Fichte's system is grounded in the *primary of practical reason*, of acting, and his first axiom or postulate is none other than that of the absolute I, which as a purely rational being is beyond all empirical self-consciousness, and as pure will is the necessary condition for it.[89]

Thus, like the axioms of logic, the axioms of geometry presuppose the activity of judging and the activity of the absolute I. In this sense Fichte's *Grundsatz* is inspired by the axioms of geometry, but does not require them. In fact, the opposite holds, as Fichte points out in §5 of *Über den Begriff der Wissenschaftslehre*. The *Grundsatz* of geometry is the »task in general of limiting space according to a rule.«[90] Or again, geometry is none other than a specific instance of the limiting of our free intellectual

87 See Andreas Schmidt's lucid exposition of Fichte's argument in *Der Grund des Wissens: Fichtes Wissenschaftslehre in den Versionen 1794/95, 1804/II und 1812* (Paderborn: Schöningh, 2004), esp. pp. 12–24.

88 Cf. Fichte, GWL (GA I/2: 256).

89 Cf. Fichte, Ibid., p. 258; and A Schmidt, *Der Grund des Wissens*, pp. 16, 23.

90 Fichte, BWL (GA I/2: 135).

activity.[91] As we saw above for the axioms of logic, this intellectual activity of ours ultimately depends on the rational faculty of judging. The activity of geometry is a determinate or determined form of the absolute activity of the »faculty of judging.« (*Urteilskraft*). This specifying or determining of the tasks of the faculty of judgement, i.e. to order specific objects in accordance with rules and laws, gives rise in turn to the natural sciences. Unlike the unrestricted freedom of the universal science called the *Wissenschaftslehre*, this freedom is restricted and determined in the particular sciences through the faculty of judgement.[92] This is why, according to Fichte, all the *Grundsätze* of the particular sciences, including geometry, are to be traced back to the *Wissenschaftslehre*, and the latter, since it is grounded in practical and not merely theoretical reason, is supposed to be a universal philosophical science or »science of all the sciences«.[93]

In summary, Fichte's *Grundsatz* takes its inspiration from and is equivalent to a geometric postulate *not* with respect to its contents – since the geometric postulate is directed at space, whereas Fichte's postulate is aimed at the self-intuiting and conscious activity of the I – but with regard to its form, universality, characteristics and summons. Both the postulate in geometry and the Fichtean postulate claim to possess the characteristics of self-evidence, transparency and indemonstrability, and both are stated at the beginning of their respective systems of thought.[94] They are both qualified as universal propositions that may be intuited by every rational being, and are stated in such a manner that they are free invitations or summons to carry out specific intellectual acts. And although Fichte seems to prefer the term »postulate« in the later Jena period, its deployment still seems to be identical with his earlier employment of the term »*Grundsatz*«, as he himself claims in the *Wissenschaftslehre nova methodo*. Indeed, an examination of Fichte's texts seems to bear this out. For just as we saw with the *Grundlage der gesammten Wissenschaftslehre*,

91 Ibid.

92 »As soon as the faculty of judgment receives the task of observing a determined object in accordance with a determined law, in order to see whether and to what extent it conforms or does not conform with this law, it is no longer free, but under a rule; and therefore we are no longer in the *Wissenschaftslehre* but in the domain of another science called natural science.« Ibid., p. 136.

93 Ibid., p. 118.

94 As we shall see this is almost exclusively the case in the Jena period for Fichte.

in a large number of other works from the Jena period – including the *Grundlage des Naturrechts* (1796/97), the *Vergleichung des von Hrn Prof. Schmid Aufgestellten Systems mit der Wissenschaftslehre* (1796), and in the *Erste Einleitung* and *Zweite Einleitung in die Wissenschaftslehre* (1797) – Fichte clearly employs the two terms *Grundsatz* and *Postulat* to simultaneously designate his first principle of philosophy.[95] Ultimately, however, the practical nature of the Fichtean first postulate forms the basis for the postulates of geometry, insofar as the latter presuppose and are particular instances of our free activity of limiting in space and the rational activity of judging.

c. *Wissenschaftslehre more geometrico?*

With its employment of postulates, corollaries, and references to mathematical-style proofs, Fichte's *Neue Bearbeitung der Wissenschaftslehre* from late 1800 appears to be a kind of »*Wissenschaftslehre more geometrico*«.[96] As a philosophical Euclid, one may understand Fichte here in two main ways: internally, with regard to his systematic thinking; and externally, with regard to the formal presentation of his thought. With some qualifications, Fichte may be said to be inspired by Spinoza in both these senses.[97] Firstly, in the *Grundlage* of 1794/95, after passing in review the philosophies of Kant, Reinhold, Descartes etc., Fichte says that the only

95 Cf. *Grundlage* (GA I/2: 390; 398f); and *Vergleichung des von Hrn Prof. Schmid Aufgestellten Systems mit der Wissenschaftslehre* (1796): »The *Grundsatz* of all philosophy that we have just established is given to the whole of philosophy itself. [...] The certain action through which it is both [subject-object], through which it is an »I«, and which everyone gives to himself through the realization of this *first postulate*, must now be analysed.« (my emphasis) (GA I/3: 255); the *Erste Einleitung* ponders the consequences of an »unproved postulate« (GA I/4: 202), while the *Zweite Einleitung* uses the term »postulate« to refer to the original *Grundsatz* of intuiting the I: »The first postulate: think yourself, construct the concept of your self and observe how you do this.« (GA I/4: 213); finally, the *Grundlage des Naturrechts* (1796/97) refers to the »postulate, I am I.« (GA I/3: 384).

96 See Daniel Breazeale, »Toward a *Wissenschaftslehre more geometrico* (1800–1801). It could be argued that both Fichte's *Foundations of Natural Right* (1795/96) and *System of Ethics* (1798) are also set out *more geometrico*, insofar as the two contain propositions, proofs and corollaries, add solutions and carry out deductions. This is true, but the two texts are *applications* of Fichte's system and not the pure system of the *Wissenschaftslehre* as such, which is our concern here.

consistent philosophical systems are the critical philosophy (i.e. Kant's philosophy and its most developed form in the *Wissenschaftslehre*) and Spinoza's. Hence, we can see from this that internal consistency was paramount for Fichte. He attributes to Spinoza, rigorousness, consistency and systematicity, three characteristics which are inherent to geometry. By adopting them and admitting their necessity for the Fichtean system, the *Wissenschaftslehre* can also be placed in the tradition of Spinoza, whom Fichte considers as an intellectual precursor: »Spinoza's system is, incidentally, a highly spiritual and profound system. [...] If he had only thought further [...] he would have discovered the critical philosophy, and would have accomplished it very well with his towering spirit. There is also such a system as the Spinozian one in the critical philosophy.« [98]

But of course the *Wissenschaftslehre* also differs from Spinoza's attempt, and Fichte tried to make philosophical advances that he believed Spinoza had not made, such as a philosophy of freedom on a transcendental basis and a rigorous and systematic theory of the I.[99] Furthermore,

97 Although Spinoza's *Ethics* is undoubtedly the most well-known, this mode of exposition did not begin with Spinoza. The *more geometrico* presentation in philosophy can be traced back to at least to the Middle Ages, e.g. to the *Ars fidei catholicae* of Alan de Lille (Alanus ab Insulis), or even further back to the fifth century to Proclus's *Elements of Theology* (The former text is sometimes attributed to Nicholas of Ameins, see: *Nikolaus von Amiens: Ars fidei catholicae – Ein Beispielwerk von axiomatischer Methode*, ed. Mechthild Dreyer, Münster: Aschendorff, 1993). And concerning specific influences on Fichte's thought in this regard, it is worth acknowledging the early impact of Lessing's *Antigöze*. The latter's mode of presentation also echoes a *more geometrico* style, and both this fact and its rigorous arguments were a revelation to the young Fichte at Schulpforta (cf. FG, vol. 1, p. 17). Finally, it is interesting to see that in March 1793 Fichte had been working on: a »Critique of Reflection Ideas (providence, miracles and revelation), written in a rigorous mathematical theoretical style (*streng mathematischer Lehrart*).« Fichte, letter to Gottlieb Hufeland, March 28, 1793 (GA III/1: 379).

98 Fichte, *Vorlesungen über Logik und Metaphysik 1797* (GA IV/1: 370). See too Fichte, *Platner Vorlesungen*: »Pl.[atner's] presentation [of Spinoza] is wholly incomprehensible. I will provide my own presentation of his system according to its spirit: for the letter especially kills here.« (GA II/ 4S: 199f.).

99 See the early *Practische Philosophie* of 1793/94, where Fichte hopes that if Spinoza were to return he would agree with these Fichtean advances in philosophy: »*I am.* – Spinoza denies this proposition and from this it appears to result that Spinoza

in Fichte's eyes, Spinoza also does not show how the transition is possible from the infinite or absolute, to its finite modes. In this respect too Fichte believed he had made progress upon the philosophy of Spinoza by attempting to address this problem of a transition.[100]

At this point it is worth asking: what exactly is a mode of exposition *more geometrico* in philosophy? Since Spinoza's masterpiece is commonly held to be the paradigmatic example in this genre, let us look at his *Ethics*, first published posthumously in 1677.[101] Spinoza's *Ethica* was originally written in Latin and bears the subtitle: *Ordine Geometrico demonstrata*, that is to say: presented according to the geometrical method. Looking at the beginning of Part One, »*De Deo*« (Concerning God), Spinoza immediately commences with six concise *Definitiones* (definitions) of cause, finite, substance, attribute, mode and God.[102] Spinoza's presentation here closely follows Euclid's, insofar as the latter also immediately begins with a list of 23 definitions in Book One of the *Elements*. Spinoza, slightly departs from Euclid by next adding a series of *Explicatio* (explications). However, by defining essential terms and meanings such as ›absolute‹, ›truth‹, ›free‹ and ›eternity‹, these first explications are similar in form and function to his initial list of definitions.[103] Spinoza next sets out seven *Axiomata* or axioms, which again closely corresponds with Euclid's list of

would not be refuted by my philosophy. What is to be done here? To hope, that if Spinoza were to come again and find our system he would give up his own.« (GA II/3: 236f.)

100 »[Spinoza] does not ask at all about these transitions: hence fundamentally there is none, his substance is not one, nor is his accident, but he only sometimes terms it so, and does not really address the issue.« Fichte, *Wissenschaftslehre 1801/1802* (GA II/6: 228). During this period Fichte also criticized Schelling's approach to Spinoza for likewise overlooking the difficulty of the transition problem: »Or again, how does the *One* first become the *Infinite*, and then a totality of the manifold? – This is the question that a consistent speculation still has to solve, and which you necessarily have to ignore because you already find this form at once in and with the absolute.« Letter of J. G. Fichte to F. W. J. Schelling, October 8, 1801 (GA III/5: 91).

101 See Benedictus de Spinoza, *Die Ethik*, Latin-German edition, trans. Jakob Stern (Stuttgart: Reclam, 1977); English translation in: *The Ethics and Selected Letters*, trans. Samuel Shirley (Indianapolis, IN: Hackett, 1982), pp. 31–225.

102 Cf. Spinoza, *Die Ethik*, p. 4; *The Ethics*, p. 31.

103 Ibid., *Die Ethik* pp. 4/6; *The Ethics*, pp. 31f.

five postulates and five common notions. Like Euclid, Spinoza's axioms aim at stating self-evident and uncontroversial truths, such as we find in his axiom number one: »I. Omnia, quae sunt, vel in se, vel in alio sunt« (All things that are, are either in themselves or in something else).[104] Similar to the axioms of geometry or logic, we can either accept or reject such an axiom, they are not subject to proof and cannot be proved. These definitions, explications and axioms constitute the introductory part of Book One of Spinoza *Ethics*, and he immediately follows these with the main text, which consists of a body of *propositio* (propositions) and *demonstratio* (demonstrations or proofs). Entirely in the manner of Euclid, and in contrast to the axioms, a »proof« of a proposition for Spinoza is said to take place when the proposition can be logically or rationally traced back to the initial list of definitions, axioms or explanations, or to a preceding proposition. For example, proposition number one in part one of the *Ethics* is: »Substantia prior est natura suis affectionibus« (substance is by nature prior to its affections). The proof or demonstration: »Patet ex Definitione 3 & 5« (This is evident from Definitions 3 & 5).[105] Even though we later find corollaries and scholia, the system or method does not change, as the former are essentially equivalent to propositions and explanations respectively. This is Spinoza's geometrical method: a step-by-step rational or metaphysical system in which the proof of a theorem or proposition is established if one is able to show that it follows from the proceeding proposition or the initial small set of definitions and axioms. Despite their obvious difference in content (metaphysical topics such as God, nature and the human being for Spinoza, spatial topics such as triangles and circles for Euclid), Spinoza's method in the *Ethics* is entirely in the spirit of Euclid's method in the *Elements*.

We can now ask: to what extent is Fichte's *Neue Bearbeitung der Wissenschaftslehre* an attempt at writing a *Wissenschaftslehre* in the geometric style favored by Euclid and Spinoza?

104 Ibid., *Die Ethik* p. 6; *The Ethics*, p. 32.
105 Ibid., *Die Ethik* p. 8; *The Ethics*, p. 32.

The Neue Bearbeitung der Wissenschaftslehre (1800)

Like the *Erlanger Logik*, the *Neue Bearbeitung der Wissenschaftslehre* is an unrevised and highly fragmentary text from Fichte's *Nachlass*. With its reflections on *Grundsätze* or first principles, the text is closer to the early Zurich and Jena writings than to the later Berlin presentations. After reworking and testing the different formulations of his first principle in the *Eigne Meditationen*, the *Grundlage* and the *Wissenschaftslehre nova methodo*, in 1800 we see Fichte again troubled by the question that has so far permeated and distinguished his entire career: what is the best starting point for philosophy? In the *Neue Bearbeitung* Fichte now chose to dispense with the term *Grundsatz* as such and to focus instead on a geometrical-like exposition replete with the classic terminology of geometry, such as theorems, postulates and corollaries etc., or as he puts it: it was attempt to present his philosophy »entirely in line with the procedure in mathematics.«[106]

In fact, the stated goal of the *Neue Bearbeitung der Wissenschaftslehre* 1800 was still one of the major themes to be found in the Fichtean corpus: to justify the employment and conditions of *intellectual intuition* in philosophy. Fichte formulates this »true objective« right at the beginning of the text in point number §1: »In fact, the goal is still only to prove that intellectual intuition is the condition of all possible consciousness.«[107]

Why does Fichte's text deserve to be labelled a »geometric« exposition of the *Wissenschaftslehre*? On the one hand, we see obvious differences with the presentations of Euclid and Spinoza, for Fichte does not at all begin with either a list of definitions or explanations, but a series of statements of intent and »preliminary remarks« in standard prose (which could also simply be a sign of the unrevised and unfinished nature of the text).[108] On the other hand, the main body of the *Neue Bearbeitung der Wissenschaftslehre* includes eight main sections in which each begins with

106 *Neue Bearbeitung der Wissenschaftslehre* (GA II/5: 335). Fichte does in fact use the term *Grundsatz* one time in the text, but it is employed more in the sense of a general principle and *not* as an absolute *first* principle of philosophy (cf. GA II/5: 350).

107 GA II/5: 331.

108 Ibid., pp. 331–334.

what Fichte terms a »theorem.«[109] These theorems are followed by various »postulates« which are in turn followed by a varying number of »corollaries.« Unlike the traditional *more geometrico*, which places the postulates at the beginning, Fichte inverts the order of postulates and theorems and distributes postulates throughout the text.

Similar to the *Eigne Meditationen*, one of the most fascinating aspects of the text is Fichte's ongoing inner dialogue with himself, especially his deliberations on his choice of geometric terminology. Fichte strives to come to an understanding with himself as to the precise role, function and interactions between the different geometrically expressed philosophical principles, and after some testing he eventually decides upon a number of conventions.[110] As mentioned, each of the eight sections begins with what is called a »theorem.« For example, theorem number one is: »1. Theorem. Every other consciousness presupposes immediate self-consciousness as a condition of its possibility.«[111] The role of the theorem in the *Neue Bearbeitung* is to formulate in concepts a proposition that asserts for itself a necessary and universal truth. Thus, unlike traditional Euclidean geometry, and unlike the *Wissenschaftslehre nova methodo*, here Fichte does not start with a self-evident postulate but with the above »theorem.« In the *Neue Bearbeitung* the postulate (or the mediating postulate) is to follow each stated theorem, and I agree with Daniel Breazeale's analysis that the postulate's function is to describe a specific intellectual act, in which the performance and observation of it are supposed to furnish »apodictic evidence for the theorem.«[112] The corollaries associated with each theorem (and sometimes there are up to six in number) are the different propositions that Fichte considers should be derived from the theorem in question by means of conceptual analysis. For example, the postu-

109 This holds for theorems one to seven, but with theorem number eight Fichte employs again »postulate.« (Cf. GA II/5: 391).

110 I am indebted to Daniel Breazeale's analysis of the *Neue Bearbeitung* in a recent article. See his »Toward a *Wissenschaftslehre more geometrico* (1800–1801)« pp. 18ff.

111 »1. Theorem. Alles andere Bewußtseyn, sezt ein unmittelbares Selbstbewußtseyn, als Möglichkeitsbedingung voraus.« Fichte, *Neue Bearbeitung der Wissenschaftslehre 1800* (GA II/5: 335).

112 See Daniel Breazeale, »Toward a *Wissenschaftslehre more geometrico* (1800–1801)« p. 18.

late following theorem number one is simply stated as: »Mediating postulate: *To think oneself.*«[113] As should be apparent from the last section, this formulation of the postulate is entirely in line with the earlier Jena versions of Fichte's *Grundsatz.* That is to say, the first principle or postulate of intellectually intuiting the I (i.e. the postulate of the *Wissenschaftslehre nova methodo*), which like Euclid's postulates cannot be proved but must be accepted. The distinction between postulate and theorem in the *Neue Bearbeitung der Wissenschaftslehre* may be summarized as follows: whereas a theorem is a *necessary and universal truth* expressed in concepts, a postulate describes a *specific intellectual act* grasped in intuition.

Several other postulates follow theorem number one, including postulate number 3 that directs one to become conscious in turn of one's own activity and type of thinking.[114] This 3rd postulate is linked to 2 corollaries of which the first asserts that immediate self-consciousness is what is eternally and invariably subjective and can never become the object of consciousness itself. This point raises Fichte's misgivings about ever consciously grasping self-consciousness, and therefore the only option available is to resort to »abstraction«, i.e. to »infer« the existence of this consciousness as the component and condition of another kind of consciousnesses.[115] The second corollary connected with postulate 3 states that it be derived in a strictly systematic and logical fashion, i.e. in a process of inference that is to be performed »entirely like a mathematical proof.«[116]

Fichte struggles with his geometrical terminology in the *Neue Bearbeitung* and at times calls into question his particular choices of terms. In the so-called »Preliminary Remarks« of the text, Fichte's procedure initially appears identical to the »self-performative« procedure of the *Wissenschaftslehre nova methodo*, when he says: »We will proceed from postulates, and the reader should forget the author and his book [...] and develop his own train of thoughts.«[117] And like the *Wissenschaftslehre nova methodo*, Fichte chooses in his projected chapter one of the *Neue Bearbeitung* a first principle that he calls a postulate: »I. – *Postulate.* Immediate

113 »Vermittelndes Postulat: *Sich selbst zu denken.*« (GA II/5: 335).

114 »3tes Postulat: Sich seines Bewußtseyns des vollzognen Denkens (u. der Art desselben) wiederum bewußt zu werden.« (GA II/5: 338).

115 Ibid.

116 Ibid.

117 *Neue Bearbeitung* (GA II/5: 333).

self-consciousness is the necessary condition of ever other consciousness.«[118] Yet Fichte soon finds himself in difficulty with his announced procedure and terminology and begins to have doubts as to how exactly the above sentence should be designated or named. This is because in line with his reasoning up to this point (and in his other writings), a postulate refers to a definite mental act or conscious intellectual activity. This is obviously not the case for this sentence. Hence, Fichte now changes his mind and chooses the term »theorem« instead. However, he then realises that although this sentence does indeed assert a necessary and universal truth, it cannot be demonstrated by means of concepts, as stipulated by his prior definition of a theorem. Moreover, the sentence is similar to a postulate insofar as it requires intuition. In consequence, Fichte is faced with a linguistic and philosophical dilemma: »What kind of proposition is the above proposition? Not actually a postulate. Yet just as little a theorem, i.e. a demonstration from concepts. – The manner of proceeding is exactly the same as in *mathematics*. It is demonstrated in intuition. I can do nothing else. Mathematical demonstrations are only signposts for leading us to the point of intuition. Hence, it is certainly a *proposition* (*Lehrsatz*). Only not from concepts. Thus, I had better retain ›*theorem*‹.«[119]

In spite of the fact that this proposition cannot be proved using concepts but only through intuition, Fichte persists with the term »theorem« for the remainder of the text of the *Neue Bearbeitung*. Later in the text Fichte could not help making a kind of mental footnote to himself to retain the term »postulate« in the sense of the *Wissenschaftslehre nova methodo*, i.e. as something self-evident and immediately clear. However, he does not choose the intellectual intuiting of the I as the content of this postulate. Instead, Fichte selects an example that will form the centrepiece and new beginning of the *Wissenschaftslehre 1801/1802* – an example based on Euclid's first postulate in the *Elements*; or in other words, *das Linienziehen* – the drawing of a line between two points: »Postulate. To draw a line. In the future I'll also give the name ›postulates‹ to propositions like this: the straight line is the shortest path between two points; (in order to better separate the way).« [120]

118 Ibid., p. 335.
119 Ibid.
120 Ibid., p. 348.

The very fact that Fichte hesitates and reflects so much on his terminology demonstrates in my opinion the flexibility and variability of his language when confronted with the reality of mental acts in conscious human experience. In Fichte's philosophical system, it is *not* a question of any fixed and set terminology, or of words or scientific »signs«, but of the immediate activities and intuitions signalled by the term. As he says of the *Wissenschaftslehre* in the *Ankündigung*: »It does not matter what sign it attaches to its object – whether it calls it ›I‹ or ›Not-I‹, X or Y. The sign itself in nothing; the *Wissenschaftslehre* is only concerned with what occurs within the immediate intuition of everyone.«[121] This holds for both Fichte's incessant striving for a perfected written presentation of the *Wissenschaftslehre*, as well as for his choice of geometrical language in the *Neue Bearbeitung*. – Despite the difficulty of correctly formulating all these terms, Fichte dismissed the view that philosophy first requires a »*meta-critique of language*«, as proposed by Hamann and others. This would be equivalent to saying we need some higher critique of »expressions such as the mathematical point and line, etc.«[122] – The *intuiting* of the *content* of these mathematical terms and not the word or concept is the key issue for Fichte. The intuition for him is a »unifying medium« (*Vereinigungsmittel*) within reason that is higher than all concepts or words, »for words are nothing but the frequently adulterated, second-hand copies of concepts.«[123]

Two further questions could be asked at the end of this section: (1) Firstly, why did Fichte think it worth *embarking* on a geometric-style presentation of his philosophy precisely at this time? The answer is not clear, but as mentioned above it might have been due in part to Fichte's study of Spinoza's *Ethics* during this same period in connection with his dispute with Schelling.[124] Moreover, we also know that Fichte wanted to incorporate or revise his lectures on the *Wissenschaftslehre nova methodo*, in or-

121 Fichte, *Ankündigung* (1801, GA I/7: 160); English translation by Daniel Breazeale in IWL, p. 198. See too Fichte's letter of July 2, 1795 to Reinhold advising him to go beyond his linguistic formulations: »Do not place so much value upon my expressions. [...] The soul of my system is the proposition: The I absolutely posits itself. Without the inner intuition of the I through itself these words are devoid of all sense and worth.« (GA III/2: 343f.).

122 Ibid., p. 159, IWL, p. 199.

123 Ibid., p. 158.

der to reply to the increasingly vocal criticisms of his philosophical contemporaries. Fichte made a preliminary reply in the *Sonnenklarer Bericht*, which constitutes a »popular« version of what he was trying to accomplish in a more »scientific« and technical manner in the *Neue Bearbeitung*.[125] (2) Secondly, why did Fichte soon *abandon* or cease working on the *Neue Beabeitung der Wissenschaftslehre* in early 1801? Again, it is difficult to say, but some have suggested that it may have been that Fichte realized that a geometric structure of his philosophy with theorems, postulates and corollaries would still cause misunderstandings, like all his earlier presentations similarly beginning with a *Grundsatz* or a postulate.[126] I would like to propose that the reason might lie in a different direction – in Fichte's fragile relationship with his soon-to-be-rival Schelling, because during this period Fichte and Schelling were in the midst of reconsidering and rupturing their intellectual friendship. More specifically – it is striking to note that at exactly when Fichte was preparing a geometrical presentation of the *Wissenschaftslehre*, Schelling was likewise preparing a mathematical exposition of his »own system«, the so-called *Darstellung meines Systems der Philosophie* that appeared in 1801, and where he acknowledged his clear debt to Spinoza's geometric exposition.[127] As Rein-

124 Fichte read Spinoza's *Ethics* in late 1800 during the compositional period of the *Neue Bearbeitung*. In the 1801 text *Zur Darstellung von Schelling's Identitätssysteme* (GA II/5: 487–508), Fichte quotes from a Latin copy of Spinoza's *Ethica* (esp. pp. 492–501). In F. Schlegel's letter to Schleiermacher from around September 1800, Schlegel mentions that Fichte possessed his own personal copy of the Latin edition of the *Ethica* (Cf. FG vol. 2, p. 395).

125 As Fichte noted in his »preliminary remarks« to the *Neue Bearbeitung*: »Although there exists no shortage of introductions to the *Wissenschaftslehre*, and because in conjunction with this text a work in the same vein has simultaneously been published under the title: *Sonnenklarer Bericht*, – I do not wish to repeat what has been said there.« (GA II/5: 333).

126 This is the view of Daniel Breazeale, see his article »Toward a *Wissenschaftslehre more geometrico* (1800–1801)«, p. 26

127 Cf. The Introduction to Schelling's *Darstellung*: »Concerning the method, which I have employed in the construction of this system. [...] I have chosen Spinoza as my model, because [...] this form permits the greatest brevity of presentation and most distinctly allows us to judge the self-evidence of the proofs.« Schelling, *Darstellung meines Systems der Philosophie*, in: *Zeitschrift für spekulative Physik*, vol. 2, 1801 (repr. Hamburg: Meiner, 2001, p. 335).

hard Lauth has pointed out[128], upon receiving a copy of Schelling's *Darstellung* from Schelling himself in 1801 Fichte was so curious that he set about comparing the respective methods of Schelling and Spinoza, writing a small commentary[129] on the first 51 paragraphs of Schelling's new work. Fichte was relatively unimpressed by the geometric presentation and could not understand how Schelling could make the transition from his supposedly first clearly defined starting point to the following theorems and corollaries.[130] Fichte finally labelled the work as »nonsense«, and »muddled-headed«, and more poetically – as a blinded Cyclops: »*Polyphemus without an eye.*«[131] And somewhat ironically, he now reapplied Jacobi's earlier criticism of the methodology of the *Wissenschaftslehre* as inverted Spinozism to Schelling's own *Darstellung*: »Isn't this method then thoroughly inverted, and can any good come of it?»[132] Thus to conclude, it is difficult to see how the discovery that Schelling had first published his own system of philosophy in a geometrical form did not play some sort of role in Fichte's decision to abandon his plan to complete and publish his own geometric exposition.

To summarize our findings: to the extent that Fichte's *Neue Bearbeitung der Wissenschaftslehre* is set out using theorems, postulates, corollaries etc., whose thoughts are supposed to be deduced or »proved« from one another »just like in mathematics«, and whose methodology above all privileges pure intuitions and construction, then the text may be labelled as an exposition of the *Wissenschaftslehre* »*more geometrico*«. However, the differences with the presentations of Euclid and Spinoza should also not be overlooked. Fichte neither employed a list of rigorous definitions

128 Cf. R. Lauth, »Fichtes Sicht der Philosophie Spinozas«, pp. 26, 31.

129 Fichte's commentary is entitled: *Zur Darstellung von Schelling's Identitätssysteme* (GA II/5: 487–508).

130 Schelling's point §1 is: »I call *reason* absolute reason, or reason insofar as it is conceived as the total indifference of the subjective and the objective« *Darstellung meines Systems der Philosophie* (Hamburg: Meiner, 2001), p. 336. Fichte commentary on this passage is as follows: »Through an explanation or real definition of this kind the defined matter is exposed and concluded as a finished object: hence, I don't see how the transition is to be made from here to the next and the subsequent thoughts«, *Zur Darstellung von Schelling's Identitätssysteme* (GA II/5: 487).

131 *Vorarbeiten Gegen Schelling* (GA II/5: 484).

132 Ibid., p. 483.

and explanations at the beginning of his text, and he also inverted the order of theorems and postulates at times, as well as interspersing his postulates throughout the text.[133]

2. Sensible, Mathematical and Intellectual Intuition

In the *Zweite Einleitung in die Wissenschaftslehre* (1797), Fichte makes the following statement about the importance of intellectual intuition for philosophy: »Intellectual intuition is the only solid standpoint for the whole of philosophy. We can explain everything that occurs in consciousness from it.«[134] Why was intellectual intuition of such importance for Fichte? And why was it so quickly rehabilitated and reintroduced into philosophy as a legitimate and indispensable principle, especially after Kant had not long before set about putting strictures on its use and virtually banished it from philosophy, as Xavier Tilliette has demonstrated?[135] In this section we will look at the role of intuition in geometrical cognition. Starting with an examination of Kant's theory of intuition in geometry, we will then move on to Fichte's radically revised conception of intellectual intuition in transcendental philosophy in general, and its role in geometry in particular. Bearing in mind the different uses of intellectual intuition in Fichte's writings, we will show that for Fichte intuition is a necessary and indis-

133 Daniel Breazeale has also highlighted a number of differences between the presentations of Spinoza and Fichte: »(There is also an interesting parallel between Fichte and Spinoza concerning their actual *method* for securing systematic unity among the propositions that make up a philosophical system: namely, their shared commitment to philosophizing ›*more geometrico*‹ and to stressing the parallels between their practice as philosophers and the practice of mathematicians. But this similarity is rather misleading, inasmuch as what Spinoza understood by the ›geometrical method‹ is something very different indeed that what Fichte understood by this same term. For Spinoza, the essence of the geometrical method lay in its rigorously deductive or analytic form, whereas for Fichte it lay in its use of postulates and constructions.)« – »Fichte und Spinoza: ›Gemeinschaftlicher Standpunkt – wesentlicher Gegensatz‹« (unpublished manuscript, 2005).

134 GA I/4: 219. Marci Ivaldo has made this passage the focus of his article: »L'intuizione intellettuale come ›punto fermo‹ della filosofia nel Saggio di una nuova esposizione della dottrina della scienza (1797–1798)« in: *Sul trascendentale moderno - Genesi, Struttura, Problemi*, ed. Roberto Perini (Naples: E. S. I., 2004): 91–109.

135 See Xavier Tilliette, *L'intuition intellectuelle de Kant à Hegel* (Paris: Vrin, 1995), esp. pp. 13–21.

pensable element in the mathematical cognition of geometric entities; however, not intellectual intuition, but geometrical or »intelligible« intuition.

a. From Kant's Sensible Intuition to Fichte's Intellectual Intuition

Kant's conception of transcendental idealism as outlined in the *Critique of Pure Reason* makes use of the argument that mathematical knowledge is synthetic *a priori*. Mathematical knowledge or cognition is synthetic *a priori* because it is able to construct its concepts in intuition. However, the Kantian theory of mathematical cognition does not confine itself to concepts alone. An intuition for Kant is an immediate and singular representation of a particular object that is provided by the senses. In contrast, a concept is an abstract and general representation formed by the intellect.[136] In Kant's eyes all mathematical cognition requires something in addition to the analysis of concepts; we require a synthesis, i.e. we have to go beyond the concept to the corresponding intuition. This is why, as we saw, all the propositions of mathematics for Kant »are therefore without exception synthetic.«[137]

One of the revolutions in the critical philosophy was to make this role of intuition in mathematical cognition absolutely indispensable. There have been many objections to Kant's view that mathematical cognition requires intuitions in addition to concepts. These objections started very early on. Eberhard, for example, famously rejected Kant's theory because he believed that we could acquire mathematical knowledge of entities for which there is no corresponding intuition, such as a polygon with a thousand sides.[138] More recent studies have examined the reasons for Kant's reliance on intuition, which has been the subject of intense debate among scholars, precisely in its relation to geometry.[139] Our aim is not to furnish

136 Kant, *Critique of Pure Reason*, p. 155 (A19-A20/B33-B34).

137 Ibid., pp. 74–75 (AA 4: 269–272).

138 Henry Allison provides an excellent overview and history of Eberhard's objections in: *The Kant-Eberhard Controversy* (Baltimore: Johns Hopkins University Press, 1973). For Eberhard's polygon objection, see p. 126.

139 For example, in »Kant on Intuition in Geometry« *Canadian Journal of Philosophy*, Vol.27 (1997), pp. 489–512, Emily Carson argues that there are two main tendencies in the interpretation of intuition in Kant's theory of geometry. On the one hand, there is the so-called »logical« interpretation advocated by Michael Fried-

another contribution to this interesting Kantian debate, but to investigate and furnish an account of Fichte's thoughts on the role of intuition in geometrical cognition. Notwithstanding, Fichte's account of intuition draws heavily on Kant's transcendental account, so this will necessitate a preliminary discussion of Kant's conception. Our discussion will demonstrate the primacy of both *intuition* and *construction* in their accounts of mathematical cognition. We will first begin with an analysis of intuition before passing on to the role of construction in the Kantian and Fichtean theories of mathematics.

In the *Critique of Pure Reason* Kant outlines his view that the only kind of intuition that is *possible* for us as human beings is »sensible« intuition. In contrast to concepts, intuition is characterized by (1) its *immediacy* and (2) its *singularity*: »An objective perception is a cognition (*cognitio*). The latter is either intuition or concept (*intuitus vel conceptus*). The former relates immediately to the object and is singular, the latter is mediate, by means of a mark, which can be common to several things.«[140] According to Kant, there are two types of sensible intuition. Either sensible intuition is »pure«, these are the intuitions of space and time, or it can be »empirical«, these are the intuitions which by means of sensation are »immediately represented as real in space and time.«[141] Consequently, it is the determinations of pure intuition that allow us to acquire *a priori* knowledge of the objects of mathematics. However, Kant qualifies this, and says we can have cognition of these objects only with respect to their forms, i.e. as appearances. For this reason the concepts of mathematics in the Kan-

man in his highly influential text: »Kant's Theory of Geometry« *Philosophical Review*, 94 (1985): 455–506. Here Friedman says that Kant adopts his view of intuition directly on account of the limitations of syllogistic logic. This interpretation considers the Kantian incorporation of intuition in geometry to be logical or *mathematical*. Carson contrasts this *logical* view with another one – advocated by herself – in which Kant's reasons for using intuition in geometry are *philosophical*. This reading, according to Carson, »emphasizes an epistemological or phenomenological role for intuition in mathematics arising out of what may be loosely called Kant's ›anti-formalism‹.« (E. Carson. »Kant on Intuition in Geometry«, p. 490). Hence Carson takes the view that Kant adopted the notion of intuition in geometry on purely philosophical and not logical or mathematical grounds.

140 Kant, *Critique of Pure Reason*, (A320/B377), pp. 398f.

141 Kant, *Critique of Pure Reason* (B 147), p. 254.

tian theory solely become cognition to the extent that we assume there exists things that can be presented according to the forms of pure sensible intuition.[142]

In the »Discipline of Pure Reason«, the first chapter in the »Transcendental Doctrine of Method« in the *Critique of Pure Reason*, Kant takes a specific example of a triangle to illustrate his view of mathematical cognition. Here he makes a distinction between pure intuition, which relies on the imagination, and empirical intuition, which is based, for example, of our drawing of the triangle on a piece of paper. However, he argues that both cases of intuition are still *a priori*. He says: »Thus I construct a triangle by exhibiting an object corresponding to this concept, either through mere imagination, in pure intuition, or on paper, in empirical intuition, but in both cases completely a priori, without having to have borrowed the pattern for it from any experience.«[143]

For Kant, an empirical intuition can be said to result in *a priori* knowledge because its »*pattern*« is still non-empirical or not derived from any experience. Although the empirical triangle is on the paper, our cognition manages to »abstract« from all its specific and particular characteristics, including the magnitude and length of its sides and angles, to arrive at its concept which is unaltered by these particular differences.[144] Thus, the particular drawn figure of the triangle on the piece of paper may be empirical, but it still »serves to express the concept, without damage to its universality.«[145]

However, it might be asked, how is the mathematical judgment *a priori* if the intuition is empirical, since »empirical« normally signifies *a pos-*

142 »Consequently, all mathematical concepts are not by themselves cognitions, except insofar as one presupposes that there are things that can be presented to us only in accordance with the form of that pure sensible intuition.« Kant, *Critique of Pure Reason* (B147), p. 254.

143 Kant, *Critique of Pure Reason* (A713/B741), p. 630.

144 Cf. Kant, *Critique of Pure Reason*: »In the case of this empirical intuition we have taken account only of the action of constructing the concept, to which many determinations, e.g., those of the magnitude of the sides and the angles, are entirely indifferent, and thus we have abstracted from these differences, which do not alter the concept of the triangle.« (A714/B742), p. 630.

145 Ibid.

teriori?[146] Kant argues that we therefore need a »pure« intuition to have a priori mathematical cognition. He agrees that empirical intuitions *as such* would only result in empirical propositions and therefore this would not suffice to account for the universality and necessity of mathematical judgments. Hence the second component of mathematical cognition requires what he calls »construction« – since constructions allows us to rise to pure intuitions by passing beyond the mere concept.[147]

Hence, there is an absolute need for *intuitions* in Kant's theory of mathematical cognition. If mathematical judgments are to be *synthetic* they require pure and not simply empirical intuitions. Empirical intuitions in mathematics of course also lead to cognition, but only *a posteriori* cognition. If we wish to acquire a synthetic mathematical judgment we must not remain at the stage of the concept but »go beyond« it to the required intuition. Remaining at the stage of the mere concept would for Kant only yield »analytic« judgments, since its explanation would already be in the thought in question. Thus intuition is vital, for only intuitions in mathematical cognition are able to provide us with necessary and absolutely certain mathematical truths: »If one is to judge synthetically about a concept, then one must go beyond this concept, and indeed go to the intuition in which it is given. For if one were to remain with that which is contained in the concept, then the judgment would be merely analytic, an explanation of which is actually contained in the thought. However, I can go from the concept to the pure or empirical intuition corresponding to it in order to assess it *in concreto* and cognize *a priori* or *a posteriori* what pertains to its object. The former is mathematical cognition through the

146 Cf. Lisa Shabel's detailed discussion of this point in *Mathematics in Kant's Critical Philosophy*, pp. 102–105.

147 »Now this is not possible in any other way but by determining my object in accordance with the conditions of either empirical or pure intuition. The former would yield only an empirical proposition (through measurement of its angles), which would contain no universality, let alone necessity, and propositions of this sort are not under discussion here. The second procedure, however, is that of mathematical and here indeed of geometrical construction, by means of which I put together in a pure intuition, just as in an empirical one, the manifold that belongs to the schema of a triangle in general and thus to its concept, through which general synthetic propositions must be constructed.« Kant, *Critique of Pure Reason* (A718/B746), p. 633.

construction of the concept, the latter merely empirical (mechanical) cognition, which can never yield necessary and apodictic propositions.«[148]

Fichte's Appropriation of »Intellektuelle Anschauung«

One of the key themes in Fichte's philosophy is one of the grand themes of metaphysics in general – the question of man's *freedom*. Following Kant, Fichte did not put forward in his early writings a conclusive »proof« of freedom but repeated Kant's claim in the second Critique that freedom rests on a »fact of reason.« We know we are free agents on account of our voice of conscience. The reality of our freedom is demonstrated by an »immediate fact« (*unmittelbare Thatsache*) – our awareness of the moral law. And thus for Fichte, the reality of freedom is an integral element of his transcendental philosophy – freedom is a necessary condition of the possibility of consciousness itself.[149]

In the manuscript *Practische Philosophie*, Fichte says that although we do not have anything like an absolute freedom, as humans we must still *strive* to attain this goal. Fichte coined a name for his philosophy at this time on the very last page of this text: »*Strebungsphilosophie*« – a philosophy of striving.[150] This striving defines our human nature. The human I is infinite for Fichte, but only in accordance with its striving – the I strives to be infinite.[151] The human I *is* striving and activity, and it is only to be understood in this sense: »The essence of the I consists in its activity.«[152] In connection with human freedom, only the attempt at striving after freedom can prove to us the reality of freedom. Like a Kantian regulative idea, we can never fully and completely attain it, but we should at least strive after this goal of freedom, a notion that would later form the basis of Fichte's views on the *Bestimmung* or destination of humanity.[153] According to many Fichte scholars, the challenge of Fichte's early Jena writings was

148 Kant, *Critique of Pure Reason* (A721/B749), p. 634.

149 Cf. Frederick Beiser, *German Idealism*, pp. 289–294.

150 GA II/3: 265.

151 Fichte, GWL (GA I/2: 404).

152 Ibid., p. 405.

153 Cf. Fichte, *Practische Philosophie*: »The state, the sublime state of man therefore is a constant striving; his joys are a mere heightening of this striving; he always needs something, and all his satisfaction consists in the heightening of his needs; and only he is the true man for whom it is so« (GA II/3: 189).

to provide a satisfactory defence of his philosophy of freedom, which is a philosophy of striving insofar as we realize and are aware of our freedom by striving to be free. The cognition of our freedom is therefore based on actions that we ourselves undertake.[154] Thus, freedom for Fichte was a practical problem, and could only be solved by practical action. However, if we ask the question as to how we recognize the essence of ourselves as self-determining and autonomous entities not solely bound to causal laws, i.e. as moral agents capable of freedom, then we cannot know this discursively but only immediately through *intuition*. However, not a cognition through sensible intuition but through an *intellectual* intuition; or as Fichte calls it in the *Sittenlehre* of 1798, it is a »real intellectual intuition« (*wirkliche intellektuelle Anschauung*).[155]

In contrast to Kant, who asserted only the reality of sensible intuition for human reason, Fichte, Schelling, the young Hegel and many of the German Romantics, spoke of the possibility of intellectual intuition for human cognition. In the words of the early Hegel: »Without intellectual intuition one cannot philosophize.«[156] Of course, Kant himself had already suggested the notion of intellectual intuition, but ultimately rejected it as lying outside the cognitive faculty of finite human beings, because for him the use of the categories cannot reach beyond the limits of the objects of sensory experience.[157] Kant still admits the usefulness of this notion as a limiting concept – it serves to conceptually define the limits of human knowledge by positing other modes of non-sensible cognition – yet ultimately rejects its possibility as in blatant contradiction with our human finitude.[158] Nevertheless, the romantics and post-Kantian philosophers rehabilitated the concept and at some stage of their careers many held it like Fichte to be the foundation of all philosophy. Accordingly, the precise

154 Frederick Beiser nicely sums this up: »Like our knowledge, then, the self-knowledge of freedom is the result of action, and there can be no guarantee of its reality except through the success of action. This was the essential spirit of Fichte's response to scepticism in the Jena years, which he applied to the problem of freedom itself.« *German Idealism*, p. 294.

155 GA I/5: 60.

156 Hegel, *Differenzschrift* (Quoted in: Besier, *German Idealism*, p. 580).

157 See Kant, *Critique of Pure Reason* (B308–309), p. 361.

158 Cf. Günter Zöller: *Fichte's Transcendental Philosophy: The Original Duplicity of Intelligence and Will*, pp. 33f.

meaning of the term »intellectual intuition« (*intellektuelle Anschauung*) has been the source of much debate and dispute. For example, important works such as Jürgen Stolzenberg's *Fichtes Begriff der intellektuellen Anschauung*, Xavier Tilliette's less well-known but broader *L'Intuition Intellectuelle de Kant à Hegel*, and Daniel Breazeale's article: »Fichte's Nova Methodo Phenomenologica: On the methodological role of ›intellectual intuition‹ in the later Jena Wissenschaftslehre«, highlight the metamorphoses that this principle underwent not just within the course of German idealism and romanticism but even within the careers of the different philosophers themselves.[159] These scholars point out how misunderstood and misapplied this concept of intellectual intuition is, even by so-called experts of the period, and of the difficulty therefore of precisely defining the term.

Defining Intellectual Intuition

Let us therefore start with one such definition or characterization of intellectual intuition provided by Fichte in the *Zweite Einleitung in die Wissenschaftslehre* (1797). As with Kant's definition of sensible intuition discussed above, here Fichte lays the accent on the immediacy and directness of intellectual intuition. However, unlike Kant's assertion that the self is unknowable, Fichte attributes an immediate consciousness to it during the performance of the act of self-intuiting in which the I comes into being. For Fichte the principle of intellectual intuition does not belong to discursive thought for it is *not* a matter of demonstrating it by means of concepts, but like a geometrical axiom or postulate its truth and reality must be discovered by each person himself: »I call *intellectual intuition* the intuiting of oneself that is ascribed to the philosopher in the carrying out of the act in which the I arises. It is the immediate consciousness that I act, and how I act: it is that through which I know something, because I do it. That such a faculty as intellectual intuition exists, cannot be demon-

159 See Jürgen Stolzenberg, *Fichtes Begriff der intellektuellen Anschauung: Die Entwickelung in den Wissenschaftslehre von 1793/94 bis 1801/02* (Stuttgart: Klett-Cotta, 1986); Xavier Tilliette, *L'Intuition Intellectuelle de Kant à Hegel* (Paris: Vrin, 1995); and Daniel Breazeale, »Fichte's Nova Methodo Phenomenologica: On the methodological role of ›intellectual intuition‹ in the later Jena Wissenschaftslehre« in: *Revue Internationale de Philosophie* 206 (1998), pp. 587–616.

strated in concepts, or even developed from concepts. Everyone must find it directly in himself or he will never know it.«[160]

Similar to Kant's apperception of the »I think« that has to accompany all my representations, the Fichtean intellectual intuiting also accompanies all my actions, and indeed, is a condition of the possibility of finite empirical self-consciousnesses.[161] It is exactly Fichte's goal to bring this activity of the mind into a scientific discussion, and we realise his distance from the Kantian standpoint. Fichte is at pains to stress that such a »direct consciousness« is possible and ultimately rests on a revised understanding of intuition that allows a self-positing and awareness of one's own I.[162] But what exactly does »intellectual« mean for Fichte in this context? In *Versuch einer neuen Darstellung der Wissenschaftslehre* (1797), Fichte draws attention to the non-empirical nature of this activity of the intelligence, in which it remains entirely in the sphere of intellectual activity: »The intelligence intuits itself, simply as pure intelligence. [...] It is for this reason »*intellectual*« in distinction to all other intuitions.«[163] And Fichte signals here what we have just mentioned above, that intellectual intuition is always linked with a »sensible« intuition because it is connected with the sensible intuition of ordinary self-consciousness.[164] Accordingly, the terms »intellectual« and »sensible« are often set in opposition, and Fichte does this from the very beginning of the *Wissenschaftslehre*, especially when trying to argue for the non-sensible character of the I and the cognition of its activity.[165] For instance, in the *Eigne Meditationen* we behold Fichte privately pondering the idea and role of intellectual intuition both within

160 GA I/4: 216f.

161 Cf. Fichte's rather dramatic words: »I can nether take a single step, nor move my hand or foot, without the intellectual intuition of my self-consciousness in these actions. [...] This intuition is the source of life, and without it there is death.« (ibid., p. 217).

162 Cf. Fichte, *Versuch einer neuen Darstellung der Wissenschaftslehre*: »Scientifically speaking, such an immediate consciousness is called an *intuition* (*Anschauung*) [...] This immediate consciousness is precisely the described intuition of the I« (GA I/4: 276f.).

163 »Die Intelligenz schaut sich selbst an, bloss als reine Intelligenz. [...] Deswegen ist es eine *intellectuelle* zum Unterschied zum andere Anschauungen« *Versuch einer neuen Darstellung der Wissenschaftslehre* (GA I/4: 278).

164 *Zweite Einleitung* (GA I/4: 217).

165 See for example the *Zurich Wissenschaftslehre* (GA IV/3: 34).

Reinhold's *Elementary Philosophy* and in philosophy in general. Fichte initially inquires into its possibility, before stating that it is the *forms* of the faculty of representation which are intuited: »Aren't there also pure intellectual intuitions? The forms of the faculty of representation, of which we are speaking, are intuited purely intellectually.«[166]

Fichte makes his first written public declaration on intellectual intuition in the small but important *Recension des Aenesidemus*.[167] The review is ostensibly concerned with the sceptical philosophy of G.E. Schulze, but more significantly Fichte makes the assertion that no empirical intuition is able to posit the I (absolute subject), but that it requires instead an intellectual intuition. This entails, Fichte further suggests, that the positing of the I be set in opposition with an absolute object, or a so-called »Not-I.«[168] Similar remarks may be found in the very first lectures of the *Zurich Wissenschaftslehre* from February 1794, where the language is even more forceful than the tentative words of the *Recension des Aenesidemus*.[169] Surprisingly, despite his public declaration of support of the theory of intellectual intuition in the *Aenesidemus* review, Fichte's first major public exposition of his philosophy – the *Grundlage der gesammten Wissenschaftslehre* – does not contain a discussion of the term and its meaning. Instead, Fichte couches it in the neologism *Thathandlung*, a so-called »deed-act« (or fact/act) to signify the same idea of the self-intuiting I that is expressed in his *Grundsatz*.[170]

Only two to three years after the *Grundlage*, Fichte not only philosophically and publicly argues in favour of intellectual intuition, but does so in explicit opposition to Kant. This especially occurs in the above mentioned *Zweite Einleitung* and *Versuch einer neuen Darstellung der Wissenschaftslehre*, and in the university lectures on the *Wissenschaftslehre*

166 Fichte, *Eigne Meditationen* (GA II/3: 24).

167 Fichte, *Recension des Aenesidemus* (GA I/2: 48)

168 »The absolute subject, the I, is not given through empirical intuition, but posited through intellectual intuition; and the absolute object, the Not-I, is posited in opposition to it.« (Ibid.).

169 Cf. GA IV/3: 34ff.

170 »[The *Grundsatz*] ought to express that *Thathandlung* that does not nor cannot occur among empirical determinations, but is rather lies at the basis of all consciousness and alone makes it possible.« In §1 of Fichte, GWL (GA I/2: 255).

nova methodo of 1796–1799.[171] If Fichte had not perhaps been confident enough to publicly challenge Kant's view of intellectual intuition in the *Grundlage*, in these later Jena texts Fichte set about making intellectual intuition into one of the cornerstones of his new system of philosophy. Some scholars even see a shift in the methodology of Fichte's *Wissenschaftslehre* during and *within* the Jena period itself, from an axiomatic and deductive structure derived from a *Grundsatz* around 1794–1795, to a more descriptive, phenomenological or »inner experimental« method in the writings running from 1796 to 1799, since the focus becomes one of consciously describing and observing the acts of positing the I.[172] I tend to agree with this broad division between the earlier and later Jena *Wissenschaftslehre*, although we must bear in mind that there exist certain early unpublished and fragmentary writings (certainly not destined for the public at large in their extant form) which are even prior to Jena – such as the *Eigene Meditationen* (1793/94) from the Zurich period. These texts likewise contain similar exhortations or descriptive »summons« to reflect upon, observe and intuit the acts of one's own self-consciousness, or in other words, a philosophy based on intellectual intuition.[173]

b. What is Geometrical Intuition?

The *Zurich Wissenschaftslehre* was Fichte's first presentation in a public setting of his newly discovered philosophy. Fichte was especially keen to show the Lavater circle in Zurich that his new model of philosophy was scientific, and he therefore compared his philosophy with geometry, as the ideal model of a science. With regard to the topic of intuition, we can

171 See especially volumes two and three of the *Kollegnachschriften* (series IV) in the Fichte *Gesamtausgabe*: GA IV/2: 17–267; Nachschrift Eschen (GA IV/3: 151–196); and Nachschrift Krause (GA IV/3: 323–535).

172 »Such a description, in turn, will constitute a ›genetic account‹ of consciousness, grounded securely upon the immediate evidence of ›inner‹ or ›intellectual‹ intuition.« Daniel Breazeale, »On Fichte's Nova Methodo Phenomenologica« p. 589. Ives Radrizzani argues that the turn of the WLnm consists in Fichte actively involving his students in his an »inner experiment« to consciously postulate their own I. See his monograph, *Vers la fondation de l'intersubjectivité chez Fichte. Des Principes à la Nova Methodo* (Paris: Vrin, 1993), pp. 81–84.

173 See Fichte's statements in the *Eigne Meditationen*: »›The I can be intuited.' »Intuit your I.« »You are conscious of your I.«(GA II/3: 26ff.).

see a distinct parallel between Kant's strategy in the *Critique of Pure Reason* and Fichte's strategy in the Zurich lectures. But with a singular difference: Kant had appealed to the pure intuitions of geometry to justify his theory of *sensible* intuition in the Critique, Fichte on the other hand appeals to the intuitions of geometry to justify his theory of *intellectual* intuition in the *Wissenschaftslehre*. Fichte was to employ this strategy throughout his career. The key aspect for Fichte is that both the intuitions of geometry and his intellectual intuition are also »pure«, that is to say, they do not rely on any empirical contribution.[174] It would therefore seem worthwhile examining the relationship between intellectual intuition and geometrical intuition or Fichte's conception of the intuitions of geometry. However, we might first ask: What is geometrical intuition?

The Nature of Geometrical Intuition

Let us first look at the remarks on geometrical intuition made by some of the greatest mathematicians in history. As we have seen, in Fichte's eyes many geometers do not reflect enough upon the philosophical underpinnings of their science and therefore remain at the naive level, or at the ordinary standpoint, as opposed to a higher transcendental standpoint. This was not the case with Henri Poincaré, however, who has furnished one of the more remarkable accounts of what happens during the process of geometrical activity. In his article »Mathematical Discovery« he provides a classic account of how mathematical intuition functions in the minds of real mathematicians.[175] Geometrical intuition has similarities to the everyday conception of intuition (or inspiration), except that it is preceded by a period of intense mental work, typically when trying to resolve mathematical problems, such as the discovery here by Poincaré of a difficult branch of mathematics called Fuchsian functions. Consequently, when one becomes more relaxed and even forgets about the problem, one is overcome by the intuition of the solution: »One is struck by these appearances of sudden illumination, obvious indications of a long course of pre-

174 Cf. Ibid., p. 34.

175 Henri Poincaré, »Mathematical Discovery« in: *Science and Method* (New York: Dover Publications, 2003) pp. 46–63. Concerning mathematical discovery and invention in general, see Jacques Hadamard, *The Psychology of Invention in the Mathematical Field* (Princeton: Princeton University Press, 1949).

vious unconscious work.«[176] This account is similar to Kant's account in the *Critique of Pure Reason* of the scientific birth of geometry in ancient Greece, or the birth of any scientific discovery or theory, where Kant employs such language as: »a new light broke upon the first person who demonstrated the isosceles triangle«, or »a light dawned on all those who study nature.«[177] Moreover, these accounts of mathematical or scientific discovery also bear an uncanny resemblance to the legendary account of Fichte's own discovery of the *Wissenschaftslehre*, which apparently occurred in November 1793 with some kind of sudden illumination or »original insight« (to borrow Dieter Henrich's term), and revealed to Fichte the primal significance of intellection intuition to philosophy.[178]

What is interesting here, however, is not the resemblance of Poincaré's species of intuition to everyday notions or to mysticism, but to the philosophical notion of intuition. This is because Poincaré furthermore characterizes geometrical intuition in a manner not unlike Fichte and Kant. Geometrical intuition is characterized by its »conciseness, suddenness and immediate certainty«, in which the »decisive idea presents itself to the mind.«[179] And his account of geometrical intuition is confirmed by other great names in mathematics, such as Kurt Gödel, who spoke of using the »eye of the spirit« that resembles our perceptual sensible eye, but beholds and discovers the »objective« entities and objects of mathematics.[180] This obviously recalls Fichte's and Goethe's notion of the eye of the spirit, which, as Eckart Förster has argued, belongs to the *Ideengut* of German idealism and is related to the philosophical notion of intellectual intuition.[181] This is not to forget the differences, for as we shall see, intuition is of course neither a concept, nor an idea nor a perception, although concepts also require intuitions.

176 Ibid., p. 55.

177 Kant, *Critique of Pure Reason* (Bxii-Bxiii), pp. 108f.

178 *Cf. Johann Gottlieb Fichte. Lichtstrahlen aus seinen Werken und Briefen nebst einem Lebenabriss* (Leipzig: Brockhaus, 1863), p. 46.

179 Henri Poincaré, »Mathematical Discovery«, pp. 54f.

180 Cf. Mark Balaguer. *Platonism and Anti-Platonism in Mathematics* (1998), pp. 25–29; and P. Davis and R. Hersch, *The Mathematical Experience*, p. 357.

181 Eckart Förster, »Goethe and the ›Auge des Geistes‹«, *Deutsches Vierteljahrschrift für Literaturwissenschaft und Geistwissenschaft* (75) 2001, pp. 87–101.

Furthermore, we are confronted with a number of problems and contradictions here. How can geometry – a science that claims to be exact and rigorous – rest on abilities and methods that seem entirely mysterious, intuitive, and non-rigorous?[182] Of course, an answer to this question presupposes explaining how mathematical discovery actually occurs, and this has yet to be fully explained. It is worth noting that with geometrical intuition we are confronted with some of the problems that we saw in the investigation of geometric axioms: there comes a limit or a point at which we cannot explain and prove any further but must accept the principles as given and self-evident. Because if intuitive is set in opposition to rigorous, we are entitled to ask what the term »rigorous« actually means, and we discover that any definition of rigorous is based in turn on an intuitive understanding of its meaning.[183]

Nevertheless, mathematicians do admit that the terms intuitive and intuition can be characterized. The studies of Davis and Hersch have shown that although mathematicians consider »intuitive« to be synonymous with something »incomplete«, or more a »visual proof«, it is still accepted as »plausible and convincing in the absence of other proof«; it ultimately relies on a »physical proof« and is holistic, i.e. non-analytic.[184] There is also no unanimity among the usage, employment and precise definition of »geometrical intuition« among geometers themselves, with many calling for it to be inadmissible in the proof of theorems.[185] Be that as it may, many other geometers consider that intuition is vital in the discovery of geometrical truths, although the correctness of these discoveries are confirmed and proved by using other means at their disposal such as logic and deductive proofs. Recalling Kant's words about the »blindness«

182 Cf. Paul Bernays's discussion of the alleged inexactness of geometric intuition in: »On Platonism in Mathematics« in *Philosophy of Mathematics: Selected Readings* (eds. P. Benacerraf and H. Putnam, 2nd ed., Cambridge: Cambridge University Press, 1983), pp. 258–272.

183 See Davis and Hersch, *The Mathematical Experience*, p. 433.

184 Ibid., pp. 433–444.

185 Hans Hahn, who was ironically the teacher of Kurt Gödel, was in opposition to the use of mathematical intuition. See his classic article, »The Crisis in Intuition« in: James R. Newman, ed. *The World of Mathematics, Vol. III, pp. 1956–1976* (reprinted in: H. Hahn, *Empiricism, Logic and Mathematics* (Dordrecht: Reidel, 1980), pp. 73–102).

of concepts without intuitions, Poincaré says, »It is by logic we prove, it is by intuition that we invent. [...] *Logic, therefore, remains barren unless fertilised by intuition.*«[186]

Perhaps the greatest appreciation for geometrical intuition is to be found in the domain of the teaching of geometry and mathematics. Many researchers and teachers consider the awakening, stimulating and training of geometrical intuition to be an indispensable aspect of learning geometry correctly. And their characterization of geometrical intuition bears a striking similarity to the philosophical notion of intuition in German idealism. For example, E. Fischbein outlines five qualities that distinguish geometrical intuition. 1) It is a species of »immediate knowledge«; 2) It possesses »self-evidence«; 3) It has a certain »intrinsic certainty«; 4) It has »coerciveness«; that is to say, it compels us to accept it as true; and 5) it has a universal or »global« character.[187] At the very least, the characteristics of nos. 1, 2 and 3 of immediacy, self-evidence and certainty are identical to Fichte's qualifications of intellectual intuition, and as we shall see in the next section; in the *Ankündigung*, no. 4 is called »necessity« by Fichte, while no. 5 is also classified as an important aspect of mathematical cognition.[188]

Moreover, the charge brought against geometrical intuition that it is a mysterious and privileged »faculty« that concerns unknowable abstract entities is precisely the same charge brought against Fichte's version of intellectual intuition. Despite their best attempts at claiming scientific certainty, are both geometrical intuition and intellectual intuition therefore nothing more than mysticism? Fichte vehemently contested the charge that intellectual intuition is just a species of mysticism or depends on some kind of special »psychological« state, as was openly suggested by Heusinger in 1799 and then tacitly repeated by Schelling in an October

186 Poincaré continues: »Thus logic and intuition have each their necessary role. Each is indispensable. Logic, which alone can give certainty, is the instrument of demonstration; intuition is the instrument of invention.« »Intuition and Logic in Mathematics«, translated by G. B. Halsted in: *The Value of Science* (New York: Cosimo, 2007), p. 23.

187 E. Fischbein, *Intuition in Science und Mathematics: An Educational Approach* (Dordrecht: Springer, 1987), pp. 13ff.

188 Cf. Fichte, *Ankündigung* (GA I/7: 153–164).

1801 letter to Fichte.[189] For Fichte, the methods and practices of both mysticism and psychology cannot and should not be conflated with the method of the *Wissenschaftslehre*. The former merely dimly »observe« *passively* discovered facts, whereas in the *Wissenschaftslehre* one is required to be fully *active* while simultaneously paying *conscious* heed to the performance of one's acts.[190] But this is also the dilemma of the *Wissenschaftslehre*: on the one hand its fundamental act cannot be proved but has to be experienced and performed by each person himself. On the other hand, its cognition is scientific like geometry *because* it is based on intuitions and not on concepts. This is why Fichte places intuition higher than concepts, because we need intuition to arrive at an understanding of concepts, not the other way round: »Precisely because all intuition lies higher than any concept, one cannot employ concepts to explain what intellectual intuition is; the only way one can become acquainted with it is by possessing it.«[191]

Furthermore, at first glance both geometrical and intellectual intuition could be mistaken for a kind of intellectual »perception.« This is a common train of thought and analogy made in geometry about geometrical intuition. As noted, Kurt Gödel describes the process of mathematical intuition by drawing upon precisely this analogy: »Despite their remoteness from sense experience, we do have something like a perception also of the objects of set theory, as seen from the fact that the axioms force themselves upon us as being true. I don't see any reason why we should have less confidence in this kind of perception, i.e. in mathematical intu-

189 This was Johann Heusinger's charge against Fichte in 1799. See Daniel Breazeale discussion of this point and the other charges against intellectual intuition from the same period in his essay: »Toward a *Wissenschaftslehre more geometrico* (1800–1801)« (especially pp. 6–16). Whereas Schelling, in a letter to Fichte dated October 3rd 1801, claimed the same of Fichte's idealism in the *Sonnenklarer Bericht*: »You forgot to enclose the *Sonnenklarer Bericht,* but I got a copy of it anyway. The idealism in it appears to me to be rather psychological, almost like in Lichtenberg's posthumous writings.« (GA III/5: 89).

190 See Fichte, *Sonnenklarer Bericht*: »Here we are not concerned with the discovery of something already finished, but rather with the discovery of something that first has to be produced by a free act of thinking. The *Wissenschaftslehre* is not psychology, which is nothing at all.« (GA I/7: 222).

191 Fichte, *Ankündigung* (GA I/7: 255). English translation by Daniel Breazeale in *Introductions to the Wissenschaftslehre and Other Writings*, p. 195.

ition, than in sense perception, [...] They too may represent an aspect of objective reality.«[192]

For Fichte, however, although both geometrical intuition and intellectual intuition *appear* to be analogous to sense perception, intellectual intuition should not be interpreted in this manner. This is despite the fact that Fichte himself used the eye's visual perception of external objects as a metaphor for the internal »seeing« of the I.[193] For unlike the act of visual perception, in which our eye is passive and does not perceive itself as such, the activity of intellectual intuition is both an active seeing and a doing, it is both the active agent *and* the acted upon, and the seeing I and the mirrored eye. Or as Fichte says in the *Wissenschaftslehre nova methodo*: »The I of philosophies hitherto is a mirror, but a mirror that does not see itself [...] The I of the *Wissenschaftslehre* is not a mirror but an eye. Everything inwardly spiritual has an external image. Whoever does not know the I, also does not know what an eye is.«[194]

c. Fichte on the Intelligible Intuitions of Mathematics

Despite being wary of conflating geometrical intuition with intellectual intuition, Fichte *did* take his inspiration from the intuitions of geometry as a concrete and accessible example to help explain the operation and functioning of intellectual intuition. Here the intuitions of mathematics are primarily viewed as a model for comprehending intuitions that have a non-sensible character. Fichte explains this very clearly in two published texts from 1801, the *Sonnenklarer Bericht* and the *Ankündigung*. In the former work Fichte states that the intuition of the pure I has to be carried out in connecting scientific steps precisely in the same manner as geometry proves each of its necessary steps through intuitions.[195] In the latter

192 Kurt Gödel, quoted in: *The Mathematical Experience* (eds. P. Davis and R. Hersch), p. 357.

193 Günter Zöller furnishes a detailed examination of this point in his book: *Fichte's Transcendental Philosophy: The Original Duplicity of Intelligence and Will*, pp. 35–38.

194 Fichte, *Wissenschaftslehre nova methodo* (GA IV/2:49). Cf. Günter Zöller, ibid., p. 36. See too Fichte's discussion on the terminology *Bild* and *bilden* in the *Neue Bearbeitung der Wissenschaftslehre 1800* (GA II/5: 395ff.).

195 »Reader. The *Wissenschaftslehre*, therefore, starts from the pure I, or from intuition in its highest purity, and with every further step of that science a new link

text, Fichte goes even further. He dismisses the view that the starting point of the *Wissenschaftslehre* is like any other previous conceptual philosophy, i.e. that it is founded on a »rational cognition *based on concepts.*«[196] Since mathematics demonstrates the possibility of a science and rational cognition that is based on intuitions, he deduces that there must exist in turn another science that itself cognizes this mathematical art of cognition. We can only immediately grasp this higher science through intuition and not concepts. In Fichte's eyes, this latter science is the *Wissenschaftslehre*, which he now terms the »*mathesis of mathesis*«, for not only is its cognition based on intuitions like those of mathematics, but it is the very »*cognition of this cognition.*«[197]

The *Ankündigung* furnishes a detailed account of the nature and qualities of mathematical intuitions and mathematical cognition, and brings to fruition some of the ideas that he had been pondering seven years before in 1793 in texts like *Bei Lesung der K. Kr.d.r. Vft.* and the *Eigne Meditationen*, especially concerning the a priori status of »mathematical« intuitions.[198] In the *Ankündigung* Fichte articulates his view that the act of mathematical cognition is distinguished by three precise characteristics:

connects with the previous links, the necessary connecting of it is then proved in that very intuition. Author. Precisely in the same manner in which geometry proceeds, where each new proposition adds something new to the proposition previously established, the necessity of which something new is also proved only in intuition. In fact, such must be the method of every real progressive science which does not turn round in circles.« Fichte, *Sonnenklarer Bericht* (GA I/7: 209).

196 GA I/7: 155.

197 »First of all, since there is certainly supposed to be a kind of rational cognition based on intuitions (for this is what one claims mathematics is), then – so long as all cognizing and thinking is not supposed to come to an end with this sort of cognition, and indeed, so long as it is supposed to be possible for us even to assert that there is such cognition – there must, in addition, be some cognition of this cognition. And since an intuition, as such, can only be intuited, this latter sort of cognition must itself be based on *intuition.*« (GA I/7: 156). English translation by Daniel Breazeale in *Introductions to the Wissenschaftslehre and Other Writings*, p. 189.

198 See Fichte, *Bei Lesung der K. Kr.d.r. Vft.* from 1792 or 1793 (GA II/2: 327f.). See Jürgen Stolzenberg's comments on this small manuscript, where he draws a link between it and the *Eigne Meditationen* (cf. *Fichtes Begriff der intellektuellen Anschauung: Die Entwickelung in den Wissenschaftslehre von 1793/94 bis 1801/02* (1986), pp. 35–40).

(1) It is immediately self-evident; (2) It possesses necessity; (3) It has universal validity. The first two of these characteristics are familiar to us already from his characterization of intellectual intuition, but Fichte elaborates more on the third characteristic of universal validity. It is not universally valid because it remains identical to itself and does not change, but because it is able to communicate this unchangeability to the concept that comprehends it.[199] He adds that such an understanding of philosophy based on intuitions of this kind will complete even further the revolution in philosophy commenced by Kant, a revolution that Kant »himself had failed to carry through to completion.«[200] Thus, it is also a matter of *breaking* with the Kantian model of cognition, and realising that *intuition* and not the concept is the »unifying medium« (*Vereinigungsmittel*) within reason itself. As already mentioned above, this is because intuition cannot be proved or demonstrated by any concepts or words, and Fichte therefore maintains that it is therefore superior to them or any philosophy taking its point of departure from them. Concepts and words must be judged by intuition and not vice-versa: »*Accordingly, philosophy would be the cognition of reason itself by means of reason itself – through intuition.*«[201] It should be clear – Fichte's theory of cognition is grounded in intuitions and not concepts.

What is the function of the intuitions of geometry within the *Wissenschaftslehre*? For Fichte, the intuitions of geometry are a *paradigmatic model* for grasping the act of intellectual intuition. – Here we have to be careful – for it is *not* a matter of extending Fichte's theory of intellectual intuition into other domains, even into the realm of geometric entities.

199 »Moreover, since mathematics already existed and was something with which we were already acquainted, everyone should have realized and been familiar with the fact that the foundation of immediate self-evidence, necessity, and universal validity is never to be found within any concept, but lies in the intuition of the very act of comprehending. [...] Such an intuition is ›universally valid‹, not because it itself always remains one and the same, but precisely because it communicates its own unalterability to every concepts that comprehends it – precisely because and precisely insofar as this concept grasps it.« (GA I/7: 157). English translation by Daniel Breazeale in *Introductions to the Wissenschaftslehre and Other Writings*, pp. 191ff.

200 Fichte, *Ankündigung* (1801, GA I/7: 159).

201 Ibid.

And it is not a matter of broadening or raising geometrical intuition up to intellectual intuition. *Intellectual intuition is not the same as geometric intuition.* Because, as Fichte points out in the *Erlanger Logik*, geometric intuition is »intelligible intuition« (*intelligible Anschauung*).[202] It is intelligible because it is pure, i.e. not based on empirical or sensible objects and therefore remains like intellectual intuition in the sphere of the intelligence, and arises through my own activity. This intuiting of the propositions and laws of geometry is a purely a priori and rational activity. It is also an *Anschauung* or intuition, because it is directly and immediately grasped, but only to the extent *that it is in movement, or pure activity* – as is the case for any *Anschauung* for Fichte. This is clear from the 1797 *Versuch einer neuen Darstellung der Wissenschaftslehre*, where Fichte also uses the term »intelligible intuition« (*intelligible Anschauung*) in this sense, as a faculty for grasping the »intelligible world« (*intelligible Welt*), i.e. a sphere that is not opposed to me but arises through *my own acting.*[203]

How does this *Anschauung* or intuition differ from the concept of the line or circle? For Fichte they are intimately related, except that only once the initial geometrical activity has »come to rest« is it then possible to grasp it as a concept. This *Anschauung-Begriff* (intuition-concept) distinction is also clearly expressed at the end of the *Versuch einer neuen Darstellung der Wissenschaftslehre* (1797/98). It also explains why Fichte talks in other texts of the idea of a pure triangle as such as an *Urbild* (archetype) or *Urbegriff* (archetypal concept) or »*geistiges Bild*« (spiritual image).[204]

For Fichte the form of geometrical intuition is therefore similar to intellectual intuition, and he had the »brilliant idea« while reading Kant's *Critique of Pure Reason* in 1793 of the basis of a cognition on which to construct his whole philosophy: »A mathematical intuition can be given *a priori*; consequently, it is scarcely different from a mere pure concept. – (I will here precisely build my idea to construct the *Grundsätze* of pure reason through empirical experience).«[205] Historically speaking, intellectual

202 Cf. Fichte, *Erlanger Logik* (GA II/9: 129).

203 See Fichte, *Versuch einer neuen Darstellung der Wissenschaftslehre* (GA I/4: 220).

204 See *Die am Ende des Winters [Sommers] 1794 von Fichten Gehaltene Schlussvorlesung als Beylage der Verantwortung* (1797) (GA I/4: 413).

205 Fichte, *Bei Lesung der K. Kr.d.r. Vft.* (GA II/2: 327f).

intuition is inspired by geometrical intuition, yet epistemologically speaking, the intuitions of geometry are grounded in and presuppose intellectual intuition: »the I is also an intuition, however, it is not a sensible intuition but an *intellectual* intuition. Therefore the *Wissenschaftslehre* is in possession of a pure intuition, just as geometry is, and all the intuitions of geometry are founded on the intuitions of the *Wissenschaftslehre.*«[206]

Importantly, although both geometrical intuition and intellectual intuition remain in the same sphere of the pure intelligence, they should not be confused or their boundaries blurred. For intellectual intuition has certain limits and borders that should not be transgressed, and it is solely limited to the realm of the activities and acts of the I (a transgression Schelling would undertake around 1800 to Fichte's consternation).[207] Hence, we should be aware of these limits of intuition and not conflate its different modes in geometry and philosophy. For Fichte the differences between geometrical and philosophical intuition are due to their own inherent laws, goals and boundaries, and this betrays a tension in their relation to necessity and freedom. The goal of geometry is to exhaust the possibility of lawfully limiting space within a single universal space. This is unlike philosophical (intellectual) intuition, which is to determine and exhaust the activity of the pure I. In the 1805 *Erlanger Logik* Fichte is clear about the differences between these two forms of intuition: »A condition of this insight is to [place] ourselves into this possibility and allow these laws to arise within ourselves, and to work directly upon our limiting of

206 Fichte, *Zurich Wissenschaftslehre* (GA II/3: 34).

207 This limitation to the I and its activities is what also separates Fichte from Schelling's attempt at expanding intellectual intuition to other domains during the years 1800–1802. Fichte rejected Schelling's suggestion of an »objective intellectual intuition« in art, saying that intellectual intuition can only be limited to the activity of the I, and could not be extended beyond this. Fichte, therefore, places limits on his notion of intellectual intuition. Ironically, just as Kant had restricted intuition to the sensible sphere, but his so-called self-proclaimed followers and students Fichte (and Reinhold) had overstepped these limits, so now in turn Fichte's limits were transgressed by his one-time friend Schelling. The latter proposed that intellectual intuition be extended to incorporate the notion of not merely a »subjective« intellectual intuition like Fichte's, i.e. one focused on the activity of the I, but also one extending to the world beyond to an »objective« intellectual intuition. (See Fichte's long letter to Schelling dated May 31 – August 7, 1801 (GA III/5: 43–53).

space, like in limiting and in construction. Hence, this is *geometrical* intuition, which has often been confused with philosophical intuition, and is something entirely different.«[208]

It is here in the text of the *Erlanger Logik* that we have what is perhaps Fichte's most explicit example of the need for such a form of intuition. Pure geometrical intuition is intelligible intuition, and precisely when it is a question of grasping the original or archetypal elements of geometry. As explained above, these archetypal elements are purely ideal, infinite, absolute and must be grasped in intuition. But what kind of intuition is therefore required for them? On the one hand, sensible intuition does not suffice, because the archetypal elements *cannot* be represented sensibly or empirically. On the other hand, intellectual intuition is also not appropriate, since as a form of intuition it is solely employed for grasping the pure activity of the I. Hence, intelligible intuition remains as the sole possibility, because Fichte's archetypal elements are inherently rational or intelligible. For Fichte, sensible and intelligible intuition are ultimately dependent on intellectual intuition and not vice versa. How are we to understand this relation?

Jules Vuillemin explains this relation between sensible and intellectual intuition by means of the example of abstract algebra. Vuillemin's original interpretation in *La philosophie de l'algèbre* takes it starting point from Fichte's notion of genesis and genetic knowledge and contrasts it with the idea of structure. In abstract algebra, for example, Vuillemin explains that with regard to the notion of genesis the equations came first, and then this led to group theory. Yet with regard to the structure of abstract algebra as a whole, groups form the basis for equations. That is to say, although discovered later, the notion of a group is more foundational than that of an equation. Applying his theory of genesis and structure to the *Wissenschaftslehre*, Vuillemin argues that with regard to notion of genesis, empirical self-consciousness comes first and then subsequently leads to the concept and necessity of intellectual intuition. – Whereas, with regard

208 Fichte, *Erlanger Logik* (GA II/9: 125). See too Fichte's remarks in the *Zurich Wissenschaftslehre* on geometrical intuition and necessity: »Finally, it will be shown that the pure intuition of space is likewise produced by means of the activity of the power of imagination, though not freely, as in geometrical intuitions, but rather necessarily, i.e. it is constructed, and that therefore the *Wissenschaftslehre* is also in possession of a construction, but which is unique.« (GA IV/3: 34).

to the structure, intellectual intuition is the condition for the possibility of all consciousness. Hence the concept of intellectual intuition is more foundational than that of empirical self-consciousness. Vuillemin's structural method is not just an inversion of the genesis, but the transition from the consideration of objects, to the consideration of forms and operations. However, we have to be careful here, for although Vuillemin calls Fichte's method genetic in his text, his employment of the word ›genetic‹ is different to what Fichte himself understood by the term.[209]

To summarize: In terms of their characteristics of immediacy, self-evidence and non-empirical character, there are striking similarities between intellectual and geometrical intuition. This is confirmed by Fichte taking geometrical intuition as *the* model of a pure intuition. Though their form and structure are similar – both are concerned with pure non-empirical intuitions in the domain of the intelligence that ultimately have a relation to the physical, empirical world – their contents, laws and functions are different. Just as we saw that the axioms or postulates of geometry are narrow or more limited examples of the subject-object postulate of the *Wissenschaftslehre*, so according to Fichte geometrical intuition should also be viewed as a more limited form of intellectual intuition. Geometrical intuition is *intelligible* intuition, and is the faculty for grasping the archetypal elements of Fichte's theory of original geometry, whereas intellectual intuition is the faculty for grasping the pure activity of the I.

3. Geometrical Construction in the *Wissenschaftslehre*

This section will examine Fichte's ideas and employment of the philosophical notion of »construction« (*Konstruktion*). I will argue that like intuition, Fichte's philosophical understanding of construction is primarily taken from the domain of geometry – i.e. that it is indebted to and modelled on the geometrical (mathematical) theory of construction.[210] However, similar to the geometric postulates and geometrical intuition, Fichte views geometrical construction as a more specific example of the univer-

209 See Jules Vuillemin, *La philosophie de l'algèbre*, pp. 112–120.

210 Among others, Michael Rath has argued for a geometric origin, saying that Fichte's terminology is inspired by contemporary theories of mathematics (e.g. Kästner, Kant and Maimon) and oriented on the methodology of geometry (cf. *Selbstbewusstsein und Wille*, pp. 34f., 137).

sal form of philosophical construction. Other influences for Fichte's knowledge of construction stem from Kant's exposition in the Transcendental Doctrine of Method in the *Critique of Pure Reason*, and Maimon's theory of construction, which is especially prominent in his *Versuch über die Transzendentalphilosophie*.[211]

a. Fichte's Understanding of Construction

It is no exaggeration to say that the theory of construction plays a seminal role in Fichte's discovery and elaboration of the *Wissenschaftslehre*. In the opening pages of the *Eigne Meditationen* from 1793/94 Fichte asserts that philosophy itself could make a huge leap forward and defeat the objections of sceptical philosophers if it could only solve the problem of construction: »How do [Reinhold's] thoughts agree in general with the actions of the spirit, and how can such an agreement be demonstrated? [...] Is construction not possible within Elementary Philosophy? Is it not possible to provide an inner intuition which would explain and prove these thoughts? Were this to occur, then Aenesidemus would be refuted.«[212] These words and other letters from the time demonstrate the extent to which Fichte was concerned with the intimate relationship between construction and intuition at the start of his philosophy.[213] Although Fichte developed a method of philosophical construction whose origin and

211 Regarding Maimon, see the discussion of construction in Gideon Freudenthal's: *Maimon's Philosophy of Geometry*, pp. 33–39; and Meir Buzaglo, *Solomon Maimon: Monism, Skepticism and Mathematics* (Pittsburgh: University of Pittsburgh Press, 2002).

212 Fichte, *Eigne Meditationen* (GA II/3: 23f.). Despite the mention of Maimon's name a few lines before this statement, Stolzenberg (1986, p. 16) claims that this passage is inspired by the objections of Schulz/Aenesidemus (against Reinhold's philosophy). However, I would agree with Paul Franks who considers that »the idea of responding to these objections by developing a method of philosophical construction is almost certainly inspired by Maimon.« (*All or Nothing*, p. 250 note).

213 For example, in a letter to Reinhard from January 15, 1794 Fichte had already questioned the relationship between construction and intuition: »However, can't philosophy, like geometry and mathematics, *construct* its concepts in *intuitions*? This is entirely right; and it would be bad if it could do this: for then we would not have any philosophy but mathematics: but philosophy can and should deduce them *through thinking* out of a single *Grundsatz* that everyone can admit.« (GA III/2: 40).

model shares many traits in common with geometrical construction, philosophical construction is ultimately the foundation for geometrical construction. It is not my intention to establish whether Fichte continued with this constructive method throughout his career or dispensed with it before returning to it again during certain periods.[214] My goal is to focus on the geometrical context of construction.

If we recall that the »construction« of a concept for Kant means »to exhibit a priori the intuition corresponding to it,« then like Maimon before him, it seems that Fichte was influenced by Kant's presentation and explicitly appropriated the method of construction from geometry and applied it to transcendental philosophy.[215] Fichte himself admits this on a number of occasions[216], but additionally argues in the *Wissenschaftslehre nova methodo* that he has transformed the method of construction in a higher transcendental manner. This is because in the Fichtean theory the philosopher is not only to construct objects so that they are empirically visible in space, as one does in geometry using a straightedge and compass, but the philosopher is supposed to consciously »construct« the *conditions* and rules of his own consciousness: »Our task is ultimately this: to bring together the conditions of consciousness according to the already known rules, and to *construct*, as it were, consciousness before our eyes; not only like the geometer does, who is not bothered by the question as to the origin of space and the ability to draw lines, for the *Wissenschaftslehre* already presupposes this.«[217]

The term »construction« has come to designate the entire method and methodology of Fichte throughout the *Wissenschaftslehre*, especially during the latter phase of Fichte's career.[218] Furthermore, in the *Erlanger*

214 This has been argued by Paul Franks: »Fichte initially considers the constructive method but then develops the dialectical in his first Jena presentation, returning decisively to the constructive method in his second Jena presentation.« (*All or Nothing*, pp. 338f.).

215 Kant, *Critique of Pure Reason*, A713/B741, p. 630.

216 See Fichte's comments from the *Eigne Meditationen* (GA II/3: 23f.) and letter to Reinhard (GA III/2: 40), both quoted and translated above.

217 Fichte, *Wissenschaftslehre nova methodo* (GA IV/2: 177f.).

218 See Michael Rath's *Selbstbewusstsein und Wille* which especially investigates construction as methodology within the framework of the *Wissenschaftslehre nova methodo*.

Logik Fichte tells his listeners that for genuine philosophizing they must first learn the art and manner of »constructing«, and that this intellectual skill is »borrowed from geometry.«[219] Of course, in geometry we also have the ordinary meaning of empirical construction, where we employ the straightedge and compass (*Zirkel und Lineal*) on the page of our book to »construct« the triangles and other objects of geometry.[220] Although related to this ordinary geometric sense Fichte has in mind here a different meaning of construction, a non-empirical and purely intelligible meaning of construction.

Fichte's distinction between these two types of construction – the purely external or empirical one and the higher intelligible or pure one – corresponds to Kant's distinction in his 1790 text *Über eine Entdeckung nach der alle neue Kritik der reinen Vernunft durch eine ältere entbehrlich gemacht werden soll.*[221] Upset by the misunderstandings (especially Eberhard's) that his treatment of construction in the 1st Critique had called forth, Kant endeavours to clarify the nature of construction by making a sharp distinction between two species of it. The first species he calls »pure« or »schematic« construction since it occurs through the imagination in accordance with an a priori concept.[222] – It corresponds to the type of construction outlined in the *Critique of Pure Reason* insofar as it is characterized as the exhibition of a concept »through the (spontaneous) production of a corresponding intuition«, and Kant furthermore equates it precisely with the one employed in the domain of (higher) mathematics: »These are the constructions which the mathematician must make use of in all his demonstrations.«[223] Kant then contrasts this type of construc-

219 Fichte, *Erlanger Logik* (GA II/9: 36). In the same text Fichte repeats that his method is »subsequently or post-constructible (*nachconstruirbar*): just like in geometry.« (GA II/9: 54).

220 Fichte discusses this ordinary sense of constructing in geometry of using straightedge (ruler) and compass in *Die am Ende des Winters [Sommers] 1794 von Fichten Gehaltene Schlussvorlesung als Beylage der Verantwortung* (1797) (GA I/4: 413).

221 Kant, *Über eine Entdeckung nach der alle neue Kritik der reinen Vernunft durch eine ältere entbehrlich gemacht werden soll* (AA VIII: 187–250). (English translation in: *The Kant-Eberhard Controversy*, translated and edited by Henry Allison, pp. 107–159).

222 Ibid. Kant, AA VIII: 192, note; English translation in: *The Kant-Eberhard Controversy*, p. 111, footnote.

223 Ibid.

tion with what he calls »empirical« or »technical« construction. This kind of construction is involved when the geometer employs mechanical instruments (such as ruler and compass) and it occurs in a material or empirical medium. Kant contends that strictly speaking this latter type of construction should not be called construction since it uses instruments and therefore does not belong to »science but to art.«[224] If we ask what *grounds* or is the basis of all construction for Kant? Then the answer seems to be: the immediate and direct activity of our own a priori *imagination* which is employed to draw the figures of geometry.[225] This had already been underscored by Kant in the *Critique of Pure Reason*, where he points out that even the drawing of a simple line requires the productive imagination.[226]

Thus, following Kant's line of reasoning in *Über eine Entdeckung*, Fichte argues on the one hand that the philosopher must take a leaf out of the »constructive« book of the geometer, yet on the other hand if the geometer wishes to explain and ground such a method he must turn to philosophy. Fichte also has recourse to the »power of the imagination.« For him, the possibility of geometric construction and the power of construction (*Konstruktionsvermögen*) are both grounded in the *Wissenschaftslehre*: »Now from where does geometry obtain this power of construction? [...] The demonstration is something that I myself, in the act of constructing, have first put there by means of the free self-activity of my own power of imagination.«[227]

What are the differences between the constructions of geometry and the constructions of philosophy? Fichte seems here to make further distinctions not made by Kant. For in the constructions of geometry Fichte claims that I am immediately certain and conscious of my free acting; whereas in philosophical constructions I have to »engage in reflection« and then furthermore *raise into consciousness my own actions* and activity

224 Ibid.

225 Cf. Michael Friedman, »Geometry, Construction and Intuition in Kant and his Successors,« in: eds., Sher, G & Tieszen. *Between Logic and Intuition: Essays in Honor of Charles Parsons* (New York: Cambridge University Press, 2000), pp. 189ff.

226 Cf. Kant, *Critique of Pure Reason* (A 162–163/B203–204), pp. 287f.

227 *Zurich Wissenschaftslehre*, lecture 4 (GA IV/3: 35).

of constructing.[228] The result is that if geometry is able to construct then why not philosophy? Or inversely, the *Wissenschaftslehre* too is able to construct *because* there exist constructions in geometry. In the *Zurich Wissenschaftslehre* construction means to give a concept a sensible underpinning. All demonstrations are grounded in this ability to construct, where *demonstration* signifies the same as it does in Kant: »showing (*vorweisen*) something in a sensible representation (*Vorstellung*) by means of taking apart (*Zerlegung*) of the same.«[229] Using the specific example of a triangle that has been constructed Fichte says that the »demonstration« of such a triangle means to establish it through intuition.[230] Fichte then asks why we should accept these demonstrations to be true, or how is it possible to furnish any kind of demonstration? For Fichte, the answer to this is once again to be sought in our *imagination* – its free activity has inserted itself into this procedure, and it is my own »free acting« that has carried out the construction.[231] On the other hand, there is necessity involved in

228 »In the first case, I am immediately conscious of my acting, since everything that occurs freely, is, in *the acting, certain* for the acting subject. / – In the second case, i.e. that of a necessary action of making, one has to engage in *reflection* in order to become conscious of this acting, inasmuch as that which necessarily occurs is ordinarily not [considered] as an acting, but rather as a state of passivity. / The actions or constructions in geometry are of the *first* kind; the actions or *concepts* in the *Wissenschaftslehre* are of the *second* kind.« (Ibid).

229 Fichte, *Zurich Wissenschaftslehre*, lecture four (GA IV/3: 33). See Kant's definition of a demonstration in the *Critique of Pure Reason*: »On demonstrations. Only an apodictic proof, insofar as it is intuitive, can be called a demonstration. Experience may well teach us what is, but not that it could not be otherwise. Hence empirical grounds of proof cannot yield apodictic proof. From *a priori* concepts (in discursive cognition), however, intuitive certainty i.e. self-evidence, can never arise, however apodictically certain the judgment may otherwise be. Thus only mathematics contains demonstrations, since it does not derive its cognition from concepts, but from their construction, i.e. from the intuition that can be given a priori corresponding to the concepts.« (A734/B762), p. 641.

230 »[These lines] form a triangle, which has come into being by means of construction. These three lines produce three angles. *To demonstrate* thus means to establish (*nachweisen*) through intuition that three angles arise from three lines.« Fichte, *Zurich Wissenschaftslehre* (GA IV/3: 34)

231 »But why then can no error have occurred in this case? Why is the demonstration true? Why is what can be established from an originally constructed pure intuition (intuition *a priori*) certainly contained in this intuition? That is, why is it

the ontological status and construction of a geometrical point (and other geometrical objects) that is in stark contrast with the freedom of the I. The activity of the geometer usually remains unconscious, and this fact is touched upon by Fichte in his *Vorlesungen über Logik und Metaphysik 1797/98* in a discussion of Maimon's criticism of Kant: »Maimon accuses the Kantian philosophy of not having any reality, for he asks: How can we apply *a priori* concepts to objects? He says, in mathematics we have insight into the reality of our concepts, and adds that we construct them ourselves. It is easy to see that this fits with the laws of our mind. However, just like it is in mathematics, so it is in the entire worldview, the sole difference is that in the constructing of the [geometric] world one is not conscious of the constructing, for it is something necessary and not free.«[232]

Another important characteristic of Fichte's theory of geometric construction has been pointed out by Werner Csech. In *Über den Begriff der Wissenschaftslehre* Fichte states that the fundamental task of geometry is to »limit« space according to certain rules and laws, and appears to equate this limiting with »construction.«[233] Here Csech says that the procedures of limiting space and construction seem to be »reciprocal concepts« (*Wechselbegriffe*) and belong together.[234] Here we also see the differences between geometrical and philosophical construction. What are the other differences? Geometric construction is to be especially distinguished from the constructions of the *Wissenschaftslehre* because of its particular limitations in space and its comprehension of them in a lawful manner; the task of the *Wissenschaftslehre*, on the other hand, is to consider the »limitability of what is capable of being limited as such.«[235] In this sense philosophical construction too is much more universal or foundational for Fichte. This geometric constructive procedure of limiting space generates

that there can be a correct demonstration? / Because this is something that I myself, in the act of constructing, have first put there by means of the free self-activity of my own power of imagination.« Ibid., p. 35.

232 Fichte, *Vorlesungen über Logik und Metaphysik 1797/98* (GA IV/I: 212).

233 »The task in general of limiting space according to a rule, or the construction in the same, is the fundamental principle of geometry, and it is thereby distinguished from the *Wissenschaftslehre*.« (GA I/2: 135).

234 See Werner Csech, *Raumlehre*, p. 69.

235 Ibid.

its individual figures, and is a free and conscious activity of the geometer. The geometer constructs these figures in his mind using the imagination and does this in accordance with the laws of the science. Thus, the possibility of geometry has various »presuppositions« – it presupposes certain limits and laws and not every arbitrary construction is possible.[236] Nevertheless, there also exist tools for aiding this construction, and these are furnished by nature herself, because according to Fichte every possible limitation in nature has a direct relationship with the *figures* of geometry.[237]

This combination of intuition and construction lies at the basis of Fichte's view of cognition, especially mathematical cognition. Fichte is especially clear on this point at the end of 1800 in the *Ankündigung* for the *Neue Bearbeitung der Wissenschaftslehre.* As already noted, here Fichte stated that his conception of cognition is not the same as Kant's. The cognition of the *Wissenschaftslehre* is *not* a »rational cognition on the *basis of concepts*«, but like in the domain of mathematics, it is a »rational cognition on the *basis of intuitions.*«[238] In this regard Fichte's *Ankündigung* is an eloquent defence of reason's ability to construct and grasp itself immediately, and that reason is self-sufficient and not based on anything else. To do this the constructions of reason have to employ *intuitions* and not simply concepts. It requires breaking with the Kantian theory of cognition, and Fichte now proposes a model of cognition based on intuitions and inspired by the constructions of mathematics. The resulting model of cognition furnishes Fichte with a *new definition of philosophy*: »It should have been obvious that reason does not grasp itself and cannot apprehend itself

236 See Fichte, *Erlanger Logik*: »*Presupposition* [of geometry]: that this possibility has its limits and laws: only a few are possible, others are not: Condition of this view: we have to place ourselves in the possibility that these laws have to arise within us, and work directly upon our limiting of space, precisely in *limiting* (*Begrenzen*), in *construction.*« (GA II/9: 125).

237 See *Die Thatsachen des Bewußtseyns* (1810): »With regard to the figure, there exists an aid for the construction, in that geometry includes every possible figure, and every possible limitation in nature can be traced back to a geometrical figure.« (GA II/12: 43).

238 Fichte, *Ankündigung* (GA I/7: 156). The importance of this text and Fichte's remarks on cognition are nicely spelled out by Daniel Breazeale in his article »Toward a *Wissenschaftslehre more geometrico* (1800–1801)«, pp. 10f.

by means of anything else (such as a concept) that has to be derived from something else and does not possess its foundation or ground within itself. Instead reason can grasp and apprehend itself only immediately – and there is nothing immediate but *intuition*. Hence, if philosophy is from now on to be synonymous with *reason's own self-produced cognition of itself*, then philosophy can by no means be considered to be cognition based upon concepts, but must instead be cognition based upon intuitions.«[239]

b. *Geometrical Construction in Fichte's Sonnenklarer Bericht (1801) and Einleitung in die Wissenschaftslehre (1813)*

In the Introductory lectures to the *Wissenschaftslehre* given to his Berlin students in late 1813 Fichte observed that he had already attempted to explain »in a childishly easy and playful manner« the essence of construction in philosophy in an 1801 text entitled *Sonnenklarer Bericht*. However, the public just did not »understand it« and could not »get beyond« their ordinary conceptions of philosophy.[240] In this section we will examine two concrete examples of construction in geometry in the *Wissenschaftslehre* that Fichte provides in these texts. We will begin with Fichte's exposition of the method of construction in the 1801 *Sonnenklarer Bericht*, since the text is earlier and dates from approximately from the same time as the *Neue Bearbeitung* and its associated *Ankündigung*.

In lesson 3 of the *Sonnenklarer Bericht* Fichte sets out his theory of construction by means of an example of the triangle. Fichte attempts to impress upon the reader that just as the ideal constructions of geometry have an application to actual life, so too do the constructions of philosophy.[241] Beginning with a rectilinear triangle, Fichte asks the reader to reflect upon its genesis or coming into being. To generate a triangle in »scientific geometry« requires that we follow certain »laws.« That is to say, if we draw two arbitrary intersecting lines and ask how we may construct a

239 *Ankündigung* (GA I/7: 157–158); translated by Daniel Breazeale in *IWL*, p. 191.

240 Fichte, »Einleitung in die WL 1813«, in: *Ultima Inquirenda*, pp. 36f.

241 The so-called »Author« of the text tells the »Reader«: »Nevertheless, you admit that geometry has an application to actual consciousness in life, and consider it, like we do the *Wissenschaftslehre*, to be an image for a part of actual consciousness.« Fichte, *Sonnenklarer Bericht* (GA I/7: 231).

triangle by finding the third side and hence a closed space, we need to add the third by first of all »closing the arbitrarily imagined angle with its two sides.«[242] Naturally, the third side or angle may be found by other means such as the law of measurement: we may determine the length of the third side by means of the included angle and the other two sides. We could measure the empirically drawn triangle to check that it angles add up to 180 degrees or two right angles. Nevertheless, it is not imperative to do this because with any triangle that is correctly »constructed« in accordance with the right procedure and laws will be of necessity 180 degrees. However, Fichte asks the question: what is the »ground or conviction« for our belief that this is so for »all possible triangles«?[243] He answers that we treat the so-called »imaginary« or ideal triangle in the same way as we treat the empirical or »actual« triangle on the piece of paper, that is to say our inner intuition of the imaginary triangle is based on the same methodology as that of the outer intuition of the actual triangle: »I treat the actual triangle according to the laws of the merely imagined one, and with the same certainty, *precisely as if it were likewise involved in that intuition*.«[244]

This original intuition is based on the process of original construction, which requires our free imagination to bring it about. And this is where Fichte draws a parallel with the intuitions and constructions of the *Wissenschaftslehre*. Just as geometry begins with an original intuition of the line or the law of the triangle, so for Fichte philosophy ought to begin with an original intuition of the »I«. As we have seen, this first, original intuition of the I is the foundation and first principle of the *Wissenschaftslehre*, its so-called *Grundsatz*. After this beginning the science of geometry is able to »construct« a finished triangle by employing the correct rule in a cognitive procedure, just like with the *Wissenschaftslehre*.[245] Taking his lead from the geometer, the philosopher of the *Wissenschaftslehre* then constructs human consciousness a priori, i.e. the pure consciousness that is common to all rational beings: »Thus, I believe, that I have given you a sufficiently clear conception not only of the goal of the *Wissenschaftslehre*

242 Ibid., p. 232.
243 Ibid.
244 Ibid.
245 Ibid.

in general, but also of its procedure and the grounds of this procedure. This science constructs the entire common consciousness of all rational beings absolutely *a priori*, in its fundamental characteristics, precisely as geometry constructs absolutely *a priori* the universal modes of limiting space on the part of all rational beings.«[246]

Against the attacks of Kant and others, Fichte is above all concerned in the *Sonnenklarer Bericht* at informing the public of the scientific nature and lawfulness of the intuitions and constructions of the *Wissenschaftslehre*. To do this he compares the example of the lawful construction of an non-empirical entity in geometry (the triangle) with the lawful construction of a non-empirical entity (consciousness) in philosophy that is based on an original intuition (of the I or self). In Fichte's eyes, just as the so-called »imaginary« construction of an ideal triangle can be demonstrated by the external empirical or »real« triangle, so too the original »I« or self is demonstrated by actual or empirical self-consciousness. Both methods are called by Fichte the »method of construction« and involve pure intuitions. Yet as we noted there is one principal difference: whereas in geometry one is focused upon the finished »products« of the construction, that is, upon the finished constructed triangles and other geometric figures, in the *Wissenschaftslehre* one is attentive to the process and pure activity of constructing itself.[247] In this sense philosophical construction is a meta-activity of geometrical construction. It is unfortunate that Fichte's *Neue Bearbeitung* remained a fragment, for one of its aims was to shed further light on the »fundamental law« of construction in the *Wissenschaftslehre* and set it out in a geometrical fashion using postulates, theorems and corollaries.[248]

246 Ibid., p. 233.

247 »In actual mathematics one only looks at the products of construction, whereas in here [in the WL] one looks at constructing itself.« Fichte, *Neue Bearbeitung der Wissenschaftslehre* (GA II/5: 344f.). Three pages later in a »corollary« Fichte says that »we are summoned (*aufgefordert*) to the concept of the I, to its construction.« (ibid. p. 348).

248 See, for example, the following passage: »Accordingly, what would be the *fundamental law* (*Grundgesetz*) *of intuition*? Wait a minute. *Description, construction* is its fundamental law: and indeed, self-construction. *To make this clearer, to illuminate and shed light on it.*« (Ibid. p. 395).

Nevertheless, there is another text where Fichte expands on the relationship between geometrical construction and philosophical construction, and his discussion occurs just a month or two before his death. This text (or texts) is the late 1813 *Einleitung in die Wissenschaftslehre* (Introduction to the *Wissenschaftslehre*).[249] In lecture six of this introductory course Fichte returns to the theme of construction in his philosophy and again uses the example of the construction of a triangle in geometry as a basis for the understanding of philosophical construction. In fact, here Fichte discusses the nature and the elements of both mathematical and philosophical cognition and tries to provide a satisfactory answer to one of the most fundamental questions of philosophy: what is the relationship between the singular, particular, empirical image I see in nature, i.e. that of a particular, specific triangle for example, and the image I have of what seems to hold for all triangles, i.e. a kind of universally valid image? We here encounter Fichte's theory of a *Bild* or image. We cannot discuss the theory of universals and particulars or Fichte's general *Bildlehre* in depth, but we will touch on Fichte's theory insofar as it concerns the role of construction in fashioning the images of mathematics and geometry.[250]

In the *Einleitung in die Wissenschaftslehre 1813* Fichte argues that construction supplies a vital element of mathematical cognition. A wholly new element or a »new sphere, aether« is added into consciousness and its origin can only be traced back to the process of »free« construction. How does Fichte define construction here? Fichte again relies on the paradigmatic model of geometric construction, but what is noticeable in this late text is that Fichte draws special attention to the »non-sensible« nature of construction. In Fichte's definition, construction is the: »free thinking and forming (*Bilden*) of something that cannot be given through the

249 Again, I say »texts« because the manuscript exists in different versions. The most important for our purposes are the original handwritten manuscript of Fichte (forthcoming in GA II/17) and the *Kollegnachschriften*: Berlin Ms.germ.qrt. 1964b, and Halle 6c (in GA IV/6: 351–472). Cf. SW IX: 1–102), because they contain more material on geometry. Both the version in the SW and in *Ultima Inquirenda* lack some of this material with the examples from geometry.

250 For an in-depth investigation into Fichte's *Bildlehre* in general, see Wolfgang Janke's study: *Vom Bilde des Absoluten. Grundzüge der Phänomenologie Fichtes* (Berlin/New York: Walter de Gruyter, 1993).

senses.«[251] It is »free« or is »positive freedom« because it is incumbent upon each person to carry out this construction himself, that is, he cannot be passive and hence requires an »inner activity that the teacher can only demand, but that the student must carry out himself.«[252]

This relates to the *origin* of an image. But how does the internal image of a triangle arises in us? Fichte makes a number of subtle distinctions and says that construction can be characterized as the process that furnishes »the element« (*das Element*) of an image. He is careful to point out that although construction does *not* furnish the image itself but only its element, it is still an essential component in the emergence of the image which is also characterized as »the specific shape or structure« (*die bestimmte Gestaltung*).[253]

How does this shape or structure of the image in turn come about? In an apparent paradoxical manner Fichte explains that despite free construction being a kind of fashioning or forming, the specific image still has to »make itself« (*selbst machen*). To do this he invokes the importance of the term »self-evidence« (*Evidenz*), which due to its force and transparency is an insight that becomes illuminated within construction itself.[254] In self-evidence the insight into the subjective and objective coincide, it is both a glimpse or a look and the solid shaping of the look (*der Blick und die feste Gestaltung des Blicks*). Fichte illustrates his thesis with an example from mathematics – once again that of a triangle. He draws three triangles like the following:

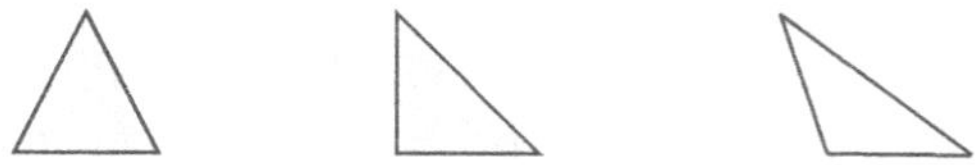

251 Berlin Ms.germ.qrt. 1964b., p. 21 (cf. GA II, 17: 258)

252 Ibid.

253 »This free constructing, however, only supplies the element of the required image; the image itself, the specific shape, is not thereby given, nevertheless the element and consequently also the free constructing are an indispensable condition if the image is to come about.« (ibid. p. 22).

254 »The insight has to absolutely and thoroughly come to us in the constructing: it is so and cannot be any other way, it is illuminated in us of itself, and the self-evidence grasps us.« Ibid.

Drawn triangles of this kind are, for Fichte, an »image of the natural sense« (*das Bild des natürlichen Sinnes*). Because they are »really present« and appear as they do and we can determine their given qualities such as lengths, angle sizes and other respective relations, it is a question of the »being« (*Seyn*) of the triangle. However, the higher (transcendental) question asks about the »*possibility* of the triangle itself« and its qualities and about the laws of the same; it poses the question about »all possible triangles«; or again: »How is a triangle possible at all?»[255] In Fichte's eyes, the intuition of this kind of triangle cannot arise through the simple study of the definite, specific image of a triangle. Ultimately, we have to instead »break free« of all external observation and raise ourselves through inner self-activity to its *construction*. This so-called »free« geometrical mode of construction is required for constructing the »triangle in itself«, or for Fichte's archetypal point or line, since it constructs a universal geometrical object that harbours the possibilities of all the specific or particular geometrical objects of the same kind. It is precisely construction that demonstrates the most intimate parallel between cognition in geometry and cognition in the *Wissenschaftslehre* and the relationship between the two. For both disciplines possess a pure inner intuition and require a construction to come about: »How did this definite intuition of the triangle come about? Certainly not through the study of the given triangle, but precisely through the breaking free of the observation, insofar as one raises oneself to the self-active formation, to the *construction* of the triangle; this initially yielded the *element* of inner intuition, and one became conscious of the construction only while carrying it out; but it did not yet come to the structure (*Gestaltung*), to the specific intuition; one came to this by elevating in freedom the possibility of constructing limited itself; the image is established only through this self-making limitation. – An intuition of this kind is absolutely not an outer one, but an inner intuition; the *triangle in itself* cannot be externally depicted or drawn, for every drawn triangle is a specific triangle, just one among the countless possible triangles.

255 »This is no longer a question about the being (*Seyn*) or about the given triangle, but the question is raised beyond it to all possible triangles, insofar as it seeks a law for reality. [...] The question is therefore posed: How is a triangle possible at all?« Berlin Ms.germ.qrt. 1964b, p 25.

[...] Thus, exactly as it is with mathematics, so it is also with the cognition in itself of which the *Wissenschaftslehre* speaks.«[256]

The examples of geometrical or »free« geometrical construction in this text also have a propedeutic role – they are to prepare the student for pure philosophical construction. It is beyond the scope of this study to discuss Fichte's conception of philosophical construction in detail[257], suffice to say that this fourth or highest mode of philosophical construction is also called by Fichte »infinite constructing«, since it concerns a mode of knowledge that contains infinite possibilities within itself.[258] Just as the pure triangle or triangle in itself is able to generate infinitely different possible triangles, so the construction of this infinite knowledge contains the possibility of all knowledge in itself: »However, our question can only be answered through an example. It is a question about thinking in general, about the genus of thought, about the One and eternal in all infinite examples. Our question about knowledge does not concern any kind of particular or specific knowledge, just as little as the question about a triangle concerns a specific and actual triangle. Rather, our question concerns knowledge in itself, beyond all its possible determinations. [...] Thus, the first requirement is to construct knowledge itself with freedom, to hover above it and to look down below, to picture knowledge in its pure possibility. This provides us with the element and the substance (*Stoff*) of all our intuitions and doctrines.«[259]

To summarize: in contrast to Kant's two kinds of construction: we could attribute four modes of construction to Fichte, the first three of which are applicable to geometry. Firstly, technical construction – schematic or by means of instruments (material). Secondly, sensible construction (in the imagination or on a blackboard). Thirdly, so-called »free« construction. Here the constructed geometrical object contains the

256 Ibid.

257 Among others, such a detailed study has already been furnished by Daniel Breazeale. See: »›Men at Work‹: Fichte and Schelling on Philosophical Construction« (paper for the Fichte Congress in Brussels in 2009, forthcoming in conference proceedings).

258 This designation of »freie und unendliche Construiren« (free and infinite constructing) appears in the *Einleitung in die Wissenschaftslehre 1813*, Berlin Ms.germ.qrt. 1964b (52r). (GA II/17: 257)

259 Fichte, *Einleitung in die Wissenschaftslehre 1813* (GA IV/6: 373, 384).

possibility of all objects in them. In Fichte's model of geometry, this would be the construction of the circle as such or the lawfulness of the pure triangle, from which one can derive all the specific triangles. Fourthly, »infinite« construction, which is no longer geometrical, since it no longer has any spatial product for its content. Instead, this mode of construction is focused on pure activity of constructing itself. Its substance consists of freedom itself, or of the free construction of the above third mode. »Infinite construction« is the construction of a mode of genetic knowledge that contains the possibility of all knowledge in itself.

Chapter Four: The Fichtean Conception of Space

1. Philosophical Images: Fichte on Fundamental Elements of Geometry

In this first section of Chapter Four we will examine Fichte's thoughts on the fundamental elements of geometry: points, lines and planes. We will begin with the classical Euclidean definitions of these basic elements and juxtapose them with their respective Fichtean versions, and then determine their philosophical role and function in the *Wissenschaftslehre*. This is perhaps one of the most original and significant aspects of Fichte's theory: he provides both *geometrical and philosophical interpretations* of the fundamental elements of geometry. In other words, from the strictly geometrical definition of the point, line and plane, Fichte draws a number of original philosophical consequences, whose aim is to facilitate an understanding of the *Wissenschaftslehre*. Fichte's philosophical interpretations shed important light on his view of the nature of the I's activity and the volitional nature of the human being. But first a few words about modern geometry's relationship to Euclidean geometry.

a. The Euclidean and Fichtean Definitions of the Point

Whereas modern geometers prefer to leave such fundamental notions of geometry as points and lines *undefined*, Fichte – following the lead of Euclid more than two millennia before him – chose to define them or at least to characterize them. Why do modern geometers not define these elements? Our modern notions are greatly influenced by the work of David Hilbert (1862–1943), who proposed a new axiomatization of the foundations of geometry in 1899, in which the point, line and plane remain as so-called »undefined primitives«. For Hilbert, modern geometry ought to leave these primitives undefined due to problems of varying interpretations and to avoid logical circularity. For example, in his seminal work *Die Grundlagen der Geometrie* (1899), Hilbert summarized the geometric point by using the letters of the alphabet, writing: »Let us consider three distinct systems of things. The things composing the first system, we will call points, and designate them by the letters A, B, C, [...].«[1] Hence these elements are simply termed »points« and designated using capitalized letters of the alphabet. Their *content* is left undefined, however, and relies on our everyday sense of intuition to form a notion of what is meant. Accordingly, as we saw earlier the term *point* may be described in a *positive* sense in modern geometry by admitting that we do at least possess some kind of intuitive notion of this term. A point may also be described *negatively*, by saying that it has no volume, length or area, or in the technical language of mathematics: a point has the status of a 0-dimensional object that can be specified in an *n*-dimensional space by means of *n* coordinates.[2] Turning back to the classical geometry of Euclid, as outlined in his masterpiece the *Elements*, points and lines are presented as the fundamental or primitive objects of the whole of geometry. They are primitive insofar as they cannot be reduced to any further elements and they cannot be proved. As we have seen, Euclid's *Elements* begins with a list of twenty-three definitions that were considered so self-evident it was not necessary (or possible) to furnish any proof of them, and where the point is Euclid's very first definition and is defined as: »A point is that which has no part.«[3] While Euclid's definition may seem vague by modern standards of logic, it was viewed as an improvement on earlier definitions abounding in ancient Greece. For example, the Pythagoreans defined a point as »a monad having position«, a definition adopted by Aristotle, whereas for Plato, a point was simply »the beginning of a line.«[4] The Euclidean definition was the classic definition of the point for over two thousand years and it was the still accepted standard definition during Fichte's lifetime. The Euclidean definition refers to an element that has no width, no depth, no length, and is zero-dimensional. If we draw a point on a piece of paper or on the whiteboard, it obviously has width, it has extension; it is an empiri-

1 David Hilbert, *The Foundations of Geometry*, trans. E. J. Townsend (Illinois: Open Court, 1950), p. 2.

2 Cf. Roberto Toretti's remarks on this modern definition in: *Philosophy of Geometry from Riemann to Poincare* (Dordrecht, Holland, D. Riedel Publishing Co., 1978) pp. 37f.

3 Euclid, *The Elements*, translated by Sir Thomas Heath, Book One, p. 1.

4 Quoted in volume II of David E. Smith's *History of Mathematics* (New York: Dover, 1958), p. 274.

cal point. The empirical point has a position in space, but lacks any notion of direction. – With the point we cannot speak of a left or right for example.

For Euclid, however, an empirical point is clearly not the same as the stated definition of the point, of an entity having no »part«. And this is crucial for Fichte. – For him, the point we picture in our mind is an ideal or intuitive point, and the sensible point merely corresponds to the former; the point on the piece of paper is a sensible image or body (*Körper*) of an idea, archetype (*Urbild*) or *spiritual* image (*geistiges Bild*).[5] It may also be termed *abstract* insofar as any reference to a sensible quality has been »abstracted away« from it. We must strip away every notion to reach a bare minimum. Before the advent of non-Euclidean geometry, Euclid's definition did not call forth any disagreement or dispute on account of its simplicity and self-evidence. We shall see that the Fichtean version of the point is (for the most part) in agreement with the Euclidean definition.

The Fichtean Point

With Fichte's lifelong interest in geometry it is not surprising to see that he had his own precise ideas and even definitions of the fundamental elements of geometry – of the point, line, plane and space. Although they are based on or related to the traditional Euclidean notions, Fichte's views do slightly differ, as we saw with his synthetic conception of the line. Yet Fichte's conceptions are entirely consistent with modern views of geometry.

Let us first consider the geometrical point. Fichte's definition of the point is similar to the Euclidean definition but perhaps more radically expressed: »A point is nothing«.[6] This is because the point for Fichte has no definite properties or genuine predicates and hence may be said to be »nothing«. On the other hand, as Fichte notes again in his *Platner Vorlesungen*, a point signifies the very essence of simplicity itself – »it is something entirely simple.«[7] Like the Euclidean definition, the pure point

5 Cf. *Die am Ende des Winters [Sommers] 1794 von Fichten Gehaltene Schlussvorlesung als Beylage der Verantwortung* (1797) (GA I/4: 413). Cf. *Die Thatsachen des Bewußtseins* (1813) (SW IX: 496).

6 »Ein Punkt ist nichts«, *Platner Vorlesungen* (GA II/4: 269).

7 »Es ist etwas ganz einfaches« Ibid., p. 269, note.

as such is an intuitive ideal entity and cannot be empirically fixed within sensible space. Fichte was often at pains to stress this difference between the ideal point and the empirical point that can be drawn on a page, but it should not lead us to conclude that the ideal point is any less real than the empirical one. In a small published text from 1800, *An das philosophische Publikum*, Fichte argues that although the ideal point is like the I, i.e. it is non-sensible in character and cannot be depicted empirically, it is not devoid of reality. Even though critics »declare that *the pure I*, from which this system departs, is a psychological illusion [...]. That would precisely be as though someone wanted to demonstrate the nullity of geometry by declaring the mathematical point and the line to be a psychological illusion, just because they cannot be presented in external perception.«[8]

Fichte also examines a number of predicates or characteristics that are often attributed to the point, especially to the empirical point. For example, he remarks that if we attribute any »roundness« to the point it immediately ceases to be an intuition (*Anschauung*) but ends up being a concept (*Begriff*). Fichte therefore sharply distinguishes between our intuition of the point and the concept of a point: »The point does not draw lines, but it *grows* (originally round); how can this roundness be described and what is its origin?/ Actually, this too is not a point; this itself is already the affair of a concept. But is itself a sphere (*Kugel*). The sphere is the most original [entity]: and through infinite diminution you always encounter a sphere. You can always contract it just as you can extend it. It is the same point, like the concretion of space in general.«[9]

Fichte makes the same argument here about the dimensions of space. Here too we must be careful about confusing the intuition of space and the concept of the dimensions of space. Like the point, the latter only arises through reflective activity: »The three dimensions only arise through thought, in precisely the same way as the mathematical point does.«[10] Another further obvious but important aspect should be recalled – the point's irreducibility and indivisibility. This is a typically Euclidean notion – the point is not a composite entity and therefore cannot be divided or

8 Fichte, *An das philosophische Publikum* (GA I/6: 458).

9 *Neue Bearbeitung der Wissenschaftslehre 1800* (GA II/5: 401). In fact, the text breaks off precisely after this discussion of the point.

10 Ibid.

broken down into any other element – it is a fundamental or primitive entity. In Fichte's conception of geometry the point is not alone in being the sole primitive entity, and as we saw this is where Fichte breaks with the Euclidean model of geometric elements. For *the line* too cannot be broken down further or reduced to any other element, including points. This notion of irreducibility and primitiveness must be kept in mind, for it is vital for comprehending Fichte's initially surprising statement in the *Erlanger Logik 1805* concerning the starting point of geometry: »The beginning of geometry is *synthetic*; neither point, nor line, but both, and both comprehensible by means of one another.[11] We cited the objection that the empirical line may be broken down into smaller pieces, and therefore into points. This is of course always possible. – But this only holds for the empirical point and line and not for the ideal point.

The Urpunkt (Archetypal Point)

Recalling our discussion from Chapter One, in his theory of geometry Fichte assumes a number of archetypal elements, that is to say, ideal and absolute entities that the Fichtean geometer and philosopher of mathematics have to simply »bring with them«. Such a point for Fichte is the *Urpunkt* – the archetypal or original point. The *Erlanger Logik* contains a specific example of this archetypal and absolute element – the so-called primal point or »*Grenzpunkt.*« It is absolutely invariable and at the centre (*Mittelpunkt*) of every line, and from which it is possible to draw an infinite number of lines. It is not in the external world, but we must »look within«, i.e. it is within ourselves and we must use geometrical intuition to find it: »Look into yourself (*Schauen Sie in sich hinein*). You have a point in yourself; from which it is a certain distance; that is, it is a specific line that you draw *unnoticed* and originally (without an arbitrary free construction), this *distance* always remains the same. Everyone as an absolute limit-point (*Grenzpunkt*) of all distance, from which infinitely many lines may be drawn, and originally drawn lines. Without this point, which is absolutely invariable in itself, and co-borne and omnipresent in infinity, nothing would have a location and we would not find anything again in

11 GA II/9: 127.

infinite space.«[12]Here Fichte underscores the necessity of this ideal point also being an infinite »limit« or boundary; it has the role in geometry of being a *limiting function* for both intuition and thinking.[13] Although as a limit the pure point cannot be perceived in space, as Csech points out it is the form of the intuition or »the form of the expression for the acting of thinking.«[14] This corresponds to Fichte's claim in the *Eigne Meditationen* that the point is not a quantity but forms the limit of all quantities that exist in space.[15]

This additionally implies that the point is not just simply one geometric limiting element among others, but that it is the limiting image of space, or more precisely: *the* limiting configuration par excellence.[16] For Fichte, the point then is the absolute limit or boundary of space. This view was also held by Kant in the *De mundi sensibilis* (*Inaugural Dissertation*, 1770), where the point is likewise considered one of the limits of space. Kant adduces the number of limits to be three because it is the same number as the dimensions of space.[17] Yet of these three limits Kant highlights the exceptional status of the point: there are three kinds of limits of space, »of which two themselves are still spaces, the third, namely the point, is the limit of all limits.«[18] In Csech's opinion the Fichtean theory of the point, especially its limiting function, may be interpreted as a theory that

12 *Erlanger Logik* (GA II/9: 129f.).

13 For example, as early as 1794 in *Über den Begriff der Wissenschaftslehre* Fichte writes: »The *Wissenschaftslehre* furnishes us with space as something necessary and with the point as an absolute limit« (GA I/2: 135).

14 »Die Form der Ausdruckshandlung des Denken.« (Csech, *Raumlehre*, p. 133).

15 »The point is not a quantity. (Footnote: If you realize the quantity, then you have infinite extension. – Negate it, then you have the point. However, I think this is the limitation, and the extension of the point (of time) would be negative quantities. […] Join the quantity with the absolute negation; there then arises the exact opposite, an infinite continuity without limits. Limit the former through the latter, then you have a continuous (quantity) limited with points. […] This all has to be better investigated.« (GA II/3: 95).

16 Cf. Csech, *Raumlehre*, p. 133.

17 See Kant, *Inaugural Dissertation*: »[…] the simple in space is not a part but a limit. […] The limit of a solid is a surface; the limit of a surface is a line; the limit of a line is a point. There are, therefore, three sorts of limits in space, just as three are three dimensions.« In: *Theoretical Philosophy 1755–1770*, p. 397 (note).

18 See Kant, *Über eine Entdeckung, nach der alle neue Kritik der reinen Vernunft durch eine ältere entbehrlich gemacht werden* (AA VIII: 220).

»intensifies the Kantian conception of the point.«[19] Apart from the 1805 *Erlanger Logik*, Fichte's intensified view of the point is especially visible in the *Neue Bearbeitung der Wissenschaftslehre 1800*, where the point is transformed into the limit and proposition of mere thought itself. Fichte here remarks that with the point we reach a boundary in thought beyond which we cannot go: »In any event, for example, the point is the *limit*, nothing else: moreover, it is a *proposition* of mere thought.«[20]

Finally, Fichte's ideas on the geometrical point are summarized in his significant letter to Schelling from November 15, 1800. Here Fichte remarks that the point is a product of abstraction in which everything that can be intuited or perceived is abstracted away, including all aspects of roundness which only arise in *intuition*.[21] Schelling's ideas on the nature of the geometric point echo Fichte's and Kant's, especially with regard to viewing the point as an absolute limit.[22] In the strict geometric sense Schelling also views the point as an abstraction of the mathematician, but still considers it an important mental aid for determining the nature and value of a limit of indivisible quantities: »The point that is assumed by the mathematician does not really exist. They too are far from conceiving it and for the purpose of construction they assume the point to be an indivisible quantity, and they have to do this because otherwise in the division of the line into halves that arise or in frequent division, the sum of the parts would not equal the whole.«[23]

To recapitulate: For Fichte the geometric point is irreducible and indivisible; it cannot be reduced to any other element. It has no properties except for an affinity with the quality of »roundness«, and it has no direction, but functions as a limiting element. Unlike a line, it has no up, down, left or right. The ideal or *Ur*-point should not to be confused with the empirical, sensible point. As we saw in Chapter Three, in contrast to the sensible or empirical points of ordinary Euclidean geometry, Fichte's archetypal point requires the faculty of pure geometrical or intelligible intu-

19 Csech, *Raumlehre*, p. 133.

20 GA II/5: 400.

21 »The point to begin with: abstraction from the infinitely many points in the sphere inclosing it (from where the *angularity* subsequently comes, since in intuition everything is *round*) The form of *positing* in general.« (GA III/4: 361).

22 Cf. Cesch, *Raumlehre*, p. 251.

23 Schelling, *Sämmtliche Werke* I, 10, 328.

ition to apprehend it. This pure geometric point can only be pictured or intuited by the mind, whereas the empirical point can be depicted in physical space. We can now turn to the philosophical consequences that Fichte draws from his geometrical definition.

b. The Point and das Bild des Ich

Fichte makes the following statement in the *Wissenschaftslehre 1801/1802*: »The point, however, is the image of the I.«[24] Or to take another example, in *Die Principien der Gottes-Sitten- u. Rechtlehre* of 1805 Fichte not only brings the point into connection with our images of the I, but also with its *genesis*: »Let's turn to the line. Precisely as we do in an intelligible light, so it also takes place in a factual light: line [...] I = point: the genesis of the I (*Genesis des Ich*): fixing of a point in a line according to the above: – this is absolute and the ground of the I.«[25]

How are we to understand these statements? In what sense is the point the manifestation or image of the I, and how can the genesis of the I be related to the fixing of a point in a line? We shall see that just as Fichte indicated a parallel between the qualities of the self-evident postulates of geometry and the *Thathandlung* of the *Wissenschaftslehre*, so he also draws a parallel between the geometrical qualities of the point and those of the pure I.

However, at the beginning of this discussion it is important to bear in mind that Fichte understands the mathematical point or the drawing of the line solely as *images* of the I and the I's activity, and not of course as the I itself. Any image presupposes an original object or activity of which it is the image. The image might assist us in understanding the original object, and this is the final goal of Fichte's philosophical interpretations of the fundamental elements of geometry. The images are borrowed or *appropriated from the domain of geometry* in order to shed light on Fichte's philosophy, especially his *theory of the I*. And as we already saw in the *Grundlage* of 1794/95, the essence of the I for Fichte consists in its striving

24 Original German: »der Punkt aber ist das Bild des Ich« (GA II/6: 288).
25 Fichte, *Die Principien der Gottes-Sitten- u. Rechtlehre* (GA II/7: 472).

and activity[26], and hence, this essence should not be confused with its image.

In the *Wissenschaftslehre 1801/02* we see that Fichte's remarks occur in the context of a transcendental argument. What are the conditions for saying that the I is the image of the point? A condition for this is that the I, like the point, is able to express its movement *directionally* (in a certain sense) and that we in turn are able to cognize this in an intuition. A fixed direction is a determination of both the I and the point. Although the point as such has no direction, in Fichte's mind it still contains within it the »genesis« of movement. The I can intuit itself in what is the genesis of directionality in the point.[27] Here we glimpse a profound connection between *freedom* and direction in Fichte's theory. As we saw in the *Zurich Wissenschaftslehre*, the actions and constructions of geometry are carried out with freedom. It is I who freely construct them and no one else can do this for me. It is likewise I myself who has the freedom to place the point *anywhere* in space, to *commence* the drawing of a line and thus to determine its specific *direction*. Freedom is one of the conditions for geometry to come into existence: »Geometry arises through the free acting of my I, by moving the point into a line in space.«[28] However, in contrast to the constructions of geometry, freedom itself *cannot* be conceptually proved, it must be assumed or intuited in pure intuition. Accordingly, freedom can be manifested, and it is manifested in acting; and all acting can be expressed by direction.[29] For Fichte, reflections such as these on the nature of the point and direction can especially illuminate our knowledge of the structure and nature of the acting of the I. After a preliminary exposition

26 Cf. Fichte, GWL : »The essence of the I (*das Wesen des Ich*) consists in its activity. [...] This is now the essence of the I [...] The I is infinite, however, merely in accordance with its striving; it strives to be infinite.« (GA I/2: 404f.).

27 See *Wissenschaftslehre 1801/02* »The self-seizing I is a point in space that is diffused everywhere. It cannot express itself, except in a direction. This is the condition of *all* expression; and a cognition of it, is the cognition of the expression. This direction itself is now everywhere, and thoroughly a determination of the point; the point, however, is the image of the I. Hence, it is to be intuited as absolutely grounded in the I; or, it itself is the I of the intuition; the I is only intuited in it and by means of it, precisely as what is directional in it.« (GA II/6: 288).

28 *Zurich Wissenschaftslehre* (GA IV/3: 24).

29 Cf. *Wissenschaftslehre 1801/02* (GA II/6: 235, 288).

in the 1794 *Grundlage*, in the *Wissenschaftslehre 1801/02* Fichte will eventually establish a number of *Grundsätze* implicating direction into the structure of the I's knowledge of itself, and asserting that the fundamental law of acting follows a linear directional path.[30]

The text of the *Practische Philosophie* (1794) contains an interesting addition to this discussion. If the point is the image of the I, then what is the image of the Not-I? Fichte draws a parallel with the notion of space and its apriority. The heterogeneous nature of the point and space consists in the different nature of their apriority. According to Fichte, space is a priori in the Not-I, whereas the point is a priori in the I.[31] As noted earlier, Fichte here believed he had hit upon a brilliant philosophical notion for the relation between the point and the I – the I generates all the forms of intuition – and it would be one of the future tasks of philosophy to demonstrate this truth: »Just as the I bears the point to the Not-I: so through synthesis the Not-I bears the line over to the I. Brilliant hypothesis. The I not only synthetically brings forth the point, but *all the forms of intuition*. [...] And philosophy also has to demonstrate this.«[32]

A further important aspect is that the point as such lacks all physical location and is non-empirical, it is an irreducible and primitive element. Yet the point is the genesis of the line and provides the line with the possibility of its direction and can therefore receive an empirical expression. Thus, exactly like the pure I compared to the empirical I of self-consciousness, the pure point does not actually enter into empirical space, yet is intimately related to it. In this sense the point and the I appear paradoxical: they can be both »beyond« space and within space.[33] And this is the implication of Fichte's view where he imagines the I as a centre with an infinite number of radii. Due to our limited or finite imagination we can assign a terminus to these radii, yet taken together these points constitute

30 See the *Grundlage* (GA I/2: 406ff.), and the *Wissenschaftslehre 1801/02* (GA II/6: 235).

31 *Practische Philosophie* (GA II/3: 217, note 1).

32 Ibid. Fichte continues his analysis here in the *Practische Philosophie* and brings the point into line with time: »time is an ever-recurring point; all time is completely equal. This is the second reason why geometry cannot begin with the point. The point is something fully heterogeneous in space.« (Ibid).

33 Cf. Csech's remarks on this paradox, *Raumlehre*, pp. 130f.

the *infinite* or original curve that is beyond our capabilities of completion.[34]

Fichte's philosophical interpretation of the point also concerns its creative and dynamic nature and its role as a limiting function. Like the I, the point may be said to be creative, because it can give rise to the drawing of lines, i.e. it is the *source* or beginning of any line. The point is also dynamic, for Fichte still insists that in spite of its abstraction the point must be conceived in a dynamic and uniform or unified manner in order for it to appear.[35] As we saw, this creative and dynamic source or starting point of a line has a crucial implication: the point is an absolute limit. Significantly, the point also has a relation to the external world, and we saw that this limiting function is the interface of the apriority of the I and Not-I, a notion echoed in the *Grundriß* of 1795: The »approximate coincidence of the effectiveness of the I and Not-I is in a […] point.«[36] In order to highlight the unique status of this limiting point, Wilhelm Metz deftly terms it a »*transcendental*« point. Metz argues that Fichte's transcendental structure of the I and the Not-I is developed with regard to both the limit of space and time. Because the real I itself and the Not-I coincide, he believes we are justified in viewing space and time as arising out of this transcendental point.[37]

34 As Fichte says in a mathematical footnote or question to mathematicians in *Über den Begriff der Wissenschaftslehre* (1794): »[…] Is the so-called curved line anything other than a stringing together of infinitely many points which are infinitely close to each other? This conclusion seems to me to be vouched for by the fact that the curved line is originally the boundary line of infinite space. (An infinite number of radii are drawn with the I as their centre. Our limited imagination, however, must assign a terminal point to each of these radii. Considered as one, these terminal points constitute the original curved line). This makes it clear that and why the task of measuring the circumference of a circle by a straight line is an infinite task, one which could only be completed if an approximation to infinity could be completed.« (GA I/2: 135); English translation in: Fichte, *Early Philosophical Writings*, p. 121. A similar argument concerning the circle line of infinite space is also to be found in the *Grundlage der gesammten Wissenschaftslehre*, cf. GA I/2: 298.

35 Cf. Fichte, *Zur Ausarbeitung der Wissenschaftslehre 1801/02* (GA II/6: 82); *Wissenschaftslehre 1804/I* (GA II/7: 195).

36 Fichte, *Grundriß des Eigenthümlichen der Wissenschaftslehre* (GA I/3: 193).

37 »The point discussed by Fichte is not a point in real space, but space and time ought to first arise *out* of it. This is why it can be designated as a ›transcendental‹

In conclusion, Fichte's philosophical interpretation of the point as the image of the I is in fact already summarized in Fichte's first publication on his new system – the programmatic text *Über den Begriff der Wissenschaftslehre*. As we just saw, Fichte explains how the point functions as the absolute limit, and how the I is the centre-point of an infinite number of radii. He subsequently shows how all these aspects are combined and depend on the freedom and creativity of the I, which manifests itself in the placing of a point in space. These philosophical interpretations belong in the domain of the *Wissenschaftslehre*. However, as soon as our freedom is limited or restricted, we are forced to enter back into the domain of geometry. Geometry for Fichte is essentially the space of the determining of our freedom within the limiting of space: »The *Wissenschaftslehre* gives as necessary space and the point as the absolute limit; however, it allows the power of the imagination complete freedom the place the point wherever it wants. As soon as this freedom is determined, e.g. to move it away against the limitation of unlimited space, and thereby to draw a line, we are no longer in the domain of the *Wissenschaftslehre* but on the ground of a particular science that is called geometry.«[38]

c. The Euclidean and Fichtean Definitions of the Geometric Line

Of all the fundamental elements of space (points, lines and planes), the line holds a privileged position in Fichte's theory of geometry. Here Fichte focuses on the *straightness* of the ideal line, and that originally it is neither composed of points nor can it be reduced to points. As we shall see, Fichte's philosophical interpretation of *the drawing of the line* becomes an image of the mind's »agility«. However, first let us recall the Euclidean definition of the line. In Book One of the *Elements*, Euclid provides the following definition of the line: »A line is breadthless length.«[39] Although the line for Euclid is a one-dimensional entity, the breadth or width of the line cannot be measured or empirically defined. That is to say, just like the point, the line may of course be depicted in physical space, may take individual form in the empirical world, but all these are merely sensible in-

point.« Wilhelm Metz, »Fichtes genetische Deduktion von Raum und Zeit in Differenz zu Kant« *Fichte-Studien* 6 (1994), p. 88.

38 GA I/2: 135.

39 Cf. Euclid, *The Elements*, Book One, p. 1.

stances of the ideal line. The Euclidean line – i.e. the one fulfilling the definition above, must satisfy the following demands: it has no breadth or thickness, it can be indefinitely extended and is perfectly straight (and not curved), and has *infinitely many points*. In contrast to the pure Fichtean line, the Euclidean line may be broken down into points. The points then constitute or make up the line. In Euclidean geometry the line is not the most primitive element, but this status is reserved for the point, for the line is reducible to the point. Moreover, only *one* line passes through two different points on a plane. Either lines are parallel or they intersect.

The Fichtean Definition of the Straight Line

Before discussing Fichte's philosophical interpretation of the line, let us recall the most important aspects of his geometrical definition of the line. We will not repeat what we have already said above, but review the key elements, as well as touching on some of the principal similarities between Fichte's and Schelling's views of the geometric straight line. As we saw, Fichte's conception of the line is a *synthetic* and not an atomistic conception. Although not strictly Euclidean, it is still geometrical and on account of its characteristics may be classified as belonging to that branch of geometry called »synthetic geometry«. The line for Fichte is a self-contained *continuous* entity, which we have to bring to geometry with us as a *whole*: »However, the line can only be constructed as a line, never, though, out of points. […] The geometer brings it with him directly and absolutely *as a line.*«[40] Furthermore, it must in no manner be thought of as a composite collection or discrete row of points, even at the most infinitesimal and minute level the line *still remains a line*: »The most infinitely small part of the line, is a line; never at all a point.«[41] The line is continuous for Fichte because the notion of *straightness* is incompatible with that of the point and therefore this predicate of straightness must already be *inherently* in the line and is essential for determining its direction, as Maimon has also

40 *Erlanger Logik 1805* (GA II/9: 127).

41 Ibid. The same view is present in: *Wissenschaftslehre 1801/02* (GA II/6: 156). Fichte continues to hold this view of the line right up until the end of his life. See his *Einleitung in die Wissenschaftslehre* from Nov./Dec. 1813, Nachschrift Halle: »Also in the line there is no indication or repose, divided into infinity the line always still remains a line« (GA IV/6: 399). Cf. Berlin Ms.germ.qrt. 1964b, p. 43.

argued.[42] For Fichte, reducing the line to points also means reduces its essential characteristics, including its straightness.[43] This has to be rejected because it would involve »transforming the very quality« of the line itself. Fichte's principal objection to the Euclidean line rests on the fact that for him the Euclidean view of the line implies that one would have to pass through an infinite number of points in order to complete the line, any line. Hence, passing from one point to another point involves an infinite process that could never be finished or terminated.[44] This highlights another difference between the geometric point and line for Fichte. One of the characteristics of the point is »repose« (*Ruhe*), whereas the essence of the line consists in motion and movement – the line has need of movement to come into existence.[45] And this reveals the line's relationship to both time and motion. – The line is the image for expressing successive motion in general: »The line is the schema or the form of forward motion.«[46]

Nevertheless, like the placing of a point anywhere in space, the line is similarly dependent on *my freedom*, for it is I myself who has the freedom to draw it, place it and direct it however I please. In the *Einleitung in die Wissenschaftslehre 1813* Fichte compares this aspect of the line to the point. Although the line is a unity of the manifold, like the point it does not possess any independent life or independence in itself. It requires the freedom of the geometer to be drawn: »I can cease drawing wherever I want, it is entirely at my whim; the life of the line depends on the arbitrariness of my drawing. All the lines of the geometer are of this kind, and

42 Compare this inherent continuity of the Fichtean definition of the line with Wolff's definition: »the parts of which are similar to the whole.« This definition is quoted by Maimon in his *Versuch über die Transzendentalphilosophie*, where Maimon agreed that this was the best definition of a straight line because it implies all its parts have the same direction and which alone for him is the distinguishing feature of a line. (see 41ff.) See too Gideon Freudenthal's discussion in *Maimon's Philosophy of Geometry* pp. 17f.

43 Fichte, *Erlanger Logik* (GA II/9: 127f.).

44 See the *Wissenschaftslehre 1801/1802*, part II, section § 2 (GA II/6: 231). Fichte's repeats this argument in the *Erlanger Logik 1805* (GA II/9: 127).

45 »[F]or the point is repose, whereas the line is constructed through movement.« Berlin Ms.germ.qrt. 1964b, p. 46f. Cf. Nachschrift Halle, GA IV/6: 399.

46 *Neue Bearbeitung der Wissenschaftslehre 1800* (GA II/5: 400).

the essence of geometry consists in this.«[47] Is there furthermore a relationship between the line and point as limits? The line is also a limit but not a space.[48] This additionally relates to the necessity of freedom with regard to the limit of the line: »The limit that is originally to be realized through freedom is the straight line.«[49] Despite its status as a limit of space like the point, the line of course is to be distinguished from the point and naturally has aspects not possessed by the point: continuity, non-discreteness, *straightness* and linearity etc. As we saw, points are only added to the line in a second or subsequent movement; originally the line is a continuous and indivisible whole.[50]

The Archetypal or Original Fichtean Line

Lastly, the Fichtean archetypal line is infinite, and may be said to be equivalent to the projective or synthetic line, which has an ideal terminal point at infinity. This is similar to the view expressed by Fichte in *Über den Begriff der Wissenschaftslehre*, in a question to the mathematician, where he says that a line is infinite. In this latter respect the projective line differs from the Euclidean, which does not admit a point or line at infinity. As we saw in previous chapters, Fichte gave a number of examples of an *Urlinie* – his so-called original diameter and perpendicular. Like the *Urpunkt*, these absolute original lines are assumed ideal elements that must be grasped in pure geometrical intuition. – The original diameter, for instance, is the originally drawn *internal line*. It possesses the invariable quality of straightness and is the absolute standard line for all direction, while the original perpendicular, is an absolute measure of distance (*Entfernungsmaas*): »But just like you did with the earlier point, so you also bring a line through itself with you, of which the first point is the centre: [this line is] therefore a *diameter*: and standing *perpendicularly* on this is the above-mentioned distance measurement (*EntfernungsMaas*), as directional measurement (*RichtungsMaas*) and bearer. [In the first case] either this original perpendicular (*ursprüngliche Perpendikel*) intersects the ex-

47 Berlin Ms.germ.qrt. 1964b, p. 46. Cf. Nachschrift Halle, GA IV/6: 399.

48 »Lines are not enclosed spaces but limits.« *Erlanger Logik* (GA II/9: 85).

49 Ibid. p. 127.

50 »What this assertion of infinite divisibility means: it is possible to infinitely limit the line that is *continuous* in itself.« *Wissenschaftslehre 1801/02* (GA II/6: 231).

ternal line, i.e. so that it makes right-angles on both sides of it [...] or [in the second case] the original perpendicular forms unequal angles, then it is convergent and divergent compared to the archetypal diameter (*Ur-diameter*), and therefore invariably determined. – And a specific direction, absolute orientation concerning the location and direction, are only possible through this absolute diameter and this perpendicular. The original diameter is the *regulator* of all direction; the perpendicular on it is the medium toward the outside.«[51]

In conclusion, we might add that unlike Fichte, Schelling did not think that geometry should begin with a straight line. Because it is a better concept for *interlinking* all the propositions in the system, Schelling considers that geometry ought to begin with the circle: »The beginning of all geometry is the line of the circle; the first proposition of Euclid, the construction of the equilateral triangle can only be mediated through the line of the circle and be understood through it and in it.«[52] Nevertheless, like with the point, Schelling's thoughts on the straight line are remarkably similar to Fichte's, since Schelling also thought that the line was not originally composed of points, but that the latter only entered into the line as a result of conceptual abstraction. According to Schelling, the line only has points »after one has presupposed that it is composed of points.«[53] Another difference between Fichte's and Schelling's view of the line is that on account of its schematic properties Schelling believed that the line belonged more to the branch of *arithmetic* than to geometry. In the important letter to Fichte from 19th November 1800 Schelling stated that the line is »the schema of arithmetic« and that the plane in fact was the »schema of geometry.«[54] This is because arithmetic deals with counting and the successive operations that play themselves out in the sphere of *time*. Hence, for Schelling: »The line – as an increasing and decreasing quantity according to one dimension – is the schema of arithmetic whose series also only has one dimension.«[55] Because of this relation to succession and to a limit or

51 The last line in the original German: »Der ursprüngl. Diameter ist das *Regulativ* aller Richtung; der Perpendikel darauf das medium nach aussen.« Fichte, *Erlanger Logik* (GA II/9: 130)

52 Schelling, *System der gesammten Philosophie* (1804) (*Sämmtliche Werke* I,6: 166).

53 Schelling, *Sämmtliche Werke* I, 10: 328.

54 Schelling, letter to Fichte, 19th November 1800 (contained in GA III/4: 367).

55 Ibid.

boundary Schelling therefore thought that the line was an intimate expression of the nature of time, which comes to external expression in the line: »Since time in and for itself originally signifies a mere boundary (*Grenze*), it can be intuited externally, i.e. brought into connection with space, and can only be intuited as a flowing point, that is, as a line. But the line is the original intuition of motion, and all motion is intuited as motion only insofar as it is intuited as a line.«[56]

Let us now summarize Fichte's geometric definition of the line: The ideal straight line is a synthetic continuous whole and not a collection of points. However, like the point, it is also a limit of space. And like the *Urpunkt* or archetypal point, there is also an *Urlinie* or archetypal line for Fichte – such as the *Ur*-perpendicular or *Ur*-diameter. The latter are infinite, absolute and invariable, and have to be intuited in intelligible intuition. Hence, the pure ideal (rational) line must not be conceived in a point-like atomistic or discrete manner. And due to its quality of straightness, the pure line is reducible to itself alone, insofar as even the most infinitely small part of a line still remains a line.

The Plane, the Solid and Mathematical Beauty

If in Fichte's theory of geometry the point and line constitute fundamental and irreducible entities or elements of space, then how does it hold with the third fundamental element, the plane? Before moving on to Fichte's philosophical interpretation of the line and the drawing of the line, we will conclude this section of Fichte's geometric definitions with a short excursus on his conceptions of the plane, figure and solid. As we shall soon see, Fichte takes the triangle to be the two-dimensional model or archetypal figure of a plane because it requires at least three lines and an angle to come into being.

Let us start with the geometric plane. Just as the line is not composed or built up of points, but is a fundamental entity, so is it the same with the plane (*Fläche*)? The answer is indeed yes. For Fichte, the essential quality of a plane – its 2-dimensional surface character – cannot be reduced to or derived from a line. And the same holds for the solid, whose essential 3-dimensionality cannot be reduced to or derived from the plane. Of course,

56 Schelling, *Sämmtliche Werke* I, 3: 487f.

like the line, they may be *subsequently* broken down into other elements such as points and lines, but not originally. As Fichte says in the *Erlanger Logik*: »[T]he plane could never be a line, a line that can be drawn as a line, and the solid never a plane, and so on. The most infinitely small part remains a *plane*, or *solid.*« [57]

How does the plane arise? It comes into being through the *drawing* or movement of the line. As Fichte again notes in the *Erlanger Logik 1805*, this is a hallmark of the progression of geometry as such: geometry »advances regularly in the form: from simple to composite forms: ›point‹ and line – line to plane etc. [...]«[58] Here it is important to note that even if the line is drawn it does not automatically describe a plane. For if with regard to the perspective of the viewer the line is drawn towards him the resulting perceptual figure is still a line, i.e. a line could be in fact a plane but simply viewed from the side, and the plane is simply fixed in a set direction. Hence, to pass from the dimension of the line requires more than one direction, or movement in more than one dimension.[59]

Thus, the point of intersection of two lines is a necessary but actually not a sufficient condition for the construction of a plane. A third line is required. In the *Sonnenklarer Bericht* Fichte describes the process of the construction of a plane out of three lines and takes as his model the example of a triangle. The latter is drawn in free imagination and is clear and certain because it arises through intuition.[60] Whereas in the *Wissenschaftslehre 1801/1802* Fichte describes the same process from a different standpoint, insofar as he asks the reader to assume in his imagination that the same angle formed by the first two lines could be closed with one or a number of other lines to generate what we call a triangle.[61] The result of all this is that Fichte takes the triangle to be the fundamental planar figure since any three points determine a plane. Again, it does not depend on what *name* we give to this figure enclosed by three lines, whether it is called a »*Dreieck*«, triangle or something else, in Fichte's eyes it is a plane

57 (GA II/9: 128).
58 Ibid., p. 135.
59 Ibid., pp. 128f.
60 Fichte, *Sonnenklarer Bericht* (GA I/7: 226, 232).
61 See *Wissenschaftslehre 1801/1802*, part I, §1 (GA II/6: 135f.).

and its representative is a triangle.[62] Hence in the *Vorlesungen über Logik und Metaphysik SS 1797* Fichte says: »A Δ is called a triangle; it ought to be called: ›a *plane* enclosed on three sides.' It is merely a matter of explicating the name.[63]

Lastly, what about the *geometric* solid? The geometric solid is an abstract three-dimensional body and not a physical or material object. If a triangle or plane requires three lines (or points) to be generated, the solid is defined by a further point that is not in this plane. The most fundamental solid is the pyramid, or to give it its geometric term, the »tetrahedron«. In the text *Practische Philosophie* Fichte talks about a kind of »*Zwischending*« (intermediate entity) between a plane and a solid: »It is the surface of objects.«[64]

Ultimately, the different figures of space for Fichte are crystallisations or spatial limitations out of an original space. Figures of this nature, including the triangle, *enclose* a certain defined area of space and are based on »genetic« laws, that is to say, we penetrate into the rules of their coming into being.[65] Here one could ask: But does this not hold for the circle? And why did Fichte, in the spirit of Schelling, not choose the circle as the model of a plane, for isn't the circle the most perfect and harmonious of all the fundamental *figures*? – Recalling the notorious footnote: »Question for a Mathematician« from *Über den Begriff der Wissenschaftslehre*[66] Fichte states his view that *only straight lines are genuine lines*, because the concept of a line implies the notion of straightness and direction. As a consequence, a curve or curved line (so-called *krumme Linie*) is simply a stringing together of infinitely many points that are infinitely close to each other, and the origin of a circle is the limiting line of infinite space

62 Csech points out that Fichte employed the German term »Triangel« for this geometric entity, which was the most common term at the time, compared to the words that are now more in use in the German language, such as »Dreieck«, »Dreiwinkler« and »Dreiseit«. (See *Raumlehre*, p. 407, note 205).

63 GA IV/1: 268.

64 GA II/3: 212.

65 »We only define where we know the genetic law: A triangle = a space enclosed by three lines. – an enclosed space.« (*Erlanger Logik*, GA II/9: 81).

66 (GA I/2: 135), footnote in §5 on the boundaries of the *Wissenschaftslehre* from particular sciences (natural science and geometry). English translation by Daniel Breazeale in: *Fichte: Early Philosophical Writings*, footnote, pp. 120 f.

(*Grenzlinie des unendlichen Raums*). If this is the case, how can we construct such an infinite, original circle? We take the I to be the centre or centre-point, and from it an infinite number of infinite radii are drawn. However, our limited, restricted imagination (*eingeschränkte Einbildungskraft*) needs to set endpoints to these radii. These endpoints are then conceived as a single unity or collection and this gives us the original circle (*ursprüngliche Kreislinie*).[67]

Thus, if we simply consider it as a collection of points, then to be consistent the circle would not belong among the planar figures. Naturally, one can still draw a circle-like figure on a plane. However, only the original circle at infinity is the *genuine* circle for Fichte; all other circles are merely finite abstractions from this original circle.

Moreover, Fichte did not share the opinion that the circle was the most attractive or harmonious geometric form for aesthetic reasons. In the *Practische Philosophie* we have a delightful intermezzo containing Fichte's reflections on the aesthetic beauty of the different mathematical figures. Any figure able to hover or pass between two other figures becomes an aesthetic figure. On account of its flexible form, the oval is therefore an example of a highly aesthetic or beautiful figure for Fichte: »It is rather correct to say that external forms are most pleasing when they approach the mathematical forms. – Yet this is not always so; among the outlines the oval is without doubt the most beautiful. The circle is far from being so. A rectangle is prettier than a regular square. [...] Thus, a hypothesis: if a given figure hovers among several mathematical figures, it will be more pleasing. The rectangle hovers between the square and the line. If it becomes too long it ends up approaching the line too closely, if it is too short, it approaches the square. [...] If anything at all can hover between the solids and the planes, and in both genus between several specific solids and planes, it becomes more beautiful, [for example]: an animated human face; a painted surface.«[68]

In conclusion, according to the Fichtean theory the *fundamental shape of space* is most adequately expressed by the *Kugel* (sphere). – Especially the flexible or mobile sphere capable of infinite contraction and expansion. In a letter to Schelling from the year 1800 Fichte says that the

67 Ibid.
68 GA II/3: 212.

spherical shape underpins all the other figures of geometry and therefore most perfectly captures the original shape of space: »Original space, or space as intuition, does not have any dimensions. Everywhere, small or large, it is a sphere.«[69] As we can see, Fichte equates the sphere with what he calls original space as an intuition. We will turn to this conception of original space after an examination of Fichte's philosophical interpretation of the line and the drawing of the line.

d. *The Drawing of the Line and das Bild des Lebens – Fichte's Transformation of Euclid's first Postulate*

> »The I intuits its mere acting as such as the drawing of a line.«
>
> J.G. Fichte.[70]

Our aim here is to examine Fichte's philosophical interpretation of the line. It hinges on what he calls: the »drawing of the line« (*das Linienziehen*). I will show that in a certain respect this act of drawing the line replaces Fichte's Jena *Grundsatz* of the intuiting of the I. In other words, the drawing of the mathematical line serves as a propedeutic or preliminary exercise for those students of the *Wissenschaftslehre* who cannot directly proceed to intellectual intuition.

In the *Critique of Pure Reason* Kant also investigates this phenomenon or ability to mentally draw a line. Yet Kant makes a distinction between this drawing of a line in thought and the motion described by an object in external space by means of a line. For Kant, the latter example does not belong in pure science and therefore not in geometry because an object that is movable cannot be cognized a priori but simply by means of experience. On the other hand, Kant considers the former activity of drawing a line in thought to be *motion as a pure act* – it is a successive synthesis of a manifold in intuition through the imagination and therefore not only belongs to geometry but even to the domain of transcenden-

69 Letter to Schelling November 19, 1800 (GA III/4: 361). Cf. the *Thatsachen des Bewußtseyns* (1810): »The straight line is [...] contrasted with the curve or circle, which is the opposite of freedom, because universal space is necessarily conceived as a sphere (*Kugel*).« (GA II/12: 60).

70 *Wissenschaftslehre nova methodo* (GA IV/2: 101).

tal philosophy.[71] – It is the *Sinnbild des Zeitflusses* (symbol of the flow of time) and there are different kinds of *Linienziehen*, depending on the kind of space etc.[72]

Kant's example is taken of course from Euclid's first postulate, which asserts what it is possible to do with two separate points in space: »To draw a straight line from any point to any point.«[73] As we saw in chapter one, already in the *Zurich Wissenschaftslehre* from February 1794 we can find Fichte philosophically interpreting Euclid's first postulate. This is the Fichtean version of it: »First of all, geometry needs a *point, by means of which* it draws; then *the space within which* it draws; and the free self-activity of the power of the imagination, which does the drawing. It obtains all three of these from the *Wissenschaftslehre*, and hence the possibility of all construction (and hence the possibility of demonstration as well) is grounded upon the *Wissenschaftslehre*.«[74]

As Knut Radbruch has remarked in an article on the significance of mathematics for Fichte, the crux of the matter here is that Fichte here poses purely philosophical questions that geometers do not typically ask.[75] For example, the philosopher poses questions like: *What* exactly is drawn? Something must be given, thus what is the real relationship between the point and line? In what *medium* does this process occur? And *who* or what *does* or carries out the act of drawing? The geometer mostly does not pose questions of this kind, for they are not of importance to him. For Fichte, however, these questions force the thinker to adopt a higher standpoint and require the imagination and freedom of the inquirer.[76] Regarding the last question, Fichte's answer is central to his philosophy: the *who* of the drawing of the line refers to the »free self-activity of the imagina-

71 Cf. Kant, *Critique of Pure Reason* (B154–155), p. 258.

72 Ibid.

73 Euclid, *The Elements*, Book One, p. 2.

74 GA IV/3: p. 34. Cf. Erich Fuchs on the drawing of the line in his introduction to the Zurich Wissenschaftslehre: *Züricher Vorlesungen über den Wissenschaftslehre Februar 1794. Nachschrift Lavater* (ed. Erich Fuchs, Neuried/Munich: Ars Una, 1996), pp. 46ff.

75 Cf. Radbruch, Knut. »Die Bedeutung der Mathematik für die Philosophie bei Fichte« *Fichte-Studien* 22 (2003), 251–267.

76 Cf. Radbruch, ibid., 251.

tion.«[77] By focusing his questions on the conditions and role of the geometer's own mental acts and activities Fichte operates in a manner that anticipates many of the operations and questions posed by mathematicians and philosophers one century later. Furthermore for Fichte, only one straight line can be drawn between two points. In the *Neue Bearbeitung* Fichte wanted to use this notion to demonstrate that the idea of necessity is only conceivable under the condition of absolute identity and invariability: »shown using the example: the straight line is the shortest [path] between two points. Actually: only one line is possible, it is always the same, and it is always identical to itself.«[78]

Of course, like Kant before him, Fichte was not alone in German idealism in establishing a relationship between the activities of doing geometry and the activity of philosophizing. As mentioned above in our discussion of dimensions, Schelling had a sophisticated set of correspondences between the dimensions of space and human faculties. In the Schelligian system self-consciousness corresponds to the first dimension, which is geometrically expressed by the straight line, sensation (*Empfindung*) corresponds to the second dimension, i.e. that of the geometric plane, and productive intuition (*productive Anschauung*) corresponds to three-dimensional space.[79] However, in contrast to Schelling's philosophical interpretations of geometry, the important point for Fichte was not to find correspondences between the human being and the dimensions of space, but to see how the fundamental elements and activities of geometry are *philosophical images* to describe the I and its activities.

In this respect we already find in the 1794 *Grundlage* that Fichte had employed a number of images to help explain the nature of the human I. Some of his most striking images, however, were not taken from geometry but from the *Naturlehre*, or what we would call today: *physics*. Here in the *Grundlage* Fichte states that it is possible to attribute *directionality* to the activities of the I. On the one hand, the outward absolute striving and activity of the I can be described using the image of *centrifugal* direction. – The I's infinite striving and activity commences from a centre within and

77 Fichte, *Zurich Wissenschaftslehre* (GA IV/3: p. 34).
78 GA II/5: 349.
79 See Csech's more detailed discussion of the topic of Schelling's theory of dimensions in *Raumlehre*, pp. 243–245.

is directed outward into the world. On the other hand, the I can be reflective and inwardly contemplate itself and its own activity. – This inwardly directed activity can be described by using the image of *centripetal* direction: »Hence, the centripetal and centrifugal direction of the activity are both grounded in the same manner in the essence of the I; they are one and the same, and the two are merely to be distinguished to the extent they are differently reflected.«[80]

Nevertheless, some of Fichte's most frequent images and analogies stem from mathematics, and in this sense he placed a huge emphasis on the drawing of the line. In the Fichtean system of philosophy the drawing of a geometrical line is intimately related to the *activity of the intelligence*, and depends on an »inner agility of your spirit.«[81] In essence: Fichte asserts that the fundamental cognitive activities of the mind may be characterized as a line, or more specially, as the *drawing* of a line: »My spiritual faculty appears to move inwardly backwards and forwards, to quickly travel from one thing to another; in short, it appears to me as the drawing of the line.«[82]

This is nuanced in the *Thatsachen des Bewußtseyns* (1810), where reminiscent of the images from physics in the *Grundlage*, Fichte indicates another affinity of the I with the straight line, insofar as it is interpreted as »the image of the immediate effectiveness of the I.«[83] – This is because all immediate effects in space appear to us as though they occurred in a linear fashion (e.g. impact and pressure). If one encounters an insurmountable object then it is avoided and the path traversed describes another straight line.

In order to avoid any misunderstanding, it must be stressed that Fichte, like Kant, is not so much laying stress upon the *empirical* drawing of a line, such as one could sketch on a piece of paper, but on the inner *generating* activity (or »drawing«) that is *freely* undertaken by the mind and within the mind. That is to say, it *only* refers to the internal activity

80 Fichte, GWL (GA I/2: 407).

81 *Bestimmung des Menschen* (GA I/6: 237). Cf. *Wissenschaftslehre nova methodo* (GA IV/2: 101).

82 Fichte, *Bestimmung des Menschen* (GA I/6: 237).

83 The original German is: »Das Bild der unmittelbaren Wirksamkeit des Ich ist eine gerade Linie« (GA II/12: 60).

and content of pure thinking.[84] In der *Platner-Vorlesungen* this activity becomes an image or the *schema* of spirituality, or of consciousness as a whole.[85] In fact, the *Wissenschaftslehre 1801/1802* begins with precisely this fact of drawing the line. Here we see another break with the strategy of the Jena period of the *Wissenschaftslehre*. That is to say, the *Wissenschaftslehre* no longer begins with the injunction »to think oneself« or with another version of Fichte's *Grundsatz* or fundamental first postulate of the self-intuiting I, but begins with an investigation of knowledge or cognition (*Wissen*), and in the Introduction (§1) Fichte uses the example of drawing a line (of the third line to enclose a triangle) to illustrate this new starting point. Here Fichte's reflection is not directed at the activity of the self-intuiting I, but at the conscious cognitive activity undertaken by the mind when drawing a line or constructing the third line of a triangle.[86]

I contend that this new starting point demonstrates *both the discontinuity and the continuity* of Fichte's different Jena and post-Jena versions of the *Wissenschaftslehre*. On the one hand, we have a discontinuity, since Fichte appears to realize that it was no longer advisable to begin directly with his standard *Grundsatz* of the self-intuiting I, but that his philosophy could be better served by a new preliminary examination of the nature of knowledge or cognition itself; and on the other hand, it demonstrates a continuity, since both starting points are directed at a conscious attentiveness to the *pure activities* of the intelligence. As a consequence, in section §3 of the Introduction to the *Wissenschaftslehre 1801/1802* Fichte's philosophical system is now described not as a science of the sciences like it was in Jena, but as a »Description of the *Wissenschaftslehre* as a Knowledge of Knowledge.«[87] Moreover, there is a further relation to the Jena de-

84 Csech makes the same point and argues that with a finished empirical line we have not really experienced or grasped the intuition of the line, and Fichte's claim therefore depends on rendering in consciousness the particular force that is present when carrying out this activity of drawing the line. Hence, our real content of this drawing is a purely mental force. (Cf. Csech, *Raumlehre*, p. 97).

85 »The schema of activity *is a line*. The effectiveness of things becomes represented as a descriptive continuous line.« (Das Schema der Thätigkeit ist *eine Linie*. Die Wirksamkeit der Dinge wird vorgestellt, als beschreibend stete Linien) (GA II/4: 85).

86 Fichte, *Wissenschaftslehre 1801/1802*, Introduction, §1 (GA II/6: 135ff.).

87 Ibid., Introduction, §3 (GA II/6: 139).

scription, insofar as Fichte again refers to the etymology of the word. In Jena he underscored that his philosophy was a theory or doctrine (*Lehre*) of what *Wissenschaft* (science) is, i.e. a *Wissenschaftslehre*, here he breaks down his neologism even further emphasizing the presence of the word »*Wissen*« in the term *Wissenschaftslehre*. Fichte therefore feels justified in stressing that his philosophy should now be viewed as a theory of cognition or knowing.[88]

Fichte argues that it is a so-called knowledge of knowledge because he differentiates between various levels of cognition. For instance, at this lower level, at the level of geometric cognition, the specific act of drawing a line can be cognized and known, it is classified as »mere knowledge« (*das blosse Wissen*) – it is a cognition of a particular and definite act. However, *we ourselves were the entity that carried out this drawing of the line*, and this aspect of the act of drawing of the line remained »unconscious« (*unbewusst*). Hence, according to Fichte, to become aware of this activity of cognition, or to genetically penetrate into its activity – to use an inherently Fichtean notion – indeed, to progress to a knowledge of this activity itself, necessitates a higher level of cognition, a knowledge of our particular and specific acts of cognition, and this mode of knowing is not to be found in geometry but only in the domain of the *Wissenschaftslehre*.[89] As we saw in Chapter Two, this kind of cognition does not have its basis in concepts but in immediate *intuition*. Thus, in this respect Fichte's mode of knowledge does not change here from the Jena conception of an intuition-based knowledge. He therefore draws the following inference: »According to the above, all knowledge is intuition. Hence, knowledge of knowledge, to the extent that it is knowledge, is intuition,

88 »The *Wissenschaftslehre* should, as the composition of the word shows, be a theory, a theory of knowledge (knowing) [...] Moreover, it is a knowledge of knowledge.« (Ibid.).

89 »Thus, in our description of knowledge (§1) we simply had as our object a mere knowing – only a specific cognition of the drawing of the line. However, what *we ourselves* – only unconsciously, precisely because the former [drawing of the line] was the focus of our consciousness – *were*, or *performed* in this describing, was a knowledge of this mere knowledge. Consequently, in the former describing we did not again stand in the domain of simple science, as we more or less did with those propositions of the line in geometry, but in the domain of the *Wissenschaftslehre*.« (Ibid.)

and to the extent that it is a knowledge of knowledge, it is the intuition of all intuition.«[90]

Moreover, this mental drawing of the line for Fichte must not be equated with any external perception (*Wahrnehmung*) or perceptual activity but is grounded in an intuition: »This consciousness of my drawing of the line lies beyond all perception is therefore without doubt what you call *intuition*.«[91] And unsurprisingly, like with the intellectual intuition of the *Tathandlung* and the validity of the *Grundsatz*, this aspect of drawing the line cannot be proved or demonstrated but depends on each person experiencing and proving it to himself.[92] This relationship between intellectual intuition and the drawing of the line is clearly characterized by Fichte in the *Ankündigung*. Again, this basic geometric act seems to serve as a *propedeutic activity* for the student of Fichte's philosophical system. If one *cannot progress directly* to the intuiting of the I, one is directed to begin with a reflection upon the simple cognitive activity of the drawing of the line: »Anyone who is not already familiar with intellectual intuition will have to await our [new] presentation. In the meantime, let such a person think about his own consciousness of *the act of drawing a line* (not his consciousness itself of the line as something *already drawn*) for such an act of drawing is also not, one would hope, something that continues to exist on its own.«[93]

In this procedure the constructing of one's own I is not the initial act of the fledgling philosopher, but we could say that the drawing of the line now becomes his *first* construction. Thus, in another sense this geometrical example of mentally drawing the line replaces, or is perhaps intended

90 Ibid. §4 *Folgerungen*. Once again in the *Wissenschaftslehre 1801/1802* (Part I, section §13) Fichte stresses the stable awareness of the participant in the cognitive act of drawing the line: »The line that is to be drawn as a line in cognition occurs only within a cognition that itself consists of a reposing and solid cognition.« (GA II/6: 159).

91 Fichte, *Sonnenklarer Bericht* (GA I/7: 228). Csech goes so far as to interpret it as *the* fundamental action of the faculty of intuition (Cf. Csech, *Raumlehre*, p. 98).

92 »That my activity is intuited as a drawing of a line, cannot be proved any further, but that everyone must go within himself and find it there.« Fichte, *Vorlesungen über Logik und Metaphysik SS 1797* (GA IV/1: 210).

93 *Ankündigung* (GA I/7: 160). (Translated by Daniel Breazeale in: *Introductions to the Wissenschaftslehre and Other Writings*, p. 195).

as a more cognitively sophisticated version of the famous »wall example«, with which the Jena *Wissenschaftslehre nova methodo* commences in §1 immediately after the statement of its »*Postulat*«.[94] The activity of consciously reflecting upon the drawing of a line does not only have a propedeutic function, but leads Fichte to a further philosophical interpretation. In the *Zurich Wissenschaftslehre* Fichte recalls the two main kinds of intuition, in which the drawing of the line presupposes space as »an originally sensible intuition«, whereas the I is »originally not a sensible but an *intellectual* intuition.«[95] In the *Grundlage des Naturrechts* (1796/7) Fichte argues that this activity carried out in the intellectual intuiting of the I can in turn be simultaneously intuited as a drawing of line, and be interpreted as the *original schema of activity* as such.[96] And not only this, in *Zur Ausarbeitung der Wissenschaftslehre von 1801/02* Fichte makes a further distinction between the sensible intuition and intellectual intuition with regard to the drawing of the line. The drawing of the line in a certain direction belongs to sensible intuition, but to grasp the destruction or annihilation of every directional aspect in this drawing, leaving simply *drawing as such*, belongs to intellectual intuition: »We receive here the line that was sought, and thoroughly devoid of all direction, and destroying all direction – but not in sensible intuition, as drawing itself, but in intellectual intuition, as mere drawing as such (*blosse Ziehbarkeit*).«[97] Hence, this focus on intuitable activity again underscores the internal continuity between the Jena and early Berlin presentations of the *Wissenschaftslehre*.

Fichte's philosophical view of the line culminates in the *Einleitung in die Wissenschaftslehre 1813* where the line is characterized as the »original

94 Cf. *Wissenschaftslehre nova methodo* (GA IV/2: 29). Although it is not at the start of the text of the *Wissenschaftslehre nova methodo*, the drawing of the line example is still present later on, occurring in sections §10 and §17 (GA IV/2: 101, 217).

95 *Zurich Wissenschaftslehre* (GA IV/3: 34).

96 Fichte here considers the resulting original line to be »*pure extension*« and constitutes the homogenous element in both time and space. – In a counter-intuitive move he declares that the line does not presuppose space but that space presupposes it. And this further illuminates his understanding of the line as a limit of original space, insofar as these original limits should not be confused with the extended sensible limits of lines in ordinary empirical space. See *Grundlage des Naturrechts* (1796/97; GA I/3: 362).

97 *Zur Ausarbeitung der Wissenschaftslehre von 1801/02* (GA II/6: 55).

image of life« (*das ursprüngliche Bild des Lebens*).[98] This is because in contrast to the »reposing« and fixed nature of the point, the line is distinguished and generated by means of *movement*, and because all life itself is expressed in movement: »[T]he point is repose, whereas the line is constructed through movement. [...] Hence, we have discovered how to express what the positing of being (*Seynsetzen*) really is! It is a line, a life, something alive that bears within it its own law of life. It is an *Act*.«[99]

2. Fichte's Conception of Space

> To comprehend space in its profundity, differently to Kant. –
>
> J.G. Fichte[100]

As we have seen, Fichte's ideas regarding the point, line and plane all have important philosophical consequences and often play a propedeutic role within the cognitive framework of the *Wissenschaftslehre*. Fichte not only possesses a theory of the fundamental elements of geometry, but he also makes an application of his ideas on geometry to sketch out a conception of space. I will allude to the unique features of Fichte's conception of space, and especially show how he departs from the Kantian conception of space.

I will above all discuss three differences with Kant. Firstly, this includes his various views on the nature of *empty* space, which is allowed by Kant but considered a fiction by Fichte. Next, in contrast to Kant, Fichte carries out a so-called »genetic« deduction of space. Finally, I will analyze Fichte's thoughts on the conventional/traditional role ascribed to geometry as the science par excellence of space, or more precisely for him: as the science of the measurement and limiting of space.

98 Berlin Ms.germ.qrt. 1964b, p. 46.

99 The original German: »So hätte sich demnach gefunden, was das Seynsetzen, das Sagen es ist, eigentlich sey! Es ist eine Linie, ein Leben, ein lebendiges Durch, in sich tragend sein eigenes Lebensgesetz. Ein *Act* ist es.« (Ibid. pp. 47f.). See the similar sentiments in the *Erlanger Logik 1805*: »He [the geometer] brings the line as a continuous and pure line with him, and the points flow into one another. This is indeed a continuous and living space: he therefore brings it with him.« (GA II/9: 127).

100 *Einleitung in die Wissenschaftslehre 1813* (SW IX: 70). (GA II,17: 292).

In order to do this, it is valuable to begin with a brief outline of Kant's theory of space. This outline does not of course aim to be exhaustive, for Kant's theory has had an enormous influence on the history of physics and mathematics, but simply to provide an overview that will help in the comprehension of the chief points of Fichte's own conception.

a. Kant's Theory of Space

Kant did not of course first begin his discussion of space in later works such as the *Critique of Pure Reason*, but treated aspects of his conception in earlier works such as the *New Elucidation* (1755) and the *Physical Monadology* (1756).[101] The former work argues for a relational theory of space in which the substances are mutually dependent: »since the determinations of substances are reciprocally related to each other [...] it follows that the concept of space is constituted by the interconnected actions of substances, reactions always being of necessity conjoined with such interconnected actions.«[102] On the other hand, the *Physical Monadology* altered this view to put forward a conception of space as appearance: »space is not a substance but a certain appearance of the external relation of substances.«[103] However, the earliest work in which Kant advanced his most detailed and coherent theory is the brief 1768 text: *Concerning the Ultimate Ground of the Differentiation of Directions in Space*. Here Kant summarized his view that space is absolute and original in which the objects of the physical world are located, and is not a sensation but a fundamental concept: »Our considerations, therefore, make it clear that differences, and true differences at that, can be found in the constitution of bodies; these differences relate exclusively to *absolute and original space*, for it is only in virtue of absolute and original space that the relation of physical things to each other is possible. Finally, our considerations make the following point clear: absolute space is not an object of outer sensation; it is

101 See Gary Hatfield's overview of the evolution of Kant's theory of space: »Kant on the Perception of Space (and Time)« in: *The Cambridge Companion to Kant and Modern Philosophy*, ed. Paul Geyer (Cambridge: Cambridge University Press, 2006), pp. 61–93.

102 Kant, *New Elucidation*, in: *Theoretical Philosophy 1755–1770*, p. 43.

103 Kant, *Physical Monadology*, in: *Theoretical Philosophy 1755–1770*, p. 57.

rather a fundamental concept which first of all makes possible all such outer sensation.«[104]

However, the *Critique of Pure Reason* provides the most graphic evidence and statements of Kant's theory, and it is also the text that Fichte most referred to and therefore helps us best bring into relief the Fichtean conception of space. In the 1st Critique Kant implicitly sets his theory of space up against two major thinkers of the past – Newton and Leibniz. Newton's view considered space to be absolute – a giant container, as it were, filled with sensible and material objects. For him, this container would exist even if it did not contain any physical objects. For Leibniz, on the other hand, space was more of a logical construction, and the objects therein interact in and with one another. Hence our statements about space are statements about this interaction.[105] Kant outlines his theory of space in the first section of the Transcendental Aesthetic of the *Critique of Pure Reason* (»*Von dem Raume*«: A22/B37ff.) Unlike the theory of Newton, Kant's goal was to show the ideality of absolute space (not just as a large container), and this was also opposed to Leibniz's view, since it was not simply a relational theory. Kant also aims to demonstrate that we must attribute a special status to space and time. That is to say, they must be distinguished from the other elements of sensible experience – since it is space and time that provide the sensible forms of all experience. In an *empirical* sense, they yield a framework to inform us how the objects outside of us interact with one another. But this is also to be understood *transcendentally*, where transcendental means: »I understand by a transcendental exposition of a concept as a principle from which insight into the possibility of other a priori cognitions can be gained.«[106] – Space too is now to be understood transcendentally, or in Kant's famous words: the original representation of space is an a priori intuition, not a concept.[107] Again, a priori here means that the representation is not derived from experience in any way, whereas an intuition signifies something that is immediate and non-conceptual. Kant furthermore says that one of the best

104 Kant, *Directions in Space*, in: *Theoretical Philosophy 1755–1770*, p. 371.

105 Cf. Kant's introductory words to his theory of space *Critique of Pure Reason* (A22-A26/B37–42), pp. 157–159; and Hatfield, »Kant on the Perception of Space (and Time)« pp. 62–69.

106 Kant, *Critique of Pure Reason* (B41), p. 176.

107 Cf. Kant, ibid. (A25/B40), p. 175.

instances of a pure intuition of space is the representation of figure and extension contained in the representation of a body.[108] Does Kant tie his theory in with the science of geometry? Yes, in fact Kant goes so far as to claim (in B40–41) that it is precisely his theory that is able to explain the possibility of geometry itself, an honour that Fichte will later claim for the *Wissenschaftslehre*: »Geometry is a science that determines the properties of space synthetically and yet a priori. What then must the representation of space be for such a cognition of it to be possible? It must originally be intuition; for from a mere concept no propositions can be drawn that go beyond the concept, which, however, happens in geometry. [...] Thus, our explanation alone makes the possibility of geometry as a synthetic a priori cognition comprehensible.«[109]

Kant furthermore argues in support of the claim that our representation of space is an a priori intuition, i.e. that it is not an empirical representation which has been derived from outer experiences, for outer experience is possible only through this representation.[110] To this extent intuitions are also termed »pure« by Kant, i.e. transcendentally they exist inside us as the framework for the conditions of knowledge. Thus, for Kant space is represented as nothing but the form of all appearances of the outer sense.[111]

b. Differences Between the Fichtean and Kantian Theories of Space

In this section we will examine how Fichte's conception of space differs from Kant's view. As with his philosophy of mathematics, Fichte did not write any detailed text on his conception of space. Materials for such a conception may be already discerned in the *Eigne Meditationen* (1793/94), they are further hinted at in the *Grundlage der gesammten Wissenschaftslehre* (1794/95), and reach a more detailed form in the 1795 *Grundriß des Eigenthümlichen der Wissenschaftslehre*, where Fichte provides a »genetic« deduction of space. My analysis is above all based on the detailed research of Hubert Horstmann[112], Wilhelm Metz[113], Paul Ziche[114] and especially

108 Cf. Kant (A20–21/B35), ibid., p. 173.
109 Kant, ibid. (A25/B41), p. 176.
110 Kant, ibid., (A22/B38) p. 157.
111 Cf. Kant, ibid., (A26/B42).

Werner Csech's monograph *Die Raumlehre Johann Gottlieb Fichtes.*[115] As noted by all these authors, Fichte's views on space are multi-sided and complex. I will chiefly limit myself to his differences with Kant, and especially to those parts of Fichte's conception that specifically relate to his theory of geometry.

What are the differences between the Fichtean and Kantian conceptions of space? Fichte himself states a crucial difference between his own and Kant's theory of space in a footnote in the 1794/95 *Grundlage*. Fichte's conception of space is based on what he calls a »genetic« deduction of space and time, i.e. a deduction able to penetrate into the inner core of space and analyze it *genesis* or how it comes into being. Fichte particularly underscores their inverse approach, or the reciprocal relationship between his and Kant's view, with the crux being the notion of *ideality*. Fichte begins with the ideality of objects, Kant with the ideality of time and space: »Kant shows the ideality of objects out of the presupposed ideality of time and space: we, inversely, will show the ideality of time and space out of the proven (*erwiesenen*) ideality of objects. He requires ideal objects to fill time and space; we require that of time and space in order to be able to place ideal objects. Hence our idealism [...] proceeds a few steps further than his.«[116]

With regard to their inverse approaches, Wilhelm Metz contends that the main difference between the Kantian and Fichtean conceptions of space is that whereas Kant passes from a mathematical to a dynamic determination of space, Fichte prefers to let mathematical space first arise out of the so-called dynamic *Urraum* (an archetypal or original space).[117]

112 H. Horstmann, »Unterschiede zwischen den Raum-Auffassungen Fichtes und Kants« in: *Transzendentalphilosophie als System: die Auseinandersetzung zwischen 1794 und 1806*, ed. Albert Mues (Hamburg: Meiner 1989), pp. 58–67.

113 W. Metz, »Fichtes genetische Deduktion von Raum und Zeit in Differenz zu Kant« *Fichte-Studien* 6 (1994), 71–94.

114 P. Ziche, »Raumkonstruktion, Deduktion der Dimensionen und idealistische Prinzipientheorie. Problemlagen im Fichte-Schelling-Briefwechsel vom November 1800«, *Fichte-Studien* 25 (2005), pp. 21–42.

115 W.V. Csech, *Die Raumlehre Johann Gottlieb Fichtes: Mit Berücksichtigung Philosophiegeschichtlicher Konstellationen* (Frankfurt am Main: Peter Lang, 1999).

116 Fichte, GWL (GA I/2: 335).

117 Cf. Wilhem Metz (1994), op. cit. p. 89.

Hence, Fichte is able to employ a genetic deduction of space in the *Grundriß*, because unlike Kant he does not conceive two separated spheres of spontaneity and receptivity, but a single line of spontaneity generating space and time. Hubert Horstmann, on the other hand, believes that there are three clear differences between the Fichtean and Kantian theories of space.[118] We will look at these three differences. The first concerns the notion of *empty space*. As mentioned above, Kant argues against Newton's idea of absolute space, and instead views space (and time) as pure *a priori* forms of intuition. For Kant, space is »not an empirical concept, derived from external experiences.«[119] Although Kant rejects the Newtonian notion of absolute space, there is still an *empty* (*leer*) space for him: space can potentially be emptied of all its physical objects. For Fichte in contrast, all perceptions are determined by affections of the external senses, which are given as states of the I, and are also determined by extension in space. The main point to bear in mind is that things and objects are not in some kind of space *as such* for Fichte, but only in a specific and definite space. As human beings we are of course intimately connected with space: »Thus, we are bound to space and time and restricted through them« and everything that we think in space has extension (*Ausdehnung*).[120] Fichte summarizes this point in the *Bestimmung des Menschen*, where no »object is in space as such, but every one is in a specific space.«[121] In Fichte's eyes, there is also such a thing as empty space, but this is a geometrical abstraction carried out by thinking, and then filled in again by the imagination. Thus, there is such a thing as the emptying of space (*Ausleerung*) and filling in of space (*Raumerfüllung*).[122] As he observes in the *Neue Bearbeitung*: »Thinking empties out space. The imagination fills it in.«[123]

For Fichte then it is erroneous to assume empty space to be a *reality*. On this account the Kantian view of »empty« space is ultimately an illusion or a fiction for him, since all spatial determination presupposes something that fills, a concrete object, and vice-versa. Here we cannot sep-

118 Horstmann (1989), op. cit. p. 58.

119 Kant, *Critique of Pure Reason* (A23/B38), p. 157.

120 Fichte, *Vorlesungen über Logik und Metaphysik SS 1797* (GA IV/I: 30, 249).

121 GA I/6: 244.

122 In the *Neue Bearbeitung* Fichte concedes that these processes still remain mysterious (cf. GA II/5: 400f.).

123 Cf. Ibid., pp. 397, 401.

arate the one from the other, they are mutually dependent. This salient difference between Kant and Fichte can be traced back to the fact that Kant assumes that space as a form of intuition comes before »all the perceptions of an object«, whereas »for Fichte space itself is *only in the act and through the act* of perception or my concrete intuiting.«[124] Hence there »is no empty space.«[125]

A second difference between the Kantian and Fichtean conceptions of space, according to Horstmann, directly relates to the certitude displayed in Euclidean geometry. It concerns Kant's belief that the qualities of space are dependent on the »phenomena« of physical processes.«[126] For Kant, »all species and determinations of space« are given *a priori* (A28/B44) and as we saw space therefore is nothing more than the form of all appearance of the outer senses.[127] That is to say, there is no genesis of the representation of space. Fichte follows Kant insofar as he also says that space »is the subjective condition of the possibility of external intuition.«[128] However, unlike Kant, Fichte does not reflect upon specific spaces that are designated by certain geometric qualities. Accordingly, we find a *genesis* or a »genetic« conception of space in Fichte, which is not the case in Kant.[129]

A third difference between Kant and Fichte concerns the relationship between space and time. For Kant, time is a form of the inner sense, and it determines the relationship of our representations in our inner state of mind.[130] For Fichte, all the objects of the senses exist in time, but only the objects of the external senses can be said to be in space. There is the limitation of the external senses, but time and space are equal in rank in Fichte: »time and space […] are intimately interlinked.«[131] This is not the case for Kant, where time has a special status. Fichte defines the sequence of time as a successive series in which every member is dependent on another, but without the same being dependent on it.[132] As a consequence:

124 Horstmann, ibid. p. 61.
125 Fichte, *Grundriss* (GA I/3: 200).
126 Kant, *Critique of Pure Reason* (A23/B38), p. 157.
127 Cf. Kant, ibid. (A26/B42), p. 159.
128 Fichte, *Grundriss* (GA I/3: 204).
129 Cf. Horstmann, ibid., p. 64.
130 Cf. Kant, *Critique of Pure Reason* (A30/B46ff.), pp. 178ff.
131 Fichte, *Platner Vorlesungen* (GA II/4: 91).
132 Cf. *Grundriß* (GA I/3: 207). Cf. Hortmann, ibid., p. 66.

»time can only form itself in space.«[133] In Fichte's conception, therefore, the relationship between time and space is conceived as an equality of status, both are indivisible from and joined with one another, and can only be reciprocally determined and measured.[134]

Hence, returning to his closing remarks about space in the *Grundriß*, Fichte argued that he had prepared the reader in his discussion of space such that he was now in a position to understand Kant's view. For Fichte believed that he had defined space and time as forms of intuition in the I, whereas Kant had simply accepted them without »the slightest proof.«[135] For the science of geometry requires a certain number of conditions to be fulfilled in order to occur. Fichte above all sees the »emptying of space« or the geometer's skill at abstracting a geometric space (as opposed to real physical space) as indispensable to the science of geometry: »*Empty out* (*ausleeren*) everything in space that is forming and limiting in it; *grasping* space *as empty and formal* is the *conditio sine qua non* of this science.«[136]

To summarize, just like for Kant, space is also an external intuition for Fichte, but no object can be in space as such, but only in a specific space. Here the object and space are both indivisible and inseparable. Because of this inseparability, Fichte considers it absurd to believe in the reality of empty space. For him empty space can exist, but only as a geometrical abstraction, not as a reality in itself.

c. Fichte on Time and Original Space

A common theme in Fichte is the distinction between everyday naïve thinking and consciousness and the reflective self-consciousness of the

133 Fichte, *Zur Ausarbeitung der WL von 1801/02* (GA II/6: 100).

134 Cf. Horstmann, ibid., p. 66.

135 *Grundriß* (GA I/3: 208). See Fichte further criticisms of Kant in the *Wissenschaftslehre nova methodo*. Here Fichte agrees with Kant's statement that space is a form of outer intuition (a priori), but Fichte brings it into connection with his theory of freedom. For Fichte, external space impacts upon our sphere of acting freely, because the material in space acts as a limitation for our freedom. (Cf. GA IV/2: 102–106).

136 Fichte, *Erlanger Logik* (GA II/9: 125). See too Fichte's comment: »In geometry, for example, by sketching a triangle, a circle and so on, empty space becomes in a certain sense restricted.« (GA I/6: 44f.).

philosopher.[137] He frequently made a distinction between these different levels of consciousness, such as when we undertake our first naïve, pre-philosophical thoughts, and place ourselves as human beings into an external world, i.e. when we consciously posit an opposition between ourselves as a thinking I and the external physical world. This external world, and we ourselves within it, are spatial objects, belonging to a world of *extended* space. These thoughts are inevitable: »An external object must be conceived in space. [...] All real objects are necessarily in space, and hence extension arises.«[138]

On the other hand, if we look at what Fichte considers the role of the philosopher and philosophy to be, then it is not to generate anything, but to explain and explicate, to instruct and to deduce. There is one exception of course, and that is that philosophy must generate itself: »Philosophy can only explain facts, and cannot bring forth any of them itself; apart from the fact that it must generate itself.«[139] And it is the same with space for Fichte. The role of the philosopher is not to generate space but to understand it, and to genetically deduce it: »The rigour of the proofs and deductions of the *Wissenschaftslehre* are not so that one inwardly acquires space for example, but that one genuinely understands, i.e. genetically penetrates into it through its principle.«[140]

Csech also makes the important point that in Fichte's view of space one must proceed from and revert back to the factual. Hence the task of philosophy is to explain the given world out of its conditions.[141] Fichte, in fact, does not simply classify one, single uniform space, but there are many spaces for him. For example, in the *Wissenschaftslehre 1801/1802* Fichte distinguishes between: pure, ›gediegene‹ ›ruhende‹ space, and constructions in space.[142] Thus, we are not just dealing with the limitations of one

137 Cf. D. Breazeale, »The ›Standpoint of Life‹ and the ›Standpoint of Philosophy‹ in the Context of the Jena Wissenschaftslehre (1794–1801)«, in: *Transzendentalphilosophie als System: die Auseinandersetzung zwischen 1794 und 1806* (ed. Albert Mues, Meiner 1989) pp. 81–104.

138 *Vorlesungen über Logik und Metaphysik SS 1797* (GA IV/1: 367f.).

139 Fichte, *Ueber den Grund unsers Glaubens an eine göttliche WeltRegierung* (GA I/5: 348).

140 *Wissenschaftslehre 1804* (3rd course) (GA II/7: 344).

141 Csech, *Raumlehre*, p. 51.

142 GA II/6: 232.

single space, but of »spaces within space.«[143] In this regard Kant also makes an important distinction between geometric spaces and metaphysical space.[144] For Kant, geometric spaces are generated by the geometer and there may be a plurality of spaces or an infinite number of spaces. Metaphysical space on the other hand, is the original space, and singular. The former geometrical spaces are grounded in this single metaphysical space.[145]

As mentioned above, Fichte's conception of space is based on a genetic deduction of space and time that is first signalled in the *Grundlage* of 1794/95, but then this genetic deduction is carried out and developed by Fichte in the *Grundriß* of 1795.[146] Here Fichte says that space is extended, interconnected and infinitely divisible, yet like with his view of the line, even the smallest part of space is always a space possessing continuity. The imagination and abstract thinking play a role in our geometrical conception of space, since they »separate space out from the things« and then subsequently project an »empty space«. Yet this empty space should only be understood as an attempt or experiment of the geometer and the physicist to penetrate and understand the nature of space, for as mentioned, empty *space as such* for Fichte, i.e. as an absolute reality, does not exist.[147]

143 *Erlanger Logik* (GA II/9: 85). See Csech, *Raumlehre*, p. 74 .

144 Cf. Michael Friedman's excellent discussion of this point: »Geometry, Construction and Intuition in Kant and his Successors,« in: eds., Sher, G & Tieszen. *Between Logic and Intuition: Essays in Honor of Charles Parsons*, pp. 188f.

145 »The geometer grounds the possibility of his problem – to increase a space (of which there are many) to infinity – on the original representation of a single, infinite, subjectively given space. This agrees very well with the fact that geometrical and objectively give space is always finite. For the latter is only given in so far as it is generated [*gemacht*]. To say, however, that the metaphysical, i.e. original but merely subjectively given space – which (because there are not many) cannot be brought under any concept capable of construction but which still contains the ground of all possible constructions – is infinite means only that it consists of the pure form of the mode of sensible representation of the subject as a priori intuition. Therefore, the possibility of all spaces, which proceeds to infinity, is given in this space as a singular representation.« Kant's handwritten notes for Schulze's review (AA XX: 419ff.); English translation by Michael Friedman in ibid., p. 188.

146 Cf. *Grundlage* (GA I/2: 362) and *Grundriß* (GA I/3: 129ff.).

147 *Grundriß* (GA I/3: 129ff.)

As outlined in the previous section. Fichte's original view of space depends on his theory of the elements that make up space: Points, lines and planes, or the so-called *Urgebilde* of space, and in this sense the *Neue Bearbeitung* even contemplates the idea of purely »mathematical spaces.«[148] Here space for Fichte is broken down into a number of basic elements, dimensions and intuitions. As we saw in the *Zurich Wissenschaftslehre*, space is originally a sensible intuition: »Within the *Wissenschaftslehre* it will become apparent that [...] mere space is an original sensible intuition, which first makes possible all pure and empirical intuitions.«[149]

Here we have a comparison between two different types of intuition in Fichte's conception of space. Space = original sensible intuition – which is the condition for the possibility of all the pure and empirical intuitions of geometry. On the other hand, the I = intellectual intuition. Moreover, we saw in the *Eigne Meditationen* that Fichte always takes care to consider space's connection with time. He observes that time can only be extended in a single dimension whereas space can be extended in all three directions, and their reciprocity forms a unity based on their different relationship to dimensionality: »Time is only extended in one dimension; space is extended in all – (Therefore, the opposition would consist in unity and totality).«[150] And he sums this up in a nice formula: »Extension is space. Time is an extension of what is not in space.«[151] For Fichte, however, though they are interrelated, time and space are in fact fundamentally opposed to each other: »Space: In space, however, we cannot discover or

148 Fichte, *Neue Bearbeitung der Wissenschaftslehre* (GA II/5: 400f.).

149 *Zurich Wissenschaftslehre* (GA IV/3: 34).

150 *Eigne Meditationen* (GA II/3:76). Fichte adds: »And there would have to be developed in advance the concept and the intuition of extension; as to what is common to both time and space. What is extension? – Doesn't this concept presuppose space? Time is clearly conceived as extended; but isn't space then already transferred to it? *Nous verrons.* – Time, space, and their common/communal extension are without doubt intuitions; and they must have arisen through the joining of the given intuitions ›reality and quantity‹. [...] A portion of the unlimitness is filled up (*angefüllt*) with reality. – on both ends non-reality. If you limit the unlimitedness on all sides, as on one side, then you still have an unlimited extension of one dimension, time. – If you limit the unlimitable as such, then you have space: a defined extension in 3 dimensions. Thus, unity and totality have to be here assumed.« Ibid., p. 73.

151 Ibid., p. 72.

conceive any opposites. Time: likewise in time we cannot invent any opposites: but time and space are themselves *opposed*. – Consequently, A (I) is opposed in time –A (Not-I) in space. [...] An internal *a priori* intuition in time equals an external one in space.«[152] Furthermore, for Fichte time has to have one dimension because it is successive, for one cannot simultaneously experience two things sensed after one another. Time is the making objective of a subjective successive ground.[153]

Finally, all the geometric elements, such as points and lines, geometric figures, like triangles and circles, and the Fichtean notion of planes and space, are all related to and interact with one another. The line has a certain direction, however, and the ability to draw lines in all possible directions gives rise to space. Not just any space, but the empty space of the geometer.[154] In the Fichtean conception of geometry, the point and line are viewed as the two limiting elements, the plane and mathematical solids are considered as types of enclosed figures, and Fichte crucially makes a distinction between pure space, empty space, and the particular limitations, restrictions or enclosures of space that make up spatial figures.[155] Thus, for Fichte we have three levels of space and objects of space, but *everything presupposes and arises from out of the Urraum or original space*:

(1) The essence of archetypal or original space (*Urraum*).[156]
(2) The archetypal elements of space (e.g. *Urpunkt*, *Urlinie* etc.).
(3) The *particular* forms and shapes of space.[157]

152 (GA II/3: 101). See too the following observation: »*Time and Space*: temporarily investigated, Species of the genre »*extension*«. Oh no: but opposed things.« (ibid., p. 72)

153 Cf. ibid. page 123ff. See also the text from 1812/13 *Vom Unterschiede zwischen der Logik und der Philosophie selbst* since it also performs a similar deduction of space. (GA II/14: 350ff).

154 Cf. *Wissenschaftslehre nova methodo* (GA IV/2: 101). See to Ives Radrizzani's discussion of empty space in: *Vers la fondation de l'intersubjectivité chez Fichte. Des Principes à la Nova Methodo*, pp. 139ff.

155 Cf. Csech, *Raumlehre* pp. 74f.

156 Reinhard Lauth also considers that the designations *Urraum* and *ursprünglicher Raum* (archetypal or original space) are synonyms for Fichte. See *Die Transzendentale Naturlehre Fichtes nach den Prinzipien der Wissenschaftslehre* (Hamburg: Meiner, 1984), pp. 57f.

157 Cf. Csech's list of levels, *Raumlehre* pp. 75f.

It is beyond the scope of this dissertation to provide a comprehensive account of the interrelations and transitions between these levels of space, including the transition to the concrete objects of the sense world. Werner Csech has already furnished an excellent contribution to this topic in his volume: *Die Raumlehre Johann Gottlieb Fichtes: Mit Berücksichtigung philosophiegeschichtlicher Konstellationen*. (Frankfurt: Peter Lang, 1999). It was our goal to concentrate on Fichte's theory of original elements and original space, because as we have seen they play a central role in Fichte's model of geometry.

To summarize: the Fichtean model of geometry is grounded in a theory of archetypal or original (*Ur*) elements. As we saw in earlier chapters, and in line with Fichte's Platonistic conception of mathematics, each of the finite or ordinary elements of geometry, has a corresponding archetypal or *Ur*-element, and all reside in *Urraum* or original space. Original space is to be conceived as an intuition, and in contrast to the fundamental elements of geometry it does not possess any dimensionality. In terms of shape, using the power of our imagination it can be represented by the shape of a self-contracting and enlarging sphere. As a result, all the dimensions of space for Fichte are *generated in geometry by a process of imaginative thinking*, and the dimensions themselves are defined as the »universal forms of thinking.« This means that the fundamental elements of geometry, such as the point, line and plane all require a similar process of abstracting from this original sphere in order to come into existence. Fichte summarizes his view and the transitions between these elements in his letter to Schelling from November 15, 1800. This letter brings together many of his thoughts on space and the elements of space. Thus, because of its importance we will translate and quote it at some length:

»For my part, I believe the following: (1) *Original* space, or space as intuition, does not have any dimensions. It is uniformly a small or large sphere (*Kugel*); and the entire work of the imagination with it is to merely *enlarge* or *contract* this sphere. This is why the deduction of the three dimensions is not at all incumbent upon the pure *Wissenschaftslehre*, but initially upon the *philosophy of mathematics*; from which *Naturphilosophie* presupposes this deduction. (2) The three dimensions arise through thought abstracting itself in space: and are nothing more than *the universal forms of thinking* itself. First and foremost, the *point*; abstraction of infinitely many points enclosed in the sphere (from which later *angularity* arises, since everything is *round* in the intuition) and the form of *positing* in general. Then the *line*: the abstraction made in the point continues; otherwise with every point of the line an infinite number of points would grow together: form: [...] The solid is then now really a space, just as the intuition would like it to be. Only the angularity betrays the work of cognition and abstraction.«[158]

d. Geometry as the Science of Space

After this discussion of Fichte's theory of original space, and to conclude this chapter, we might now ask: how does Fichte understand the connection between his conception of space and the discipline of geometry? If there is one discipline that deals with the science of space, it is of course the science of geometry. As is well-known, the word »geometry« comes from the Greek, *geōmetriā*, meaning to measure earth or land (*geo* = earth, *metronthen* = to measure). For Fichte the role of geometry is not simply to be understood as a science of deductive reasoning, but also in its traditional sense as a science of the measurement of space. According to Fichte, geometry is the science of the systematic and lawful »limiting« of space. Already in *Über den Begriff der Wissenschaftslehre* from 1794 we find a telling reference concerning the *Grundsatz* of geometry, and how it is to be distinguished from the *Grundsatz* of the *Wissenschaftslehre*: »The first principle (*Grundsatz*) of geometry is the overall task of limiting space in accordance with a rule, or the task of spatial construction. In this manner geometry is sharply separated from the *Wissenschaftslehre*.«[159]

This is true for even specific instances of geometrical figures, where the geometer restricts or limits »empty« space: »In geometry, for example, empty space becomes restricted in a certain manner through the projecting (*Entwerfung*) of a triangle and a circle, and so on.«[160] This conventional role for geometry had naturally been alluded to by Kant, who consid-

158 Fichte, letter to Schelling, November 15, 1800 (GA III/4: 361).

159 GA I/2: 135; translated by D. Breazeale in: *Fichte: Early Philosophical Writings*, p. 121. In the *Zurich Wissenschaftslehre* Fichte also draws a delimitation between the activity of geometry and that of the *Wissenschaftslehre*. He explicitly says that the limitation of the construction of the intuition of space is not the task of the *Wissenschaftslehre* but of geometry. (Cf. GA IV/3: 34ff).

160 Fichte, *Verantwortungsschriften gegen die Anklage des Atheismus* (1799). (GA I/6: 44f.).

ered that all figures are possible only as different ways of restricting infinite space.[161] And this also holds for Johann Schultz, whose definition of geometry follows the Kantian line: »Geometry is the science of the extension of space,« and it is based on the axiomatic principle of: »a single infinitely extended space.«[162] Whereas Fichte, in contrast, focuses on the abstract aspect of the geometer's ability of emptying out space.[163]

Thus, geometry is both a process of *abstraction* for Fichte, and a process of *construction*. On the one hand, Fichte starts from the abstract notion of empty space and sees the role of geometry in methodically abstracting all the material, with the goal of investigating the possibilities and laws of the limits of space. Although this empty space as such for Fichte is an abstraction or a fiction, it is still a *lawful* process in which the geometer has to generate or construct these configurations. This is an important point that separates geometry from the *Wissenschaftslehre*. In fact, as we saw, Csech argues that the notions of the construction of space and the limiting of space are reciprocal concepts, or the two sides of the same coin. – This is because the *Wissenschaftslehre* takes into consideration the limitability of what is limitable *as such*.[164] This coincides with Fichte's belief that his philosophy and geometry ultimately aim at different kinds of limitation: the former seeks to explore the limitations of the mind, the latter those of space. Not in an arbitrary fashion, but both in a systematic manner according to a rules: »Just as, for instance, geometry involves the entire system of the limitations of space, so the *Wissenschaftslehre* involves the system of the whole of reason.«[165]

However, Fichte considers the *Wissenschaftslehre* to be higher than the science of space. This is for two reasons: Unlike geometry, which investigates the constructions and *necessity* of space, the *Wissenschaftslehre* is placed at a higher standpoint, investigating both the activity of the constructing agent and the laws of constructing itself.[166] Hence, whereas geometry deals with fixed and necessary laws, we need another science to

161 Cf. Kant, *Critique of Pure Reason* (B 606), p. 557.

162 Johann Schulz, *Anfangsgründe der reinen Mathesis* (Königsberg: Hartung, 1790), p. 245.

163 Cf. *Erlanger Logik 1805* (GA II/9: 125).

164 See Cesch, *Raumlehre*, pp. 68–71.

165 *Ankündigung* (GA I/7: 160).

explicate the nature and laws of human freedom. For Fichte, this science is the *Wissenschaftslehre*. As he famously remarked in a letter to Reinhold in January 1800: »From the beginning to end my system is simply an analysis of the concept of freedom.«[167] And naturally, just as we saw with Fichte's philosophical interpretations of the point and the drawing of a line, Fichte also joins the notion of freedom with his conception of space. The *Grundriß* states that when freedom arises it manifests itself in connection with the spatial realm: »thereby that [...] freedom becomes posited, it first shows itself as space.«[168] This highlights the eternal tension in our human nature – the tension between our corporal body bound to natural laws and anchored in a spatial world, and our philosophizing rational mind, seeking to make sense of this by means of its own strict and systematic laws: »It is however also true; in another respect philosophy is not brought about like experience (*Erfahrung*). In experience the spirit is compelled and bound; everyone, for example, has to posit his world in space; this rational law imposes itself, for here we are dealing with the lower cognitive faculty: however, philosophy does not impose itself, like experience, but is called into being through freedom.«[169]

Geometry cannot account for the phenomenon of freedom and the laws of the mind, it confines itself to laws of space. In the next Chapter we will now examine in more detail five further similarities and differences between the science of the *Wissenschaftslehre* and the science of geometry.

166 Regarding this constructing agent, see Csech, *Raumlehre*, pp. 70f. Here he refers to Maimon's views on the relationship between freedom and necessity in the construction of geometric figures.

167 Letter January 8, 1800 (GA III/4: 182). See too his remarks from 1795: »My system is the first system of freedom.« (GA III/2: 300).

168 *Grundriß* (GA I/3: 202).

169 Fichte, *Vorlesungen über Logik und Metaphysik SS 1797* (GA IV/1: 351).

Chapter Five: The Relationship between Geometry and the »Wissenschaftslehre«

1. Geometry as a Propedeutic for the Wissenschaftslehre

> I wish people had some knowledge of mathematics before embarking on a study of the *Wissenschaftslehre*.
>
> J. G. Fichte.[1]

In the first section of this final chapter I will address the question of the pedagogical value of geometry with respect to the *Wissenschaftslehre*. Fichte was acutely aware of the difficulties faced by his students in their attempts to grasp the significance of his words and directives. He knew his system was complex and innovative, and he attempted to articulate it in many different ways: in lectures, addresses, letters, unfinished fragments and texts, popular presentations, in particular applications – such as the *Sittenlehre, Rechtslehre, Staatslehre* and theory of religion – in highly abstract and technical expositions, and in numerous introductory lectures. He was frequently disappointed and sometimes exasperated at the failure of his students, colleagues and readers to understand the basics of his system, at times wanting to even »*force* them to understand.«[2] The problem was twofold: on the one hand, he often attributed this failure to himself, to his inability to communicate his philosophy in a satisfactory *written* form;[3] this dissatisfaction helps us to explain the over 15 (!) different unpublished manuscript versions of the *Wissenschaftslehre*. – On the other hand, it sheds light on Fichte's preference for oral communication,

1 *Ankündigung* (GA I/7: 160).

2 Cf. the notorious subtitle of the *Sonnenklarer Bericht*, a volume written in an attempt »to force the reader to understand« (*die Leser zum Verstehen zu zwingen*« (GA I/7: 183).

3 Cf. *Ankündigung* (GA I/7: 154f.).

where he could »immediately eradicate« any misunderstandings.[4] In this sense, the numerous introductory university lectures and so-called »popular« (writings) lecture courses in Berlin are an implicit admission of the *Wissenschaftslehre's* need for some kind of propedeutic.

However, Fichte passionately believed that there *already* existed a perfectly good introduction and propedeutic to his system – *the study of geometry*. As we have seen in earlier chapters, Fichte directs his students to reflect upon the basic act of mentally drawing a straight line if they are unable to directly progress to the intuiting of their own I. This cognitive act of pure geometry thus serves as a preliminary activity for the more strenuous and abstract cognitive acts of the *Wissenschaftslehre*.[5] In a very real pedagogic sense this again places Fichte squarely in the Platonic tradition. Our aim in this section is to show the extent to which Fichte considered the study of geometry to be an indispensable tool and preparation for developing philosophical thinking. As should be evident from Chapter Three above, this includes the ability to have *non-empirical intuitions and constructions*, which form the basis of philosophical cognition for Fichte.

a. Pestalozzi's Influence and the Teaching of Mathematics

Like many other philosophers before him, Fichte stressed the importance of the study of mathematics to help train the mind to think complex and abstract thoughts. Fichte's views on the value of mathematics for the mind and the imagination began early. In a »*Plan anzustellender Rede-Übungen*« from Zurich 1789 Fichte lists two ways for acquiring the capacity to »think coherently« (*zusammenhängendes Denken*) – composing one's own essays and pure mathematics.[6] In Switzerland Fichte also came into contact with an exceptionally gifted educationalist: Johann Heinrich

4 In a *pro memoria* to the Royal Cabinet in 1804 Fichte expressed his desire to renounce all written publications of the *Wissenschaftslehre* to solely confine himself to »oral communications« (GA III/5: 223).

5 For example, see *Ankündigung* (GA I/7: 160), in: *Introductions to the Wissenschaftslehre and Other Writings*, p. 195.

6 »Apart from learning pure mathematics, which, however, occupies less the imagination, the sole way to practise coherent thinking is *to compose one's own essays*. It is the best and surest training in thinking for oneself, in attention and in interconnected reasoning; the most useful and sole practical learning of logic.« (GA II/1: 130).

Pestalozzi (1746–1827).[7] The two immediately found a like-minded thinker in the other, and engaged in »intimate philosophical, moral and cosmopolitan conversations of the most Socratic, theistic and Kantian kind.«[8] Pestalozzi's ideas and methods would not only leave a lasting impression on Fichte's didactic thoughts but also on Fichte's student, the philosopher Johann Friedrich Herbart.[9] Both Fichte and Herbart were impressed by Pestalozzi's method of stimulating the rational and imaginative faculties of man. The foundation of Pestalozzian cognition was *Anschauung* (intuition), and his system placed little value on mechanical rote learning.[10] With regard to geometrical intuition, Pestalozzi proposed, among others, the construction of a table depicting right-angled triangles of various shapes and sizes. In his eyes, the *constant* viewing and exposure to such a table stimulates the conception of the right-angled triangle in the child. Moreover, Pestalozzi's method is distinctly Socratic in spirit, in which the pupil is to produce and discover for himself all knowledge. Here the teacher is foremost a guide to direct and foster the free and independent

7 Regarding Fichte's relationship to Pestalozzi, see Peter Paret, *Clausewitz and the State* (Oxford: Oxford University Press, 1976), especially chapter 8: »Fichte, Machiavelli, Pestalozzi«, pp. 169–208.

8 See Jens Baggesen's account to Sophie Baggesen on December 10, 1793 of the meeting between Fichte and Pestalozzi: »Apart from clearing up many things with our Lavater – I have become close, inward and trusted friends with two of the greatest, most sublime and admirable mortals: Fichte, the author of the *Critique of Revelation* and the *Contributions*, and of a work even more important than these two – and Pestalozzi, the author of the magnificent folk book *Leonhard and Gertrud*, an inexpressibly interesting, worthy, free, active and humanly pleasant man. [...] Here [in Richterswyl] we three: Fichte, Pestalozzi and I passed the evening in intimate philosophical, moral and cosmopolitan conversations of the most Socratic, theistic and Kantian kind.« FG, ed. Erich Fuchs (vol. 6.1, p. 28).

9 Johann Friedrich Herbart (1776–1841) was Fichte's student in Jena in 1794, and as we saw in the Introduction above, had replied to Fichte's »Question for a Mathematician« from the *Über den Begriff der Wissenschaftslehre* about the nature of straight and curved lines (cf. Herbart's letter to Fichte, October 1, 1795 (GA III/2: 411–415).

10 After a visit to Pestalozzi in 1798 Herbart was inspired to write his book on the importance of Pestalozzi's theory of intuition: *Pestalozzi's Idee eines ABC der Anschauung* (Göttingen: J.F. Rower, 1804). Herbart later became Kant's successor in Königsberg in 1809.

thinking of the student.[11] Indeed, Fichte ultimately believed Pestalozzi's system was even more significant than its creator had realized, and that generations of students educated in accordance with this method would later be in a better position to understand the aims of the *Wissenschaftslehre*.[12]

However, what was education (*Erziehung*) for Fichte? In the 1796 *Naturrecht* Fichte explains that human beings can only develop self-consciousness properly if they continually receive external »incitements« (*Aufforderungen*) to act freely and develop conscious relationships. Precisely this »incitement to self-activity [...] is what we call *education*.«[13] Like many other thinkers at the time, including Goethe, Fichte considered that the best method and textbook on mathematics was still the *Elements* of Euclid, and even believed that following his exact deductive method could aid in the comprehending of philosophy itself.[14] Accordingly, Fichte was convinced that learning mathematics and geometry were some of the best subjects for developing rational and logical thinking. His unpublished *Aphorismen über Erziehung aus dem Jahre 1804* are to be read in

11 Cf. Felix Klein and Robert Hermann, *Development of Mathematics in the 19th Century*, pp. 115f.

12 See Fichte's comments on Pestalozzi in *Der Patriotismus und sein Gegentheil* (1807): »Pestalozzi's thought is infinitely more and greater than Pestalozzi himself. [...] The soul of Pestalozzi's life was love to impoverished and neglected people: his love was so blessed that he found more than he sought, the sole remedy for the whole of humanity. That he likewise found the sole means for educating a generation that is capable of understanding rational science [...] reflected light from him can be shed on the actual goal of the *Wissenschaftslehre*.« (SW XI: 267). Cf. Fichte's similar comments about the *Wissenschaftslehre* and Pestalozzi in the *Staatslehre 1813* (SW IV: 590).

13 *Grundlage des Naturrecht* (GA I/3: 347). Cf. Alois K. Soller. »Nationale Erziehung und sittliche Bestimmung« *Fichte-Studien* 2 (1990): 89–110. We know from Fichte's *Nachlass* that he had already penned ideas on the correct method of educating children as early as August 1789 in his so-called *Tagebuch zur Erziehung*. Here we find scattered educational thoughts on the children of the Ott family in his care as well as the »most striking mistakes in education.« (cf. GA II/1: 141–198).

14 For example, Goethe praised Euclid's *Elements* as a preparation for philosophy in his *Maximen und Reflexionen*: »Geometry is here conceived in its first elements, as presented by Euclid [...] And therefore it is the most perfect preparation, indeed, introduction, to philosophy« (no. 354). (München: C.H. Beck, 1989), p. 412.

this light. According to Fichte, after instruction in the ancient languages, Euclidean geometry and its method are the next best tools for developing thought, and therefore constitute the second main component of the curriculum for children: »The pupil should actually incessantly work upon the inner spirit and character of his intellectual education, without which all external skill and knowledge has no value. Acquire everything through one's own spiritual force, never learn it mechanically. [...] Hence the study of mathematics – preferably according to Euclid himself or his method – should be the second main branch of the actual instruction.«[15]

In these *Aphorismen* Fichte suggests that early childhood education should above all focus on two things: (1) Strengthening the imagination and making it free and independent, and (2) Training the intellect to follow a lawful and logical course. Yet Fichte also warns against the dangers of commencing too early with a speculative and systematic study of the sciences.[16]

Berlin University and the Reden an die deutsche Nation

These aphoristic reflections gained a concrete theatre of action when Fichte was asked in 1807 to draw up a curriculum or pedagogical plan for the new Berlin University.[17] Fichte's plan also discusses the teaching of mathematics, and he was of the opinion that pure mathematics should be already taught at the lower school. However, at the university level he thought it beneficial to create a »*Comité*« of the most able mathematicians and then to select one of the mathematicians themselves to be responsible for the plan and direction of the course of study.[18]

In 1807 we likewise find what is perhaps Fichte's most detailed thoughts on Pestalozzi's method and the teaching of mathematics. This is

15 Aphorism no. 5 (GA II/7: 19f.).

16 »In other words: transcendentalism of every kind, even the slightest indications, does not belong to the subject of instruction« (ibid., p. 20). In Aphorism no. 7 of the same text Fichte also briefly mentions the possible benefits of arithmetic in helping to stimulate the imagination of children.

17 For an overview of Fichte and his connection to the mathematical education and reform of the time in Berlin, see Hans Niels Jahnke: *Mathematik und Bildung in der Humboldtschen Reform* (Göttingen: Vandenhoeck & Ruprecht, 1990).

18 See Fichte remarks in point §24 of the 1807 *Deducirter Plan einer zu Berlin zu errichtenden höhern Lehranstalt* (GA II/11: 111).

because at this time Fichte procured a copy of Pestalozzi's masterpiece *Wie Gertrud ihre Kinder lehrt*[19] and made an in-depth study and commentary on it.[20] Fichte was impressed with the educator's instructions for how language relates to concrete objects, and noted that this was also »an essential task for the teacher of the *Wissenschaftslehre* himself; the relationship of language to things.«[21] In his commentary, Fichte paid especial attention to Pestalozzi's examples for teaching mathematics, and agreed that the art of measuring presupposes a so-called »ABC of intuition«; this was a fresh and innovative art: »the art of intuition as a new art. / highly fortunate [expression]!»[22] Although Fichte initially took issue with a number of indications, particularly that the square and not the curve should be the foundation of learning geometry, after further study he eventually came around to Pestalozzi's view.[23] Like Pestalozzi, Fichte above all emphasized that counting should be first learnt in a practical manner, since it »facilitates the *free drawing of the line* in empty space.«[24] But for his own stu-

19 Pestalozzi's book *Wie Gertrud ihre Kinder lehrt* was first published in Bern and Zurich in 1801.

20 See Fichte, *Bei Lektüre von Pestalozzis Buch ›Wie Gertrud ihre Kinder Lehrt‹* (GA II/10: 431–457).

21 Ibid., p. 431.

22 Fichte, Ibid., p. 450. Cf. Widmann's thoughts on Pestalozzi's view of intuition: *Die Grundstruktur*, p. 10.

23 »My above criticism has now completely disappeared.« Fichte, Ibid. Cf. Fichte's further comments in GA II/10: 447–450. Fichte's student Herbart also disagreed with Pestalozzi that the square (quadrate) should be the foundation of the mathematical theory of intuition. Instead, in his *Pestalozzi's Idee eines ABC der Anschauung*, Herbart wrote that he thought that »the triangle [...] lies at the basis of the whole of geometry« (p. 108).

24 »*Grundsatz*: counting has to be learnt with objects that can be freely joined and calculated so that it becomes something *practical*: Here the spirit already elevates itself higher. It occupies itself early with multiplying and *dividing*. What do I say about this? It is entirely correct. However, the pure analysis of number must be learnt through the analysis that I outlined above. N.B. / In the table with strokes which disappear or which make itself. Is there another exercise for the power of intuition for the creation out of nothing, for filling empty space? It has to be carried out early; and probably even before measuring: – It facilitates the *free drawing of the line* in empty space, since it has what is simpler, the *point*. I therefore suggest also first counting *points*.« Fichte, *Bei Lektüre von Pestalozzis Buch ›Wie Gertrud ihre Kinder Lehrt‹* (GA II/10: 453).

dents Fichte wanted to develop Pestalozzi's indications on intuition even further: »For higher education I have to find a rational art (dialectic) of this kind so that I can furnish a similar basis for it as the square.«[25]

It was this study that formed the basis of Fichte's positive remarks on Pestalozzi in the ninth and tenth addresses of his *Reden an die deutsche Nation* (1808). Here Fichte repeated his view of the importance of training the child's faculty of intuition and even proposed setting up a national education on the basis of Pestalozzi's conceptions.[26] Yet Pestalozzi had only supplied a beginning, and it is incumbent upon the teacher of this new national education to surpass this »procedure in which we only perceive empty geometrical forms«. That is to say, to pass from »obscure intuitions to clear concepts«, i.e. initially into the higher sense-free domain of mathematics, and then ultimately to the teachings of the *Wissenschaftslehre*.[27]

Finally, in the great Swiss geometer Jacob Steiner (1796–1863)[28] we seem to have a remarkable example of the effectiveness of Pestalozzi's method of cultivating geometric intuition, though this is contested in some quarters.[29] Jacob Steiner expressly followed Pestalozzi's intuitive method and Socratic emphasis on the self-generation of all geometrical knowledge to such a extent that he forbid the use of geometric figures in his classes. The student was supposed to be so active in his listening that no material or empirical image was needed to produce a clear mental picture.[30] Fichte raises this same difficultly for the philosophical student in

25 Ibid., p. 451.

26 »We have to clearly render the guidance of the pupil, first his sensations, then his intuitions [...] this is the first central part of the new German national education. With regard to the training of intuition, we have a useful introduction from Pestalozzi.« Fichte, *Reden an die deutsche Nation*, Tenth address (GA I/10: 227) Cf. the ninth address as well (ibid., p. 217).

27 Ibid., pp. 217–220.

28 A poor Swiss farm boy, Jacob Steiner became one of the greatest geometers in history. Jacob Steiner had attended one of Pestalozzi's schools. Cf. Felix Klein and Robert Hermann, *Development of Mathematics in the 19th Century*, p. 115.

29 Felix Klein and Robert Hermann find it difficult to imagine that Jacob Steiner's faculty of intuition could have benefited from Pestalozzi's method, even though Jacob Steiner had been a pupil of Pestalozzi. Yet they do attribute some effectiveness to Pestalozzi's emphasis on the Socratic method. (Cf. *Development of Mathematics in the 19th Century*, pp. 115f.)

30 Ibid.

the fourth lesson of the *Sonnenklarer Bericht*. However, unlike in geometry, the philosophical novice does not even have any recourse to external aids: »The geometer has, at least, his lines and figures on the blackboard, and thus concentrates his intuition; but the student of the *Wissenschaftslehre* has nothing whatsoever except himself and his free reflection.«[31]

b. The Path from Geometry to the Fichtean System of Philosophy

We have seen how Fichte was impressed by Pestalozzi's method for developing intuition. But how exactly is the activity of *doing geometry* a preparation for Fichte's system of philosophy? In Fichte's eyes, carrying out geometric exercises helps to train two essential rational faculties of the mind: 1) the force or power of the imagination (*Einbildungskraft*), and 2) the power of intuition (*Anschauungsvermögen*). In his 1794/95 Jena lectures on the duties of the scholar Fichte outlines the different methods for developing independent thinking. The emphasis is on abstract thoughts, i.e. philosophical thoughts, in order to elevate oneself to »clear consciousness« and spirituality (*Geistigkeit*). Geometry is reckoned among these methods: »Among the rules of reason I can recommend geometry, which if studied with diligence is the most excellent education for the imagination.«[32]

What is the imagination? For Fichte, everything that occurs in our consciousness must have at basis a spiritual or mental image (*Bild*) that is projected by the I. In this way it becomes an image occurring within our consciousness: »We call the ability to project these images the power of the imagination.«[33] For according to Fichte, this imaginative activity of »producing for oneself« the figures and constructions of geometry trains us in sense-free thinking. We learn to abstract from the external body

31 Fichte, *Sonnenklarer Bericht* (GA I/7: 252f.).

32 *Die am Ende des Winters [Sommers] 1794 von Fichten Gehaltene Schlussvorlesung als Beylage der Verantwortung* (1797) (GA I/4: 413). Fichte develops his theory of the force of the imagination further in the ninth lecture of the *Die Thatsachen des Bewußtseins* (1813) (SW IX: 490–503).

33 »Das Vermögen diese Bilder zu entwerfen nennen wir Einbildungskraft.« *Über Geist, u. Buchstaben in der Philosophie* (GA II/3: 309). See too Csech's comments on imagination and geometry in: »Von der Reichweite mathematischen Denkens bei Fichte und Novalis«, *Theologie und Philosophie. Vierteljahrschrift* 70 (1995) pp. 53f.

(*Körper*) of geometric objects such as empirical mathematical triangles that are constructed with a straightedge and compass (*Zirkel und Lineal*), up to the spiritual image (*geistiges Bild*) of the triangle or line that is universally valid and holds for all cases. This spiritual image can only be inwardly intuited (*innerlich anschauen*), and the power of the imagination assists us in its creation.[34] Fichte defines the power of the imagination as an intermediate member between what is real and above the real; it is the member that conditions the »objectified empirical intuition itself.« – However, to arrive at this spiritual or original image we have to recognize the importance of human *freedom*, for the »empirical image of the imagination first expresses its original image (*Urbild*) when it is represented through freedom.«[35] All the constructions of geometry, therefore, require the imagination, even the simplest ones such as drawing a line: for it is »the free self-activity of the imagination that does the drawing.«[36]

In lesson 6 of the *Sonnenklarer Bericht* from 1801 Fichte especially speaks about the educative value of mathematics and geometry for creative thinking. He recommends mathematics as a worthwhile subject for the »training of the mind, and geometry for the development of the imagination«, on account of its focus on direct and immediate intuitions. Its qualities of exactness, attention, dexterity (*Gewandtheit*) and firmness (*Festigkeit*) all furnish the mind with a solid foundation for leading the student to the *Wissenschaftslehre*.[37] For it is only once we have acquired the ability to abstract from all empirical sensations and qualities and to properly intuit and construct the objects of geometry that we are prepared for the task of inwardly intuiting our own I. Not the sensible intuition of the I in ordinary self-consciousness, but the pure intuition of the I in *intellectual* intuition. For Fichte, the intuitions of geometry and the *Wissenschaftslehre* are pure, i.e. non-empirical, and for this reason the attaining of pure intuitions in geometry helps us to arrive at the transcendental standpoint in philosophy. That is to say, at the difficult starting point of

34 Cf. *Die am Ende des Winters [Sommers] 1794 von Fichten Gehaltene Schlussvorlesung als Beylage der Verantwortung* (1797) (GA I/4: 412f).

35 *Die Thatsachen des Bewußtseins* (1813) (SW IX: 496).

36 *Zurich Wissenschaftslehre* (GA IV/3: 34).

37 GA I/7: 253. See too Fichte's letter to Joseph Rückert, Sept. 1802 where Fichte gives Rückert three small mathematical examples to help him »ascend to the highest *Evidenz*.« (GA III/5: 148f.).

intellectually intuiting the I of the *Wissenschaftslehre*: »It will furthermore become evident that the *I* is originally an intuition as well, though not a sensible intuition, but an *intellectual* intuition. Thus, the *Wissenschaftslehre* is in possession of a pure intuition from the start, just as much as geometry, and all the intuitions of geometry are grounded in the intuitions of *Wissenschaftslehre*.«[38]

The Privatissimum für G.D. (1803)

The idea that the geometrical examples in the *Sonnenklarer Bericht* were to be used as a preparation for the *Wissenschaftslehre* and for philosophy in general was not an abstract or utopian theory to Fichte. For in April 1803 he used this text as the basis of some private lessons to an adult student called G.D. (hence the name of the text: *Privatissimum für G.D.*).[39] Here Fichte takes up again the idea of geometry as a basis for philosophy and expounds quite a large array of examples. His method is entirely in the Socratic spirit of Pestalozzi, since philosophy is *not* a matter of »learning things« but the formation and training of a »certain mode of thinking« *(Übung einer gewissen Denkart)*.[40] The teacher's primary role is to »stimulate the scientific spirit«, yet nobody can carry out this thinking in his stead. Someone has only attained the domain of science »to the extent he has inwardly acquired this insight.«[41] Fichte strove to point out the types (or »forms«) of knowledge and cognition employed by empiricism, mathematics and philosophy, using the former to awaken and »stimulate« the latter. For example, empiricism or empirical knowledge is chiefly concerned with what is »variable« and changeable. The empiricist can study a whole host of trees to learn what a tree is, and of course »not a nominal tree, but a *real* one, which [has] a certain construction and internal organisation in space.« – What is imperative for Fichte is invariable and unchangeable knowledge, i.e. how the empiricist could have known with his

38 *Zurich Wissenschaftslehre* (GA IV/3: 34).

39 The text is published in: GA II/6: 325–373. The student was most likely Count Dohna (Graf Dohna = G.D.), but his identity has not been fully confirmed (see the editors' remarks in the Fichte *Gesamtausgabe*: GA II/6: 327f.).

40 Ibid., p. 329.

41 Ibid., pp. 332, 353.

»very first tree that it was a tree and not his *nose*.«[42] Likewise with the discipline of mathematics. – Mathematics is focused on »particular« forms such as a triangle, which is composed of two right angles; whereas philosophy is concerned with universal forms of knowledge, with the form of cognition *as such*. Yet mathematics also seems to have a certain »invariable« knowledge of postulates and constructions, such as the *rules and knowledge* of the triangle: »For what is invariably identical is not the triangle itself, but my knowledge of the triangle.«[43] Hence, learning to attain this mode of invariable knowledge in mathematics will assist us in raising ourselves to the cognitive standpoint of the *Wissenschaftslehre*. Here we see again Fichte's mathematical Platonism – for the absolute and unchangeable is the foundation for the changeable and empirical: at this higher standpoint the »variable, i.e. differing forms, views and shapes of cognition are explained out of absolute invariability and identity.«[44]

As proof that the geometric examples from the *Sonnenklarer Bericht* were not just reserved for beginner students of philosophy like G. D., Fichte included this same example of invariably identical knowledge in his *Antwortschreiben an Herrn Professor Reinhold* (1801). Fichte tries to explain to Reinhold that the path leading to the *Wissenschaftslehre* first traverses an understanding of the geometric and scientific notions of self-evidence and construction, and that all certitude in a philosophical science rests on a self-aware thinking that »knows that it knows.« Choosing again the specific example of constructing the third side of a triangle, Fichte illustrates how our invariable *knowledge* of an infinitely repeatable (universal) triangle relates to the particular and empirical instance of a triangle, and hence yields a valuable lesson for philosophy.[45] In a text from

42 Ibid., p. 331.

43 Ibid., p. 334.

44 Ibid., p. 335.

45 Fichte, *Antwortschreiben an Herrn Professor Reinhold* (1801): »The ground of all certitude, of all knowledge in life, and all self-evidence in science, is the following: We posit the totality (*Allheit*) in and with the particular. [...] I have my reasons to prove this by means of an example from geometry. Construct an angle, that is doubtlessly an included angle, with determined straight sides. [...] You close this angle through a third side. [...] whereby in thinking as thinking the infinitely repeatable triangle becomes *kat exochin this* triangle.« (GA I/7: 292). See too the examples of geometry on pages 292ff.

the same time Fichte summed up his belief that students of the *Wissenschaftslehre* would profit from a prior study of geometry: »With regard to its contents, mathematics is the only completely scientific undertaking that exists. In other words, I would hope that no one begins a study of the *Wissenschaftslehre* without first obtaining a clear insight into the foundation of the *immediate self-evidence* and *universality* of mathematical postulates and theorems.«[46]

2. Further Similarities and Differences between the *Wissenschaftslehre* and Geometry

> Because the *Wissenschaftslehre* is mathematics, it also has the distinctions of mathematics.
>
> J.G. Fichte.[47]

What are the distinctions or advantages of mathematics that Fichte has in mind here? Or to put it in another way: what exactly is the relationship between the *Wissenschaftslehre* and geometry? In this section we will investigate this relationship by looking at five specific similarities and differences between these disciplines. We will principally concentrate on two texts: the fourth lecture of the *Zurich Wissenschaftslehre* from 27 February 1794 (GA IV/3: 32–37) and the *Ankündigung* from January 1801 (GA I/7: 153–164).

Naturally, in the preceding chapters we have *already* seen a large number of significant similarities and differences between geometry and the *Wissenschaftslehre*. For example, in Chapter Three we discussed intuition and construction and outlined their role in the cognition of both these disciplines. In addition, we showed the extent to which Fichte's basic *Grundsatz* is modelled on and inspired by the postulates of geometry. Furthermore, Chapter Four underscored the manner in which the fundamental elements of geometry, such as points, lines and especially the activity of drawing the line, become philosophically appropriated by the *Wissenschaftslehre*. And finally, in the above section we detailed the pedagogic value of geometry to Fichte – how geometry teaches us to think complex thoughts and therefore aids and stimulates the power of the

46 *Ankündigung* (GA I/7: 160).

47 *Ankündigung*, January 1801 (GA I/7: 160).

imagination. In all truth, all of the above topics – especially the notions of postulates, construction and intuition – illustrate the deepest and most crucial similarities and differences between the *Wissenschaftslehre* and geometry. We will not repeat these considerations here, but focus instead on another five specific or distinctive *characteristics* singled out by Fichte in the *Zurich Wissenschaftslehre* and the *Ankündigung*.

Accordingly, in these texts we shall see that that there exist three similar characteristics between geometry and the *Wissenschaftslehre* for Fichte: (1) Self-evidence; (2) Determinacy, and (3) irrefutability. On the other hand, the differences between them are due to (4) the more formal nature of geometry, and (5) their unequal relationship to freedom.

a. Similarities: Evidenz, Determinacy and Irrefutability

Evidenz (Self-Evidence)

Somewhat surprisingly perhaps, the *Zurich Wissenschaftslehre* has little to say about »self-evidence« *as such* in its discussion of geometry, but lays more stress on the related concept of »*Gewissheit*« or certainty, particularly in the context of Fichte's newly discovered *Grundsatz* of philosophy. As mentioned in Chapter Two, in the *Zurich Wissenschaftslehre* Fichte stated that the *Grundsatz* of philosophy should be: »certain in itself, *immediately* and without any proof; it should be clear without any explanation.«[48]

In contrast, the *Ankündigung* is much clearer about the importance of self-evidence. In fact, in its treatment of mathematics it declares that the *first* advantage or distinction that the *Wissenschaftslehre* shares in common with geometry is its *Evidenz* (self-evidence); they both possess the »same sort of immediate self-evidence.«[49] However, what is *Evidenz*? For Fichte, self-evidence is an immediate kind of *mental clarity*. Like other rigorous thinkers before him[50], Fichte recognized that when one grasps

48 Lecture 3 (GA IV/3: 30). The term *Evidenz* also plays a negligible role in the other texts from the same period: the *Eigne Meditationen* and *Practische Philosophie* (1793/94).

49 Fichte, *Ankündigung* (GA I/7: 160).

50 For example, Pascal too saw one of the intellectual benefits of geometry for philosophy in a comprehension of the transparent self-evidence of first principles: »D'où l'on voit que la géométrie ne peut définir les objets ni prouver les principes; mais par cette seule et avantageuse raison, que les uns et les autres sont dans une

something with cognitive self-evidence, then one is free of all deliberating and wavering. Self-evidence grants a surety and certitude in one's philosophical convictions, and one is struck by an obviousness that »forces« itself upon one and does not require any proof. We also saw in Chapter Three that self-evidence is one of the hallmarks of the axioms and postulates of geometry. Fichte is convinced that this quality of self-evidence also applies to the first principles of his system: Anyone who examines the *Wissenschaftslehre* will be »surprised by its immediate clarity and necessity. He will be unable to see things in any other way than the way in which he does see them.«[51]

This quality of self-evidence has a strong rapport with the *written* and *oral* elaboration of a system. In 1800 Fichte was particularly concerned about the hitherto misunderstood reception of his philosophy on account of his earlier published texts, especially the *Grundlage*.[52] Unlike his oral lectures where any misunderstandings could be detected and eliminated on the spot, Fichte agreed that his earlier over-technical expositions perhaps lacked lucidity and sometimes precluded comprehension.[53] Fichte even conjectured that with regard to the *Sittenlehre* and *Naturrecht* it might simply be impossible to present the applied conclusions of his system »with the same degree of *self-evidence* that one can easily give to the first principles themselves.«[54] In the *Ankündigung* he therefore proclaimed that his forthcoming presentation would be his most lucid ever, one that »does not require any further assistance, and I hope that it will not need any other more lucid presentation.«[55] This announced presentation was the *Neue Bearbeitung der Wissenschaftslehre 1800*. Hence at this

extrême clarté naturelle, qui convainc la raison plus puissamment que le discours.« Blaise Pascal, »De l'esprit géométrique«, essay from 1657/58, in: *De l'esprit géométrique – Ecrits sur la Grâce et autres textes* (Paris: Flammarion, 1999), p. 77.

51 Fichte, *Ankündigung* (GA I,7: 160)

52 Ibid., pp. 153f.

53 See Fichte's *pro memoria* to the Royal Cabinet, January 3, 1804 (GA III/5: 223).

54 Fichte suggested that the incomprehension of the *Sittenlehre* and *Naturrecht* might be attributed to people »skipping the introductions.« He also says that only the two Introductions from 1797 and their accompanying *Neue Darstellung* from the same time seem to have been better understood. See the *Ankündigung* (GA I/7: 154).

55 *Neue Bearbeitung* (GA II/5: 325). Cf. the *Ankündigung* for the same sentiment (GA I/7: 164).

time Fichte was still under the impression that geometry and the *Wissenschaftslehre* could have *similar presentations* and methodologies, one with the lucidity and self-evidence of Euclid's *Elements*. As he writes in the first chapter of the *Neue Bearbeitung*: »The procedure is entirely the same as in *mathematics*.«[56] However, as we saw, Fichte soon abandoned the writing of a new version of his system based on an explicitly geometric theorem, and turned to other presentations, such as an exposition of the *Wissenschaftslehre 1801/02* that did not start with any postulate or *Grundsatz*, but with a description of »*Wissen vom Wissen*« (cognition of cognition).[57]

Despite abandoning the written exposition of the *Neue Bearbeitung*, the theme of *Evidenz* did not disappear from Fichte's system, but assumes even greater importance in the post-Jena lectures.[58] For now Fichte makes a distinction between two different types of self-evidence: factual self-evidence (*faktische Evidenz*) and genetic self-evidence (*genetische Evidenz*). As the names suggest, the factual is the self-evidence associated with the world of external facts and traditional sciences, whereas genetic self-evidence purely refers to the internal rational domain of cognition and the activity of knowing itself. This latter type of self-evidence is associated with *genetic* cognition, because we not only penetrate into and possess knowledge of the origin of the object itself, but also of the attendant mode of knowledge. Moreover, there is a definite priority to the two difference types – the factual has to be derived from the genetic and not vice-versa: »Everywhere there are none other than factual self-evident principles in the real-world sciences, and the *Wissenschaftslehre* in contrast wishes to introduce a thoroughly genetic kind of self-evidence, and from which the factual self-evidence is then to be derived.«[59]

These words are from the second series of lectures on the *Wissenschaftslehre 1804*. Fichte had posted an announcement for this series of

56 GA II/5: 335.

57 *Wissenschaftslehre 1801/02*, §3 (GA II/6: 139).

58 See Fichte, *Wissenschaftslehre 1804/II* (GA II/8: 2–421); see lecture 3 in particular, 19th April 1804 (GA II/8: 37–53, esp. pp. 46ff.); and the *Erlanger Logik* (GA II/9: 125). In *Ueber das Verhältniss d. Logik zur Philosophie oder tr. Logik* (1812), Fichte also mentions how truth is not made but is presented in genuine self-evidence, just like in the *Wissenschaftslehre* (SW IX: 188).

59 Fichte, *Wissenschaftslehre 1804/II* (GA II/8: 47).

lectures using the following words: »The undersigned person proposes a continuation of the oral lectures on the *Wissenschaftslehre*, i.e. of the complete solution to the riddle of the world and consciousness with mathematical self-evidence.«[60] Echoing his words in the *Antwortschreiben an Herrn Professor Reinhold* (1801), in the *Wissenschaftslehre 1804/II* Fichte describes his »genetic« theory of self-evidence as a knowledge that *knows itself and is based on itself* – an archetypal-essential cognition (*das urwesentliche Wissen*).[61] – According to Fichte, Kant's critical philosophy also only possesses »factual self-evidence«, i.e. only the »appearance« of self-evidence and not true genetic self-evidence.[62] This is because the Kantian critiques are merely based on a transition between the sensible and supersensible words.[63] Thus, for Fichte, this new kind of genetic self-evidence is what distinguishes the Fichtean theory of cognition from all other philosophies.[64] And since it belongs among the other sciences, even the self-evidence of geometrical and mathematical principles is still only a kind of factual self-evidence for Fichte. This is because the arithmetician and geometer initially do not possess the same degree of insight into the origin and *genesis* of the elements of their sciences (e.g. numbers and space respectively) as the teacher of the *Wissenschaftslehre*. However, as we seen, Fichte notes that this may change as the former gain a greater insight into the nature of mathematical cognition: »[Mathematics] must confess that its principles admit nothing more than factual self-evidence, regardless of the fact that they will become actively genetic as we proceed.

60 »Der Unterschriebene erbietet sich zu einem fortgesetzeten mündlichen Vortrage der *Wissenschaftslehre*, d.h. der vollständigen Lösung des Räthsels der Welt und des Bewußtseins mit mathematischer Evidenz.« Fichte, »Ankündigungen zu den Vorlesungen in Berlin« (GA I/8: 17). Cf. Fichte, *Die Wissenschaftslehre. Zweiter Vortrag im Jahre 1804*, eds. R. Lauth, J. Widmann and P. Schneider (Hamburg: Meiner, 1986), p. 2.

61 Fichte, *Die Wissenschaftslehre 1804/II* (GA II/8: 48).

62 Ibid.

63 Ibid., pp. 44f. In 1813 Fichte also remarked that he did not think much of Kant's and Mendelsohnn's earlier essays on mathematical evidence in metaphysics, which won first and second place respectively in the 1764 competition set by the Prussian Academy of Sciences. Fichte said: »There has recently been a lot of talk in vain about self-evidence. Academies have even given prizes for it.« In: Fichte, *Ultima Inquierenda*, p. 33.

64 See Vuillemin, *La philosophie de l'algèbre*, pp. 112–120.

For let the arithmetician as mere arithmetician simply tell me how he is able to elicit a solid and permanent number »one«; or let the geometer explain what fixes and holds his space for him while he draws a continuous line through it; and whether these and all the other ingredients that he requires for the possibility of his genesis are given in any other way than through factual intuition? Of course, this does not in any way constitute an objection to mathematics. For mathematics as such can and should not be anything else.«[65]

Determinacy and Irrefutability

According to the *Ankündigung*, a second distinction that mathematics and the *Wissenschaftslehre* share in common is *determinacy* (*Bestimmtheit*). What Fichte means here is that the names, »signs« or technical terms of his system are of secondary importance, compared to the primary fact that everyone can have exactly the *same* immediate intuition. This relates to the universal human nature of Fichte's system, and his genuine belief that the *Wissenschaftslehre* relates to reality. We may alter or modify the terms of this philosophy, but if it has its basis in reality then the elements of the system will be accessible and identical for all healthy rational beings – especially the theory of intellectual intuition. This strikes at the core of Fichte's philosophy, and is one of the reasons why he was unperturbed about producing so many different versions of the *Wissenschaftslehre*: »The *Wissenschaftslehre* has the same sort of thoroughgoing *determinacy* as mathematics. It does not matter what sign it attaches to its object – whether it calls it »I« or »Not-I«, X or Y. The sign itself is nothing; the *Wissenschaftslehre* is concerned only with what occurs within the immediate intuition of everyone. [...] What is encountered within intellectual intuition is the same for all reason, and its remains unalterably the same for every rational being, just as long as he remains rational.«[66]

Thus, like mathematics, the objectivity and immediacy of the *Wissenschaftslehre* permits it to change its languages and signs, but ultimately it describes experiences and constructions that must be freely *generated and produced by everyone themselves*. Because we have produced them, we

65 Fichte, *Die Wissenschaftslehre 1804/II* (GA II/8: 48f.). Fichte, *The Science of Knowing*, English translation by W. Wright, p. 38 (translation modified).
66 GA I/7: 160f. English translation by Daniel Breazele in: IWL, pp. 196f.

have participated in their genesis, and therefore we know and are convinced of their objective reality. Paradoxically perhaps for many people, the form is subjective because it is produced by ourselves, but the content is objective because it is the same for every rational being. The intuitions of geometry and the intellectual intuition of the I cannot be presented *empirically as such*, yet they are not some kind of internal mental illusion, but realities for the philosopher. In the *Ankündigung*, Fichte underscores the universally rational and non-delusional nature of his claims. This emphasis on healthy human reason has its background in a dispute from the time, and is further explained by Fichte in a text entitled: *An das philosophische Publikum*.[67] It was a public reply to criticisms that Fichte's theory of self-consciousness was nothing more than a kind of »psychology«, because it apparently reduces mental acts into »facts of consciousness«, and consequently the *Tathandlung* of the pure I is a »psychological illusion.«[68] Fichte dismisses this charge of »psychology« or the negative subjectivism of his theory, and argues instead that his system's closest parallel is the wholly »objective« science of geometry, because both are ultimately grounded in non-empirical and non-perceptual elements.[69]

Finally, the third advantage or distinction mentioned by Fichte in the *Ankündigung* that the *Wissenschaftslehre* shares in common with mathematics is *irrefutability* (*Unwiderlegbarkeit*). That is to say, the *Wissenschaftslehre* is incapable of proof, and as a result we cannot argue or dispute its contents. Like with the axioms of geometry, this applies to Fichte's first principles and to intellectual intuition. Why is the *Wissenschaftslehre* irrefutable? For Fichte, all refutations rest on conceptual principles and assertions. However, like geometry and mathematics, the cognition of the *Wissenschaftslehre* is grounded in intuitions. One could simply deny the intuitions of the *Wissenschaftslehre*. But as Fichte further argues, denying is not the same as refuting, and thus this would be equivalent to denying geometry. Hence, ultimately both geometry and the *Wissenschaftslehre* cannot be disputed and are irrefutable in this sense.[70]

67 In a public letter from November 1, 1800 to the *Intelligenzblatt der Litteratur-Zeitung* (GA I/6: 457–460).

68 Ibid., p. 458.

69 Ibid. Fichte likewise dismissed the charge of psychology in the *Sonnenklarer Bericht* (GA I/7: 222).

70 *Ankündigung* (GA I/7: 156, 160).

These three claims of self-evidence, determinacy and irrefutability are summed up in a passage in lesson three of the *Sonnenklarer Bericht* (1801), which as we know was conceived parallel to the *Ankündigung*. Here Fichte stresses that both the deductive method of geometry and the *Wissenschaftslehre* progress step by step in clear and lucid knowledge, and both apply and are common to all rational beings: »Author. Precisely in the same manner in which geometry proceeds, where each new proposition adds something new to the proposition previously established, the necessity of which something new, is proved also, only in intuition. In fact, such must be the method of every real progressive science which does not turn around in a circle. [...] Thus, I believe, that I have given you a sufficiently clear conception not only of the object of the *Wissenschaftslehre* in general, but also of its procedure and the grounds of this procedure. This science constructs the entire common consciousness of all rational beings absolutely *a priori*, in its fundamental characteristics, precisely as geometry constructs absolutely *a priori* the universal modes of limiting space on the part of all rational beings.«[71]

b. Differences: Formal Language and Freiheit

As we have seen in the preceding chapters, there are numerous differences between geometry and the *Wissenschaftslehre*. The first postulates of geometry are more limited than Fichte's first postulate and are grounded in the latter; the figures of geometry require sensible and intelligible intuition, the *Wissenschaftslehre* requires intellectual intuition; geometry employs technical and free constructions, Fichtean system generates free infinite constructions. Thus, although Fichte saw profound similarities between the disciplines of the *Wissenschaftslehre* and mathematics, he also warned against confusing or conflating them, or trying to aim at an absolutely mathematical philosophy. The two disciplines may share the same certainty and aim at self-evidence, but ultimately they have different goals, symbols and subject matter. Hence, as we saw in Chapter One, anyone who literally seeks to turn the *Wissenschaftslehre* into geometry or the philosopher into a geometer, would be either a »pedant« or a »confused

71 GA I/7: 223.

and confusing mind.«[72] Hence, the two sciences are different, and for Fichte the first principles of mathematics are in fact derived from the *Wissenschaftslehre* and not vice-versa. They require a genuine philosophy of mathematics to explain them in their entirety. As he says in the *Wissenschaftslehre 1804/II*: »Of course, it is not our business to blur the boundaries of the sciences; but it has to be simply recognized that [...] the principles of the possibility of [mathematics] lie in another higher science.«[73]

In the last chapter we also saw that one of the major differences between geometry and the *Wissenschaftslehre* concerns the purpose and scope of each science. The »purpose« (*Zweck*) of geometry is systematic limiting, it is »to exhaust the possibility of limiting of space within One universal space (*Einen allgemeinen Raumes*).«[74] This notion of geometry as a science of rules for the limiting of space was already stated by Fichte ten years earlier in his *Über den Begriff der Wissenschaftslehre*.[75] However, our task is not to examine the functions of each science again, but to outline some more differences between the two.

In the *Ankündigung* Fichte briefly remarks upon another one of these differences. According to Fichte, the *Wissenschaftslehre* still does not possess the same *authority* as geometry.[76] That is to say, it is still not universally accepted as being on a par with the scientific exactness of geometry. The *Ankündigung*, however, does not dwell too much on the differences between mathematics and the *Wissenschaftslehre*, as its function was obviously to underscore more their affinities. However, in the *Zurich Wissenschaftslehre*, Fichte states that there is especially a difference in the *products* and *action* of geometry and the *Wissenschaftslehre*. The methods, or as he terms them here – the product and activity – are reversed or inverted in geometry and philosophy. In geometry we are presented with a method of acting, and we seek the products of this acting, i.e. the constructions of geometry. Geometry is therefore concerned with facts (*quid*

72 *Erlanger Logik* (GA II/9: 125).

73 Fichte, *Wissenschaftslehre 1804/II* (GA II/8: 47f.).

74 *Erlanger Logik* (GA II/9: 125).

75 »The task in general of limiting space according to a rule, or the construction in the same, is the first principle of geometry, and through this it is distinguished from the *Wissenschaftslehre*.« (GA I/2: 135).

76 Cf. GA I/7: 161.

facti), or factual self-evidence. In the *Wissenschaftslehre*, the product – the subject or empirical self-consciousness is given – and we have to seek the right lawful action (*quid juris*) to arrive at the transcendental standpoint. Thus, in this sense the methods in the two disciplines are different: »In *geometry*, the action is *given* and what has to be sought is the *product*. In the *Wissenschaftslehre*, the *product* is given, and what has to be sought is the action by virtue of which it is there. Geometry asks, *Quid Facti*? The *Wissenschaftslehre* asks *Quid Juris*? Thus, with respect to its inner essence, the method of proof employed in the *Wissenschaftslehre* is precisely the opposite of the one employed in geometry, and the external differences between geometry and the *Wissenschaftslehre* follow from this.«[77]

In the *Neue Bearbeitung der Wissenschaftslehre 1800* Fichte again highlights precisely this difference with regard to the activity of constructing in the two sciences: »In actual mathematics one only looks at the products of construction, whereas here [in the *Wissenschaftslehre*] one looks at constructing itself.«[78] This relates to another specific difference between geometry and the *Wissenschaftslehre*. In geometry there is *Notwendigkeit*, necessity, there are strict laws and rules for spaces and the elements of space, we cannot arbitrarily change them and invent our own rules. However, by means of the power of our imagination we are free to chose where to put a point, when and where to draw a line in space. Thus, philosophically speaking, in the *Wissenschaftslehre* there is pure *freedom*, whereas in geometry there is a limit to our freedom.[79] And of course, as a system of transcendental philosophy the *Wissenschaftslehre* was to be a philosophy of freedom *par excellence*.[80]

Moreover, geometry and mathematics in general, are expressed in formal or symbolic systems. They are expressed using figures and signs, and furnish us with »symbolic knowledge« (*symbolische Erkenntniß*)[81] or a *Ze-*

77 GA IV/3: 36. On Quid Facti and Quid Juris, see Kant, *Critique of Pure Reason* (B116–117/A 84), pp. 219ff.

78 GA II/5: 344.

79 Cf. Fichte, BWL (GA I/2: 135).

80 See Andreas Schmidt's 2004 study, which argues for the continuity of the *Wissenschaftslehre* as a »Philosophie der Freiheit«. – Andreas Schmidt, *Der Grund des Wissens: Fichtes Wissenschaftslehre in den Versionen 1794/95, 1804/II und 1812* (Paderborn: Schöningh, 2004, pp. 7-10.

81 *Platner Vorlesungen* 1796 (GA II/4: 123).

ichensystem, a system of signs. Here Fichte often compares this with the *Wissenschaftslehre*, with his famous distinction between the spirit and the letter: The living internal spirit is found in the philosophical activity of the *Wissenschaftslehre*, the dead letter in the formal external system. Of course it is a matter of degrees, the external system of geometry is more formal than the external system of the *Wissenschaftslehre*. Both of them try to express *Denkregel* and *Denkgesetze*. These conceptual laws or rules can be formulated in a precise system of formal signs. This formal system helps us to have an overview of the conceptual system, whose goal is the attainment of intuitive self-evidence.[82] Geometry has such a system, and Fichte thought that this should also be the case for philosophy. If we compare the *Eigene Meditationen* and *Practische Philosophie*, we see that his thoughts are sometimes articulated in a logical, algebraic or geometric fashion, i.e. in a formal system: »Attempt to represent everything hitherto in letters.«[83] And as Fichte says in the *Erlanger Logik* 1805: – in geometry our freedom is only *formal*, and does »not condition the content«, whereas in the *Wissenschaftslehre* the content is »living.«[84] However, in the *Eigne Meditationen* Fichte finally realises that his first principle (*Grundsatz*) must go beyond all formal principles in order to become an *absolute principle*: »The first principle remains the first principle. [...] A spark of light! I cannot make any progress with the purely *formal* grounding principle (*Grundsatz*); thus, it drives me endlessly in a circle and does not make possible the fact (*das Faktum*). – Therefore, I require an unconditioned, absolute [*Grundsatz*], a supreme unity.«[85]

And as Fichte stressed often enough, for example in the *Grundlage des Naturrechts* from 1796/97, the goal of the *Wissenschaftslehre* was *not* to be a mere formal science like logic or algebra, a so-called *Formular-Philosophie*, but a living or »*Realphilosophie*«, a system of real and genuine thinking, that is vastly different to the mere abstractness and emptiness of modern logic.[86] This is one of the dangers of a mere formalized language and science that is a characteristic of mathematics, and especial-

82 Cf. Csech, *Reichweite*, p. 45.
83 »Versuch das Ganze bis hie her in Buchstaben vorzustellen« (GA II/3: 63).
84 *Erlanger Logik* (GA II/9: 45).
85 *Eigne Meditationen* (GA II/3: 48).
86 Cf. Fichte, GNR (GA I/3: 316 ff.).

ly in branches such as algebra and arithmetic. But the same can also be said of geometry, if we focus only on the outer external signs and symbols of the points and lines, and neglect to pass to their inner intuitions. We saw that the drawing of the line can be grasped as an *image* of the inner agility of the mind. However, the image as such, the geometric symbol, is formal and unreal, compared to the animated force of the mind itself. As Fichte says in the *Neue Bearbeitung*, the »inner character of thinking« is alive, and should not be »grasped as an empty form.« Thus, »the line is only the path that the ray of force describes; it is not its intension, it is not *real*.«[87]

But just as we saw with certain passages in the *Eigne Meditationen* and the *Practische Philosophie*, the idea of a formal symbolic presentation of his philosophy also intrigued and was cherished by Fichte. This gave rise to one of his more cryptic comments at the very end the *Ankündigung* for the *Neue Bearbeitung der Wissenschaftslehre 1800* – where in the spirit of Leibniz's mathematically-inspired universal system of symbolic notation or *universal characteristic* – Fichte declared that there could also be a future version of the *Wissenschaftslehre* based entirely on formal and symbolic notation.[88] – Yet only after the age had become receptive to his *Neue Bearbeitung*. Unfortunately, Fichte neither completed the *Neue Bearbeitung*, nor lived to carry out his new plan of a formal symbolic system of pure concepts: »I will concern myself later with such matters as scientific elegance, the strictly systematic arrangement of the parts, the exclusion of any foreign elements, the adoption of a precise terminology, and the creation of a symbolic system of pure concepts (such as that »universal characteristic« which was already sought by Leibniz and which first becomes possible only subsequent to the *Wissenschaftslehre*). That is to say, I will attend to these matters only after I have found that the age is making some use of this new presentation of the *Wissenschaftslehre* and has thereby made itself receptive to a purely scientific presentation of the same.«[89]

87 *Neue Bearbeitung der Wissenschaftslehre* (GA II/5: 400).

88 Regarding Fichte's plan, see Joachim Widmann's article: »Exact Concepts: Fichte's Contribution on a Problem of Tomorrow« *Idealistic Studies* 11 (1981), pp. 41–47.

89 *Ankündigung* (GA I/7: 164). Translated by Daniel Breazeale, in: IWL, p. 201.

Conclusion: The »Wissenschaftslehre« as »Mathesis of the Mind«

a. Fichte's New Philosophical Mathesis

As stated in the Introduction, our primary objective in this study was twofold – to explain on the one hand Fichte's philosophy of mathematics and model of geometry, and to demonstrate on the other the internal cognitive affinities between the science of geometry and the science of the *Wissenschaftslehre.*

With regard to the first objective, we believe we have demonstrated that Fichte has a coherent and consistent philosophy of mathematics, which is visible in writings such as the *Zurich Wissenschaftslehre* (1794), the *Ankündigung* (1801) and the *Einleitung in die Wissenschaftslehre* (1813), but most fully articulated in the *Erlanger Logik* (1805), albeit in a fragmentary and incomplete form.

Fichte's philosophy of mathematics contains nine principal elements: 1) *geometry is viewed as the foundation* of the whole of mathematics; 2) *arithmetic* is grounded in geometry, and the aim of arithmetic is to terminate the process of division to arrive at a final solution or unity; 3) *algebra* is both a system of formal and symbolic notation, and philosophically beneficial in training the mind to think abstractly; 4) Fichte's *conception of infinity* is twofold: he makes a distinction between mathematical infinity – which is essentially a rule, procedure or task that we can carry out *ad infinitum* – and Infinity as such, which grounds mathematical infinity but is beyond our conceptual grasp as a totality; 5) Fichte's philosophy of mathematics assumes *original or archetypal elements* to provide fixed absolute points and to avoid circularity; 6) it is a *transcendental* science, because it investigates the conditions and possibility of the various branches of mathematics, which in themselves are only »factual«; 7) a solution to the *parallel problem* is sought, including Fichte's own proposal for an archetypal diameter to yield an invariable and standard normalizing direction against which all other parallel lines can be compared; 8) it treats of the *deduction of the dimensions of space*, where the latter are considered as »universal forms of thought« generated by a process of abstraction; and 9) the system of signs or the *language of mathematics* here finds its final arbiter.

In addition, Fichte's philosophy of mathematics can be classified as being *Platonistic* in four ways: Firstly, Fichte sees mathematical *knowledge* and the propositions and laws of mathematics as invariable, eternal and universal. Secondly, Fichte grants a higher existence to ideal elements than sense elements, and assumes archetypal or original entities, such as the archetypal point and lines. Thirdly, mathematics has a divine origin for Fichte. And lastly, mathematics and geometry serve as an excellent intellectual preparation for philosophy.

Consequently, we saw that Fichte's model of geometry contains a number of features that differ from both the Kantian and Euclidean models. For example, Fichte suggests an entirely *new starting point* for geometry. Instead of the »ordinary« Euclidean beginning of a line joining two points, he proposes starting with his »original« diameter and perpendicular. This gives rise to what he calls »true, original« geometry – a system of geometry having its basis in absolute, infinite and invariable elements that can only be grasped in geometrical intuition, or what he calls in the *Erlanger Logik*: *intelligible* intuition. Moreover, Fichte ascribes a circularity assumption to Euclid's fifth or so-called *parallel* postulate, and as mentioned, he additionally posits his archetypal diameter as an absolute standard or normal direction to remedy the parallel problem. Furthermore, Fichte's conception of the *straight line* is at odds with the Euclidean straight line. Fichte possesses a synthetic and *continuous* conception of the geometrical line as opposed to the Euclidean version of a line as a *discrete* set of points. Though Fichte's model of geometry differs from the Euclidean model in a number of fundamental respects, his model is perhaps not strong enough to be classified as non-Euclidean geometry. He held reservations about the assumptions of Euclid's fifth postulate, but did not replace or find a substitute as such for this postulate. Fichte is indebted to the Kantian conception insofar as he follows Kant in distinguishing between synthetic and analytic methods and propositions in geometry. Nevertheless, Fichte's model of original geometry also follows the synthetic method in the classic sense of synthetic or projective geometry; that is to say, it privileges the exclusion of algebraic proofs and relies instead on the method of pure construction, and posits ideal entities at infinity.

With regard to our second objective, we saw that Fichte has a twofold conception of the fundamental elements of geometry: a purely *geometric* conception, as just outlined, and a *philosophical* conception, in which the fundamental elements of geometry become *images* (*Bilder*) in the *Wissenschaftslehre* for the purely intellectual activities of the human being. – The point is an image of the I, the line is an image of the effectiveness of the I, and the drawing of the line is an image of »life« itself, because like life it is expressed in motion. And, we noted how Fichte's theory of space differs from the Kantian theory, insofar as Fichte's view is grounded in an original space or *Urraum* from out of which all the other elements of space and geometry arise.

Moreover, we investigated the relationship between Fichte's *Grundsatz* and the axioms of geometry. We saw that on the one hand Fichte's *Grundsatz* takes its inspiration from the postulates of geometry and is equivalent to a postulate not with respect to its contents – since the geometric postulate is directed at space whereas Fichte's postulate is aimed at the self-intuiting activity of the I – but with regard to its form and qualities. On the other hand, the postulates of geometry are grounded in Fichte's more universal *Grundsatz* or philosophical postulate. Both the geometric postulate and the Fichtean postulate claim to possess the characteristics of self-evidence, transparency and indemonstrability, and both are stated at the beginning of their respective systems of thought. Fichte prefers the term »postulate« in the later Jena period, yet its deployment appears to be identical with his earlier use of the term »*Grundsatz*«, as he himself says in the *Wissenschaftslehre nova methodo*. And in fact, as we saw, in a number of major texts, including the *Grundlage der gesammten Wissenschaftslehre* (1794), the *Grundlage des Naturrechts* (1796/97), and the *Zweite Einleitung in die Wissenschaftslehre* (1797) – Fichte employs the terms *Grundsatz* and *Postulat* synonymously to designate his first principle of philosophy.

We next examined geometrical (mathematical) and intellectual intuition to try and unravel the relationship between the two in Fichte's system. We concluded that the role of geometric intuition within the *Wissenschaftslehre* is to function as a paradigmatic model for the act of intellectual intuition, because of its immediacy, self-evidence and intelligible character. This is confirmed by Fichte taking geometrical intuition as *the* model of a pure (non-sensible) intuition. Yet it is not a matter of extending Fichte's theory of intellectual intuition into the realm of geometry.

This is because intellectual intuition has its limits and boundaries which should not be transgressed, and is solely limited to the realm of the free activities and intuiting of the I. Thus, for Fichte, geometrical intuition is not intellectual intuition but *intelligible* intuition. This is because both are non-sensible intuitions within the sphere of the pure intelligence. Hence, we should be aware of the limits of intuition and not conflate its different modes in geometry and philosophy.

Furthermore, we saw that there are essentially *four types of construction* for Fichte. The first three belong to the sphere of geometry: technical, sensible and free construction – with the third mode of free construction required for Fichte's original or archetypal elements. Fichte's distinction between the first two types of construction – the purely external or empirical one and the higher internal or sensible one – shares parallels with Kant's distinction in his 1790 text *Über eine Entdeckung nach der alle neue Kritik der reinen Vernunft durch eine ältere entbehrlich gemacht werden soll.* However, Fichte argues in the *Wissenschaftslehre nova methodo* that he has transformed the method of construction in geometry in a transcendental manner. Since in the Fichtean theory the philosopher not only has to construct objects so that they are empirically visible, as one does in geometry using a straightedge and compass for example, but the philosopher is supposed to consciously »construct« the *conditions* and rules of his own consciousness. In the constructions of geometry Fichte claims that I am immediately certain and conscious of my free acting; whereas in purely philosophical or »free infinite« construction I have to »engage in reflection«, and furthermore *raise into consciousness my own actions* and activity of constructing. Thus in this sense the constructions of the *Wissenschaftslehre* are like the constructions of mathematics but of a higher cognitive order. For in mathematics and geometry we look at the products of the construction, whereas in the *Wissenschaftslehre* we are concerned with the activity of construction itself.[1]

Finally, we considered a further number of differences and similarities between geometry and the *Wissenschaftslehre*, including their common self-evidence, irrefutability and determinacy (universality). They differed with regard to their unequal relations to freedom and formalized language. We prefaced this with some of Fichte's thoughts on the teaching

1 Cf. *Neue Bearbeitung der Wissenschaftslehre 1800* (GA II/5: 344).

and learning of mathematics, and noted the influence on him of the Swiss educationalist Johann Heinrich Pestalozzi. We saw that on account of its ability to stimulate the imagination and self-independence in thinking, Fichte views Euclidean or synthetic geometry as an excellent propedeutic for the *Wissenschaftslehre*.

To conclude this study we can say: although the *Wissenschaftslehre* was conceived in the spirit of the Kantian critical philosophy in 1793/94, and draws scientific support from the methods and elements of geometry and mathematics, it was still viewed by Fichte as an altogether *new* philosophical *science*. Fichte himself only seemed to fully realize this fact after his public break with Kant in mid-1799. During the genesis of the *Wissenschaftslehre* in Zurich Fichte was aware of the new philosophical *discovery* he had made, i.e. the discovery of the *Grundsatz* or *Tathandlung* on which the whole of his philosophy could be based. Yet on account of the Copernican revolution wrought by Kant's critical philosophy, Fichte was privately convinced that Kant had *already* made the same discovery, but for some unknown reason would not make it public. He reasoned that Kant must be some kind of modern-day Socrates, whose philosophical conclusions are inspired by a *daimon* or genius but whose initial premises remain unknown to him.[2] Thus in the *Ankündigung* and in the *Neue Bearbeitung der Wissenschaftslehre 1800*, it finally seemed to dawn on Fichte that he had not just discovered a new starting point for philosophy, but that the *Wissenschaftslehre* as a whole was a *totally new science* that had *never existed before*.[3] Why was it brand new? It was new because it was the first science to undertake a systematic philosophical investigation

2 Cf. Schelling's letter in Tübingen, to Hegel, January 5, 1795: »*Fichte* said the last time he was here that one has to have the genius of Socrates to penetrate into Kant. I find it more true everyday.« (Quoted in FG vol. 1, p. 225); and Baggesen's account in his diary of his conversations with Fichte in Zurich, December 7, 1793: »We soon came into philosophical conversation about [...] ›Kant – does this immortal man have a genius like Socrates, who tells him things that he does not know? Or is it due to his poor opinion of the age, which for him is not mature enough, or is it due to modesty? – Everything he writes is true – but all his proofs are false.'« (FG, vol. 1, p. 67). See too Fichte's similar remarks in the letter (draft) to Flatt, November or December 1793 (GA III/2: 18); and Fichte, letter to Niethammer December 6, 1793 (GA III/2: 20f.).

3 See *Ankündigung* (GA I/7: 153), and *Neue Bearbeitung der Wissenschaftslehre* (GA II/5: 344).

of the rational mind using the strict methodology of mathematics, a philosophical science whose mode of intellectual cognition was based on the *cognition of intuitions* and not on concepts, and that followed a methodology as lucid and precise as the procedures of the geometers.

One of the chief goals of Fichte's career therefore remained constant: to demonstrate and convince his philosophical contemporaries and public that the *Wissenschaftslehre* was a science of the same rigour and self-evidence as mathematics, or what he also called around 1800, a philosophical *mathesis*: »The *Wissenschaftslehre* is *mathesis*, not merely in accordance with its *external* form, but also in accordance with its *content*. It describes a continuous series of intuitions; and proves all of its propositions in intuition. [...] Thus, because the *Wissenschaftslehre* is mathematics, it also has the distinctions of mathematics.«[4] As we see, Fichte could above all draw this parallel on account of the cognitive affinities between the two sciences: for like the *Wissenschaftslehre*, mathematics or ordinary *mathesis* too is a science of rational cognition based on pure intuition. However, Fichte rightfully asks: where is the science or discipline that examines not just the content of mathematics, but also determines whether its mode of intuitive cognition is legitimate? Such a meta-science would in turn have to employ a form of rational cognition based on intuition: »And since an intuition as such can only be intuited, there must be cognition *based on intuition*. However, where is this *mathesis of mathesis* realized?»[5] This latter science is realized in the *Wissenschaftslehre* – it aims to be a certain philosophical system examining all the various modes of cognition based on pure intuition.

Throughout his career Fichte never ceased qualifying his undertaking as a geometrically or mathematically-inspired philosophy of *pure reason*, in stark contrast to works on ordinary *mathesis*, such as the *Anfangsgründe der reinen Mathesis* by his old friend the Königsberg mathematician Johann Schultz. Despite Schultz's use of the term »*reine*« (pure) in his title, his understanding of *mathesis* is exclusively confined to measurement and quantities in geometry and spatial areas, and is therefore not to be confused with Fichte's purely *philosophical* project.[6] Instead, Fichte haboured the same lofty aspiration that was present in thinkers of the cal-

4 *Ankündigung* (GA I/7: 160).

5 Ibid., p. 156.

ibre of Descartes and Leibniz[7], whose *mathesis universalis* and *characteristica universalis* respectively, illustrate the search for a universal philosophical science possessing the precision and certainty of mathematics.[8] Thus, Fichte's philosophical project of *mathesis* is to be distinguished from ordinary *mathesis*. Ordinary *mathesis* focuses on the laws and limiting of space, whereas Fichte's art of *mathesis* is concerned with the laws and compass of reason or pure thought: »The *Wissenschaftslehre* is *mathesis* [...] it is the *mathesis* of reason itself. Just as, for instance, geometry includes the entire system of the limiting of space, so the *Wissenschaftslehre* includes the entire system of reason.« [9]

Both the science of mathematics and the new science of the *Wissenschaftslehre* strive to be absolutely rigorous and systematic and concern the domain of the pure intelligence, and both are grounded in ideal elements and non-sensible intuitions and constructions, and it is for this reason that Fichte viewed his system as a higher version of geometry or mathematics. The *Wissenschaftslehre* is not a philosophical system researching the laws of the outer physical and material world, but one above all concerned with the lawful inner world of the mind or spirit: »The *Wissenschaftslehre* is *mathesis* of the mind.«[10]

6 For example: »§6. *Mathesis* or mathematics is called the science of quantity (*scientia quantorium*). §7. The part of *mathesis* that is abstracted from all real objects, is called pure (*pura*) *mathesis*; the part that is occupied with real objects is called applied *mathesis* (*applicata*)«, Johann Schultz, *Anfangsgründe der reinen Mathesis* (Königsberg: Hartung, 1790), p. 2.

7 Regarding Fichte's hopes for continuing Leibniz's *characteristica universalis*, see the concluding words of the *Ankündigung*: »[...] a symbolic system of pure concepts (which was already sought by Leibniz with his universal characteristic, but is only possible since the *Wissenschaftslehre*) [...]« (GA I/7: 164).

8 On *mathesis universalis* as a philosophical tradition, see Jules Vuillemin, *La philosophie de l'algèbre*, esp. pp. 465- 518 on »La Mathématique universelle«.

9 Fichte, *Ankündigung* (GA I/7: 160).

10 Original German: »Die W.L. ist *Mathesis* des Geistes.« *Neue Bearbeitung der Wissenschaftslehre 1800* (GA II/5: 344).

Appendix: Johann Gottlieb Fichte: »Original Geometry«

Translated excerpt from: *Erlanger Logik* (1805)

»Original Geometry«[1]

Johann Gottlieb Fichte

[124] [...]

Let us explain the established propositions by means of an example.[2] (What these [propositions] are and can do will be recalled in due course). *Geometry*, as the ground of all mathematics. – Plato: we people of modern times are required to be more modest.[3] Moreover, it is philosophy. – Phil. [osophy] of mathematics[4], and new things might emerge; it is not as-

1 This is a translated excerpt from J. G. Fichte's *Erlanger Logik 1805* (GA II/9: 124–137). The numbers in square brackets in the text refer to the pagination in the *Gesamtausgabe* edition. All editorial additions are also indicated by square brackets. It is worth bearing in mind that the original German text is extremely dense and not entirely clear in places. Particular translation difficulties are highlighted in the footnotes. The *Erlanger Logik* was published in German for the first time in 1993 (in the *J. G. Fichte Gesamtausgabe*). I would like to thank the editors of the Fichte *Gesamtausgabe* and the Frommann-Holzboog Verlag in Stuttgart-Bad Cannstatt for kindly allowing me to translate this excerpt into English.

2 The propositions Fichte is referring to are the rules and elements of a genetic science. Cf. GA II/9: 122–124.

3 Plato (427–347 B.C.). See *The Republic*, pp. 331–337.

4 *Ph. der Mathematik.*

sumed that you know them already, but they should be first learned here. –

Do not be alarmed if I spend a few hours on this. It will serve as an example in this connection: however, you could also consider it as an important fragment from metaphysics, to which, at any rate, I will proceed later on.

It is all the more significant as an example in that I will not only indicate the rules of the external form, but at the same time show the inner penetration of the concept, [125] through which science alone receives its true scientific content and inner spirit.

I). Geometry, on account of its essence and the determination from its principle, *necessarily* only has factual self-evidence[5]; because it is a fragment from the domain of knowledge. Therefore – as a fragment – it cannot have insight into its own possibility within this domain, but only realises its possibility factually. (By saying and wanting to prove this I have to go beyond [geometry] into philosophy: how deeply, depends on the present aim).

Its purpose: to exhaust the possibility of the limitation of space within One universal space: – *Presupposition*: that this possibility has its limits and laws: only a few are possible, others are not: – A condition of this insight: [transplanting] oneself into the possibility, so that these laws reveal themselves to us, and directly work upon our limiting of space, precisely *in* the *limiting*, in the *construction*: (hence *geometrical* intuition, which is often confused with philosophical intuition, is something entirely different). Thus *NB. start by freely limiting*, so that nothing except the internal *impossibility of limiting* itself (according to its inherent laws), hinders the free limiting.

Therefore (1) *Empty out*[6] everything in space that is forming and limiting in it; *grasping* space *as empty and formal* is the *conditio sine qua non* of this science. It does this absolutely through the act of its existence: fact; without justification, and without wondering about it: (and it is entirely

5 *faktische Evidenz*. This expression is sometimes rendered into English as »factical self-evidence« or »manifestness«.

6 *ausleeren*.

right to do this: no geometry is possible without this unexplained fact). Question: does the *Wissenschaftslehre* explain this [fact]? Certainly. Why does the geometer not do this? Answer: It does not take place as geometry but as *Wissenschaftslehre*: *It*, the *Wissenschaftslehre*, can never become geometry within its [domain]. If the teacher of the *Wissenschaftslehre* were to become a geometer, then it would no longer be valid to establish this fact from out of itself, and to reason about its possibility, but he would have to decidedly carry it out: become the fact, and become rooted in it. If he then continues to be concerned about the question of the possibility, he is either (1) A pedant (2) [or] what is worse, a confused and confusing mind. …

(N.B. – Here you have a striking example of the above expounded proposition that the *Wissenschaftslehre* cannot replace the other sciences, [or] integrate them into itself, but that it leaves them alone as independent [sciences]: It is and does not immediately have a fact; but is only the genesis of all possible pure and *a priori* facts: hence the particular science now situates itself in the One [science], and does this really and in reality; [126] and this is the characteristic and eternally persisting and ineradicable difference between the two).

In any event, the *Wissenschaftslehre* inquires into this emptying of space, for original space is not empty, but thoroughly filled; it *is* matter[7] / I say this carefully: matter is filled space; and space is absolutely filled, immediately self-filling, i.e. force solves all the difficulty, and the whole of atomism has only arisen from the empty space of geometry. – The real space of the physicist, who would certainly not grant any empty space, and therefore neither any infinite space. The geometer grants the latter, renouncing the means for holding it together. – The geometer is initially in contradiction with the physicist, whose science is older: after the emergence of the latter [science], the second [the physicist] is in turn [in contradiction] with the first [the geometer]. Only the philosopher can reconcile the two, because he inwardly (not just in appearance) draws the boundaries between both by proving the possibility of the two standpoints. (Clearly everyone ought to do this insofar as it is in his own person: in addition, the attempted arbitrating expression of the philosopher will not help the two parties. (because they do not arrive at it).*

7 *Materie.*

*[*noted at the bottom of the page*] which is the correct space: geometric or physical [space]? The answer of the philosopher: neither of them, and above all space is not [fully explained by them]. Thus, it is obvious that the philosopher imperiously combines them by wholly taking away the soil on which they carry out their battle; and he only allows this soil to the extent that they do not come too close to one another on it.

– Proof of empty space, by *stretching out one's hands*: There must be *something there*: and hence *space* for it!! To the extent that you think it, you can always proceed [in] it. Thinking about it signifies an entrance into metaphysics and actual physics.

(2) Empty out, I said: and this is clearly evident from the proof; however, examined in more detail it seems that the possibility of geometry itself requires that space not to be emptied out entirely: or what amounts to the same thing, that for the possibility of its free construction geometry has not emptied itself out from *space* and its [127] own connection within itself. – I will prove this as follows:

The straight line is the original limit to be realized through freedom. (Whoever doubted this proposition, and considered for instance the point to be this [original limit], will be soon struck by the opposite). However, the line can only be constructed as a line, never from points (as it is often clumsily expressed). The geometer brings it with him directly and absolutely *as a line.* If it were otherwise, one would have to traverse and *complete* an infinity of points in order to pass from one point to another point, which is absurd. By means of points I can never pass from one point to its infinitely close [neighbour], but have to admit a piece of line, however small it would have to be: and for me, given over to only myself, this [piece of line] would always be *straight*, and this is precisely why I add it, and so on. He [the geometer] brings the line as a continuous[8] and pure line with him, and the points flow into one another. This is indeed a continuous and living space: he therefore brings it with him. He does not bring with him the infinite divisibility of space (which he subsequently and correctly asserts to be point-wise in the already generated line) for the line itself and for the possibility of the assertion. (This assertion of infinite

8 *zusammenhaltend* – or ›interconnected‹; literally: ›holding together‹.

divisibility means: that it is possible to limit the *interconnected* line *ad infinitum*).

1) I believe this is clear, and rather profound. To express it: The beginning of geometry is synthetic; neither point, nor line, but both, and the two are only comprehensible by means of one another.

2) The infinite divisibility of space has often been contested by philosophers and physicists: and certain ›spatial points‹[9] have attained a sort of renown in the literary world, and do not appear to have completely faded into obscurity. When a decisively good thinker says something decisively false, it is always to be assumed that he simply chose the wrong formulation, and meant something true; and thus I have never believed what was supposedly proven above about the line: this can be expressed as follows: the most infinitely small part of the line, is a line; never at all a point. Wanting to break down the line into points through an infinite division does not signify an increase in the *quantity*, where only geometry is unlimited, but signifies a transformation in the *quality*, [128] and I hope the good genius of this science may protect it from this.[10] It follows from this that the plane[11] could never be a line, a line that can be drawn as a line, and the solid never a plane, and so on. The most infinitely small part remains a *plane*, or *solid*; you could leave it undrawn: but *drawn* not broken down into a line and so on. For you are then no longer breaking down *your work*, with which you can do whatever you want, but the work of the interconnected space.

[This is] once again a fact that geometry does not need to grasp in its possibility; I have also not explained it. ... However, you should know it in its inner essence, just as we have established it in each of its representative [elements]; so that you will be able to counter the same unreasonable demands as the ones cited above.

9 *Raum-Punkte.*

10 As the editors of the Fichte GA point out, this is possibly a reference to certain ideas of the mathematician K. C. Langsdorf. – Cf. Friederich Hildebrandt: *Anfangsgründe der dynamischen Naturlehre* (Erlangen, 1807), p. 31.

11 *Fläche.*

(3) Space in space: holding space and making it stationary insofar as one dissolves and doubles it from itself, as it were: – with the line this has already been remarked upon about space that holds itself together. Does geometry share something in common with physics? – The difference is that there are *forces of nature* in physics, whereas in geometry it is one's own freedom that deposits stationary spaces in the universal space[12] ... Hence, this is actually the connection between both sciences in ph[ilosophy]. This is the point where philosophy mediates between the two. It is *factual self-evidence*[13], which geometry cannot get out of, which is imposed upon it by its own essence.

I say.

II). Geometry in the form known and taught to me has factual self-evidence, which it should not have, but which it should genetically penetrate; [129] and for this reason it is excluded from the rank of a deductive science, and is a mere *determinate* [science], with which it does not have to be content.

(I said with good reason ›known to me‹, as I learned it in the past. Because I am not sure if this is the case in higher domains. If a geometer were to say: that is not true in itself, and so on, and he should say that he knows this, and that would be good: if he would only concede to me that not everyone who carries out geometry knows this).

There is in fact an original mode, determined by free constructing in space itself, as to how the geometer holds his space together and gets his bearings in it without becoming disorientated (not only this holding together in general, which I have just mentioned, and whose fathoming I leave to the geometer.). The geometer should factually and genetically recognize this determined mode, and his science will thereby acquire higher domains, and become transformed from a determined [science] into a deducing (*quantum*) one.

12 *welche in dem allgemein Raume stehende Räume absetz.*

13 *faktische Evidenz.*

(a) Proving my assertion in intelligible intuition.[14] Draw a straight line: I maintain that this line has for you a determined lawful invariable *location* (*Ort*) in universal space on account of its construction. (I can say: draw another line to this line, so that an included angle arises: How do you find again the location of this [first] line, and what holds and bears it for you, while you are occupied with the second one? Its *location*, its relation: to what? – to infinite space? a relationship to the infinite? Absolutely impossible. Look into yourself. You have a point in yourself; from which it is a certain distance; that is, it is a specific line that you draw *unnoticed*[15] and originally (without an arbitrary free construction), this *distance* always remains the same.

Everyone has an absolute limit-point[16] of all distance, from which infinitely many lines may be drawn, and originally drawn lines. Without this point, which is absolutely invariable in itself, and co-borne and omnipresent in infinity[17], nothing would have a location and we would not find anything again in infinite space.

I furthermore maintain: that the line you have drawn also has a determined direction due to the fact that its construction is invariable and set.

You then draw another line to the same line, and there arises a determined (or it is not) included angle.[18] This could not occur if the direction of the first line were not fixed.

[130] What maintains this direction? In the intuition: nothing can maintain it except the invariable side-angle[19], which it makes with a line drawn to this same line. Where is this line? … Answer: The line already indicated above that determines the location drawn from yourself as the centre.[20] – Good: – But how can this line make a determined angle on the external line, if it itself is not drawn in a specific direction, and on a line [making]

14 *meine Behauptung in der intelligiblen Anschauung beweisen.*
15 *unvermerkt.*
16 *Grenzpunkt.*
17 *u. allgegewärtig in der Unendlichkeit <mit>getragen.*
18 *eingeschloßner Winkel.*
19 *SeitenWinkel.*
20 *Mittelpunkt.*

a determined angle with it, but from a mere point, as it must appear up to now?

My answer is: it is not like this at all. But just like you did with the earlier point, so you also bring a line through itself with you, of which the first point is the centre: [this line is] therefore a *diameter*: and standing *perpendicularly* on this is the above-mentioned distance measurement[21], as directional measurement[22] and bearer. [In the first case] either this original perpendicular[23] intersects the external line, i.e. so that it makes right-angles on both sides of it, then this line is parallel to the original diameter, and thereby, as its absolute and invariable quality[24], its direction is immutably determined into infinity[25]: this is because the principle for judging[26], the original diameter[27], is immutably determined; or [in the second case] the original perpendicular forms unequal angles, then it is convergent and divergent compared to the archetypal diameter[28], and therefore invariably determined. – And a specific direction, absolute orientation concerning the location and direction, are only possible through this absolute diameter and this perpendicular. The original diameter is the *regulator* of all direction; the perpendicular on it is the medium toward the outside.[29]

We are still not finished with the description of the original mode: however, let us pause for a moment to try and understand this.

(1) Ordinary geometry commences with a first straight line that is drawn with absolute freedom externally to the subject – this is how it ap-

21 *EntfernungsMaas.*
22 *RichtungsMaas.*
23 *ursprüngliche Perpendikel.*
24 *unveränderliche Qualität.*
25 *ist ihre Richtung ins unendliche unabänderlich bestimmt.*
26 *Beurtheilungsprincip.*
27 *ursprüngliche Diameter.*
28 *Urdiameter.*
29 *Der ursprüngl. Diameter ist das Regulativ aller Richtung; der Perpendikel darauf das medium nach aussen.*

pears to this geometry. – We know it to be different: Every externally drawn line presupposes a line in ourselves, the former can of course be drawn with absolute freedom, but it is not at all an original line. [The latter] has a perpendicular on it at its centre that can be infinitely extended or contracted, and which immutably determines the location and direction of the first freely drawn line, [131] and which provides the orientation, and according to which all the other lines can be easily determined.

Any geometry that is not aware of this has to *determine* its first free line by means of other likewise freely drawn lines. [Their] location and direction are estimated accordingly. I will not ask how [ordinary geometry] can deal with the second [line] when it could not deal with the first one: it will be shown that it carries it out through the first [line], and the first [line] in turn through the second [line]. Thereby in its principle it now falls prey to *arbitrariness* (it is like this, for I have done it in *this* way, since originally in general to make itself, but thoroughly not let itself be made like *this*, but must be accepted as we find it) – there is then a never-ending circle in the proof: precisely, because after failing to appreciate the original lawfulness, arbitrariness [ends up] becoming the principle.

Here the theory of parallel lines immediately springs to mind. I don't know whether this dubious blemish of geometry (which it again fully admits) has been solved. Schulz's solution does not accomplish a thing, and the entire theory is improbable.[30] The part equals the whole: arithmetic and counting, which fail[31] to orient themselves on geometry … *Parallel, relation*; on two arbitrarily drawn externalities[32]: which are not determined and directed by anything except their own *parallelity*.[33] What is now the direction of the two [parallel lines] and what is the normal direction?[34] It can be known by starting from the third arbitrarily *drawn angle*.

30 Cf. Johann Friedrich Schultz: *Entdekte Theorie der Parallelen nebst einer Untersuchung über den Ursprung ihrer bisherigen Schwierigkeit* Königsberg 1784. - Schultz theory states that »the size (Größe) of a plane angle is determined by the size of the plane lying between its infinitely extended side.« (From the preface).

31 *unterläßen.*

32 *Aeussern.*

33 *Parallelität.*

34 *NormalRichtung.*

Now, what about the *angle* itself? On parallelity. – Direction through the angle: angle through the direction of the lines: this runs through the whole of geometry, and you cannot arrive at the correct view with a πρώτον ψευδος.[35] – What then is this [correct view]? I can tell you in a few words: the direction is primary and this is absolute: and it is the direction of the original diameter that everyone brings with him: this is the straight line and straight direction *per se*.[36] In order to measure all the other directions against it, the space above it is divided into two equal parts, and this yields the original *right-angle*. This in turn is not determined by the direction of the perpendicular, but is determined by the direction itself, as [something] secondary, and it is not the primary and immediate direction; but it is determined by the equal parts of space: this perpendicular with its original right-angle now determines all other right-angles or non right-angles, i.e. obtuse or acute angles, and measures them through itself. The first parallel [132] runs as a second [line] parallel to the archetypal diameter[37], to the primary or normal direction; and this parallelity is equal to the two right angles on it: All [lines] drawn with this parallel line are parallel among themselves, not because they are as such, but because they are parallel to the normal line.[38] Now if an oblique line intersects this whole through the centre of the archetypal diameter, then it is easy to arrive at the known relation, everywhere in the same manner, from the quality of its line and the lines intersecting it, in any case not only relatively, but absolutely and therefore qualitatively oblique lines: and you won't be tempted, in alternating proofs, to make the result again into the principle of its own principle.

– In the above series the absolute direction of the archetypal diameter is what is primary. The angle, and indeed, the right-angle first arises in the synthesis of *other* non-original lines with this [archetypal diameter], and its direction is estimated according to the first one, as the first and absolute [angle] at the centre of the archetypal diameter. The angle of the ex-

35 *proton pseudos.*

36 Here Fichte uses the term ›*rechte Linie*‹ (literally: right line), which is an older term for straight line.

37 *Urdiameter.*

38 *Normal Linie.*

ternal line and its direction have to be judged in turn according to the same. The archetypal direction[39] therefore is what determines this angle; however, it is only determinate for derived and relative directions. This is the course for avoiding the above circularity and arbitrariness.

And now entirely clearly: the fact with which ordinary geometry begins is the freely constructible line. It draws another line to this one, and this is its first angle; this intersection results in either a right, an obtuse or an acute-angle respectively. This angle is determined by the slope of the lines to one another. The slope of the lines is determined in turn by this angle, and this [determination] is circular.- It is simply a reciprocal determination, a *g* and *x*, two terms determine [each other]. – and you cannot derive your own procedure.

However, I say: true original geometry, i.e. the geometry posited through pure reason, begins with an entirely different fact. Ordinary geometry cannot dispense with this fact, for ordinary geometry is only possible through this fact and within this fact. It simply does not penetrate it with knowledge, but it simply carries it out mechanically, like a law of nature. We have just penetrated this fact and thereby furnished geometry with its true principle. It is the original diameter, with the perpendicular at its centre, which everyone brings with them to geometry. This [fact] is creatively present everywhere in geometry: to begin with for the first freely drawn line, which for true geometry is not the first line: [this fact] conditions the possibility of drawing this line, and sets [133] that with which geometry is solely concerned, its *location and its direction*. This first [line] now stands solidly on the original perpendicular, in the angle that it makes with it, and does not require any *second* external [line] in order to be held by means of it. Of course, I can draw a second [line] to it, whose location and slope can be determined in turn with respect to the first, by the inherent direction of the first in relation to the archetypal perpendicular. And if I want, and it serves the purpose, through mediation I can always [proceed] from the immediate on into infinity. In this way science becomes *deductive*, purely deriving, not turning round in circles. (*Where*: it is a certain distance from the first, and makes an angle with it. – Good: but where is

39 *UrRichtung.*

the first? – it is a certain distance from the second, and it makes the *same* angle with it. And you could continue and draw a third or fourth [line] to it, and so on into infinity. A person would never become clever through this; and geometry could never ascertain the determination of a location; unless everyone, in addition to what he articulates, does something else unobserved that he does not articulate). Namely, the first line, or also surely both [lines] with the included angle, are at such and such a distance from *he himself* [as the archetypal diameter], and make a determined angle with the perpendicular standing on it: and then, however, he does not need to ask any more *where* it is: then he is absolutely where he is through absolute reason; and through this, his absolute ›Where‹, he first understands space, and this ›Where‹ is the creator of space.

And now what has been said about the particular science in general will have been made clear to you by means of the example. a). A fact that is not made, but which *absolutely makes* itself, in a material sense[40], and a qualitatively determined rational law: the life of reason and so forth. Every science that proceeds from a free act (like ordinary geometry) thereby betrays its arbitrariness. Here is the mode of the original determination of space.[41] b). It is to be *constructed afterwards*, enlivened and formed; this precisely immediately. – In order to know it correctly, one has to somehow make it in examples; I have carried it out together with you: examine this example, and take note of the mode.

I was not quite finished with the original fact of geometry: but interrupted myself, as soon as I had acquired sufficient clarity concerning the case in question. For the sake of completeness, and so that I do not instruct you using [134] a half or one-sided example, and therefore [teach] something partly untrue about another different science, I will elaborate on it, although the goal of the example has been reached.

Like before, I will remain with plane geometry: – I say: if a determined location and direction are to be possible in space, then this [space] may

40 *Materialiter.*

41 *Hier der modus der ursprüngl. Raumbestimmung.*

not expand itself through itself (because then the location would change). The space of geometry is necessarily a stationary space[42] – Non extending = it has to hold itself together, *externally* (not internally, which belongs to the line) merging together[43] / but we are here speaking of an actual *extension*[44], which is at rest there: a space that is *limited* through itself. (The space of geometric determinations is necessarily a limited [space]). This limit is *one*, and it has a specific distance from the centre: i.e. it is the line of a circle described using the original line of distance[45], which is a perpendicular, (and is the absolute radius here) about the centre, which everyone brings with them.

[This] is the original construction of the circle line that is posited absolutely through reason. From the standpoint of intuition, it is actually only a semi-circle; we are not concerned with what is behind us; but ordinarily only with the half of this semi-circle, or an original quadrant.

The semi-circle limits space through the diameter[46]; a perpendicular [limits] the two right-angles, and thus the *right-angle* becomes the measure of the quadrant.[47] (I was careful to say this: for it is originally so. Only in the relative determination of the angle does the circle curve[48] inversely become the measure of the angle).

The extension of space is infinite: this simply means that geometry can *unconditionally* enlarge this circle or semi-circle, and must be able to do this (unconditionally enlarge); however, in every real operation it necessarily has a restricted space[49], otherwise all determinations of location would cease.

Corollary. All arithmetic is based on geometry and is in fact nothing more than a presenting of geometric spatial relations in examples. (Do not elaborate on this; just sketch it out. However, I hope that those here would have understood and grasped this proposition; it is inherently true and decisive for a certain empty activity in arithmetic).

42 *ein stehender Raum.*
43 *concresciren.*
44 *Ausdehnung.*
45 *ursprüngl. Entfernungslinie.*
46 *über den Diameter.*
47 *das Maas des Quadranten.*
48 *Cirkelbogen.*
49 *beschränkter Raum.*

With regard to its extension, it necessarily has a restricted space; therefore arithmetic too only really exists on this basis.

[135] The essence of arithmetic is to posit a unity (which incidentally has nothing to do with geometry, which resides entirely in relations, and through which indeed arithmetic becomes a mere example[50] of geometry). However, arithmetic only arrives at a unity when one ceases to divide. The space of arithmetic therefore, in accordance with its own existence, is also not inwardly infinite or divisible into infinity (as geometrical space is of course), but the division is brought to a close, at the point where arithmetic starts. – These two last sentences could shed light on certain employments of infinity among mathematicians and give them food for thought.

A philosopher: *right*, and *left*. (1) The endpoints of the original diameter. *Right*, that quadrant of the semi-circle, in which everyone who is left to their own devices directly measures. Conventional *right* and *left* is something different. For the conventional left-handed person, left is his right.

From here onwards geometry now proceeds in a regular form: from the simple, to the composite: point and line. – Line to plane, and so on, from the general laws of the figure to the particular [laws]: and again in each of them, e.g. in the triangle. This formal perfection, which exists nowhere else like this, was its triumph. Everyone could do it; among other things, it was only lacking somewhat in language: but this belongs to philosophy. To create its *language*. – *Philosophy*, above all to bring about understanding.

Higher logic. *Rigorous science: non rigorous [science]*. An absolute difference, not a relative one. Therefore the first: [rigorous for] those who at least could and should genetically penetrate its factual principle; [it is] not *rigorous* for those who *cannot* penetrate it; absolutely cannot: and do not

50 *Exempliren.*

have it? If not: at least have *oneself* or *one part* of it themselves: how are we to understand this? Answer: by means of the above; not in clear insight, but mechanically, and effective as a law of nature. Thus, *it seeks its principle and* [136] *perhaps up to the still remaining part of itself, starting from those parts of itself that it already has*: and this indeed not through individual restriction, but through the diremption[51] lying absolutely in the thing itself. First and foremost (1) In general: it is knowledge, as a non-rigorous science: (That the non-rigorous is not situated in the imposing[52] of the strict form: then this is not imposed, as will shortly be shown – but concerns the matter). Science = combination of the coincidental under the unity and universality of a law. It *has* to be so, otherwise it would not be science. It has laws. Perhaps [it has] laws of laws, and so on. How have they become part of it? Not from above downwards: therefore, from below upwards: searching for the *same*, in turn for what is considered lawful in the particular: again, *searching*, I said, for the same in this specific [case] under the differing lawful aspects, using comparison, not absolutely positing and creating it through the law.

It ascends synthetically: and not like the strict science that it could also be, for the sake of an external goal of popularity, but necessitated *by its* own nature. Now this ascension also has to occur *rigorously* and systematically: without a leap, without a circle, without standing still, with rigorous determinacy: and if it does not have this, then just like in the rigorous [science] it has to keep quiet and does not have any right to conduct superficial chatter under the pretext of non-rigorousness.

Its essence, as we certainly see, is that it starts in the perception[53]; and forever remains in the perception. We could justifiably call it – not an empirical science, for this is a contradiction – but a science of the empirical world[54], within the empirical, and in the domain of the empirical.

And now to thoroughly and *lucidly* instruct you in this point (insofar as in φ [philosophy] there is no other lucidity but thoroughness), I will have to guide you into a deeper distinction between absolute *a priori* sci-

51 *Diremtion.*
52 *Erlassung.*
53 *Wahrnehmung.*
54 *Wissenschaft der Empirie.*

ence and empirical science. As preliminary remarks I might add 1). That this subject will contain the most essential elements of the philosophy of physics and its history; just as the immediately preceding contains the most essential aspects of the philosophy of mathematics: [they are] extremely essential parts of metaphysics, whose latter foundations coincide with the plan of this course. (2) That this subject supplements, rectifies, and protects from misunderstandings what has just been treated [137] concerning the rigorous science a priori, and that without this comparison you would have received something one-sided and therefore something partly false. (3) And that whoever takes the trouble to correctly grasp and understand it, will on the one hand without noticing it overcome the confusion that so rightly characterizes the spirit of our time and is virtually universal in it, and on the other hand he will come away with a considerable gain for the rest of his study. [...]

Bibliography

a. Works of Johann Gottlieb Fichte

J. G. Fichte Gesamtausgabe:

Johann Gottlieb Fichte Gesamtausgabe der Bayerischen Akademie der Wissenschaften, eds. Reinhard Lauth, Hans Gliwitzky, Erich Fuchs, Hans Georg von Manz, Ives Radrizzani, Peter K. Schneider, Martin Siegel, Günter Zöller *et al.* (Stuttgart/Bad Cannstatt: Frommann-Holzboog, 42 vols., 1962–2012)
[= GA, followed by series, volume and page numbers. Series I: Published Works; Series II: Unpublished Writings; Series III: Correspondence; Series IV: Lecture Transcripts].

The following texts are especially important for this study:

Eigne Meditationen/Practische Philosophie, 1793/94 (GA II/3: 21–266).
Züricher Vorlesungen über den Begriff der Wissenschaftslehre, February 1794 (GA IV/3: 19–41).
Über den Begriff der Wissenschaftslehre, 1794 (GA I/2: 107–172).
Grundlage der gesammten Wissenschaftslehre, 1794/95 (GA I/2: 249–451).
Neue Bearbeitung der Wissenschaftslehre, 1800 (GA II/5: 331–401).
Ankündigung [*Seit sechs Jahren*], 1800 (GA I/7: 153–164).
Letter of J. G. Fichte to F. W. J. Schelling, November 15, 1800 (GA III/4: 359–361).
Sonnenklarer Bericht, 1801 (GA I/7: 183–268).
Wissenschaftslehre 1801/02 (GA II/6: 105–324).
Wissenschaftslehre 1804/II (GA II/8: 2–421).
Erlanger Logik, 1805 (GA II/9: 57–171; esp. 124–137).
Einleitung in die Wissenschaftslehre, November-December 1813 (forthcoming in: GA II/17: 233-314) [Halle *Kollegnachschrift* version in: GA IV/6: 351–472; and version: Berlin Ms.germ.qrt. 1964b].

Fichtes Werke

Fichtes Werke, ed. Immanuel Hermann Fichte, 11 vols. (Berlin: Walter de Gruyter & Co., 1971). [= SW]

Other Single Volume Editions:

Fichte, J.G. *Züricher Vorlesungen über den Wissenschaftslehre Februar 1794. Nachschrift Lavater. Beilage aus Jens Baggesens Nachlass: Exzerptseite aus der Abschrift von Fichtes Züricher Vorlesungen über die Wissenschaftslehre*, ed. Erich Fuchs (Neuried/Munich: Ars Una, 1996).

›Ultima Inquirenda‹. J. G. Fichtes letzte Bearbeitungen der Wissenschaftslehre Ende 1813/ Anfang 1814, ed. R. Lauth (Stuttgart/Bad Cannstatt: Frommann-Holzboog, 2001).

Fichte Translations:

The Science of Knowledge (1794), ed. and trans. P. Heath and J. Lachs (Cambridge: Cambridge University Press, 1982).

Early Philosophical Writings, trans. and ed. Daniel Breazeale (Ithaca: Cornell University Press, 1988, 2nd ed., 1993) [= EPW]

Foundations of Transcendental Philosophy (Wissenschaftslehre) nova methodo, trans. and ed. Daniel Breazeale (Ithaca: Cornell University Press, 1992).

Introductions to the Wissenschaftslehre and Other Writings (1797–1800), ed. and trans. Daniel Breazeale (Indianapolis: Hackett, 1994). [= IW]

*Foundations of Natural Right, ed. F. Neuhouser, trans. M. Baur (*Cambridge: Cambridge University Press, 2000*). [= FNR]*

The System of Ethics in accordance with the Principles of the Wissenschaftslehre, eds. and trans. Daniel Breazeale and Günter Zöller (Cambridge: Cambridge University Press, 2005).

The Science of Knowing: Fichte's 1804 Lectures on the Wissenschaftslehre, ed. and trans. Walter E. Wright (Albany: State University of New York Press, 2005).

The Philosophical Rupture Between Fichte and Schelling: Selected Texts and Correspondence (1800–1802), eds. and trans. Michael G. Vater and David W. Wood (Albany: State University of New York Press, 2012).

b. Classical and Contemporary Sources

Aristotle. *The Metaphysics*, trans. Hugh Lawson-Tancred (London: Penguin, 1999).

Beck, Jakob Sigismund. *Einzig-möglicher Standpunkt, aus welchem die kritische Philosophie beurteilt werden muß* (Riga: Hartknoch,1796).

Cusa, Nicholai de. *De docta ignorantia*, ed. H. G. Senger (Hamburg: Meiner, 1994).

– . *Complete Philosophical and Theological Treatises of Nicholas of Cusa*, 2 vols. ed. Jasper Hopkins (Minneapolis: Arthur J. Banning Press, 2001).

Desargues, Girard. *The Geometrical Work of Girard Desargues*, eds. J.V. Field and J. Gray (Berlin: Springer, 1987).

Euclid. *The Elements*, translated by Sir Thomas Heath (Illinois: University of Chicago, 1952).

Goethe, J.W. *Maximen und Reflexionen* (München: C.H. Beck, 1989).
Herbart, Johann Friedrich. *Pestalozzi's Idee eines ABC der Anschauung* (Göttingen: J.F. Rower, 1804).
Kant, Immanuel. *Gesammelte Schriften*, Königlich Preußischen Akademie der Wissenschaften (Berlin and Leipzig: De Gruyter, 1922ff.) [= AA]
– . *Kritik der reinen Vernunft*, ed. Jens Timmermann, 2nd ed. (Hamburg: Felix Meiner Verlag, 1998).
– . *Über eine Entdeckung nach der alle neue Kritik der reinen Vernunft durch eine ältere entbehrlich gemacht werden soll* (AA VIII: 187–250). English translation in: *The Kant-Eberhard Controversy*, translated and edited by Henry Allison (Baltimore: Johns Hopkins University Press, 1973), pp. 107–159.
– . *The Critique of Pure Reason*, trans. and eds. Paul Guyer and Allen Wood, (Cambridge: Cambridge University Press, 1998).
– . *Theoretical Philosophy (1755–1770)*, ed. David Walford, with Ralf Meerbote (Cambridge: Cambridge University Press, 1992).
– . *Critique of Practical Reason*, trans. Thomas Kingsmill Abott (New York: Dover, 2004)
– . *Prolegomena to Any Future Metaphysics*, ed. G. Zöller (Oxford: Oxford University Press, 2004).
Langsdorf, Karl Christian. *Grundlehren der mechanischen Wissenschaften: welche die Statik und Mechanik, die Hydrostatik, Aerometrie, Hydraulik und die Maschinenlehre enthalten: mit besonderer Rucksicht auf Physiker und Praktiker* (Erlangen: J.J. Palm, 1802).
Maimon, Salomon. *Versuch über die Transzendentalphilosophie*, ed. Florian Ehrensperger (Hamburg: Felix Meiner Verlag, 2004).
Mendelssohn, Moses. *Gesammelte Schriften Jubiläumsausgabe*, eds. A. Altmann, *et al.* (Stuttgart-Bad Cannstatt: Frommann-Holzboog, 1971ff.).
Nikolaus von Amiens. *Ars fidei catholicae – Ein Beispielwerk von axiomatischer Methode*, ed. Mechthild Dreyer (Münster: Aschendorff, 1993).
Novalis. *Schriften. Historisch-kritische Ausgabe*, eds. R. Samuel, H-J. Mähl and G. Schulz (Stuttgart: W. Kohlhammer Verlag, 1960-...) [=HKA].
Pascal, Blaise. *De l'esprit géométrique – Ecrits sur la Grâce et autres textes* (Paris: Flammarion, 1999).
Plato. *The Republic*, trans. Desmond Lee (New York: Penguin Books, 1980).
– . *Protagoras and Meno*, trans. W. K. C. Guthrie (New York: Penguin Books, 1956).
– . *Timaeus. Critias. Cleitophon. Menexenus. Epistles* (Cambridge MA: Harvard University Press, 1999).
Schelling, F.W.J. *Friedrich Wilhelm Joseph Schelling's Sämmtliche Werke*, ed. K.F.A. Schelling, 14 vols. (Stuttgart: Cotta, 1856–61).
– . *Historisch-kritische Ausgabe, im Auftrag der Schelling-Kommission der Bayerischen Akademie der Wissenschaften*, edited by H. M. Baumgartner, W.G. Jacobs, H. Krings *et al.* (Stuttgart/Bad Cannstatt: Frommann-Holzboog, 1976-...).
– . *Idealism and the Endgame of Theory: Three Essays by F.W.J. Schelling*, trans. Thomas Pfau (Albany, N.Y.: SUNY Press, 1994).
– . *Darstellung meines Systems der Philosophie*, in: *Zeitschrift für spekulative Physik*,

vol. 2 (1801), (reprint Hamburg: Felix Meiner Verlag, 2001, pp. 329–436). [English translation: *Presentation of My System of Philosophy*, trans. Michael Vater in: *The Philosophical Rupture Between Fichte and Schelling* (Albany: SUNY, 2011).

– . *Further Presentations from the System of Philosophy*, (selections) trans. Michael Vater in: *The Philosophical Rupture Between Fichte and Schelling* (Albany: SUNY, 2011).

Schultz, Johann Friedrich: *Entdekte Theorie der Parallelen nebst einer Untersuchung über den Ursprung ihrer bisherigen Schwierigkeit* (Königsberg: Hartung, 1784).

– . *Anfangsgründe der reinen Mathesis* (Königsberg: Hartung 1790).

– . *Prüfung der Kantischen Critik der reinen Vernunft* (Königsberg: Nicolovius, 1792).

Spinoza, Benedictus de. *Die Ethik* (Latin/German), trans. Jakob Stern (Stuttgart: Reclam, 1977).

– . *The Ethics and Selected Letters*, trans. S. Shirley (Indianapolis, IN: Hackett, 1982).

Staudt, Karl Georg Christian von. *Geometrie der Lage* (Nuremberg: Bauer & Raspe, 1847)

– . *Beiträge zur Geometrie der Lage* (Nuremberg: Bauer & Raspe, 1856–60).

Steiner, Jacob. *Systematische Entwickelung der Abhängigkeit geometrischer Gestalten von einander* (Berlin: G. Fincke, 1832).

Wolff, Christian. *Gesammelte Werke*, eds. Jean École *et al.* (Hildesheim: Georg Olms, 1962-…).

– . *Anfangsgründe aller mathematischen Wissenschaften (Halle: Rieger,* 1710).

c. Other Authors

Asmuth, Christoph. *Interpretation – Transformation. Das Platonbild bei Fichte, Schelling, Hegel, Schleiermacher und Schopenhauer und das Legitimationsproblem der Philosophiegeschichte* (Göttingen: Vandenhoeck & Ruprecht, 2006).

Balaguer, Mark. *Platonism and Anti-Platonism in Mathematics* (Oxford: Oxford University Press, 1998).

Beierwaltes, Werner. *Platonismus und Idealismus* (Frankfurt am Main: Klostermann, 2004).

Beiser, Frederick C. *German Idealism: The Struggle against Subjectivism, 1781–1801* (Cambridge MA: Harvard University Press, 2002).

– . *The Fate of Reason: German Philosophy from Kant to Fichte* (Cambridge MA: Harvard University Press, 1987).

Benacerraf, Paul and Putnam, Hilary (eds.). *Philosophy of Mathematics: Selected Readings* (Cambridge: Cambridge University Press, 2nd edition 1983).

Bernays, Paul. »On Platonism in Mathematics« in *Philosophy of Mathematics: Selected Readings*, eds. P. Benacerraf and H. Putnam, 2nd ed. (Cambridge: Cambridge University Press, 1983), pp. 258–272.

Bolzano, Bernard. *The Mathematical Works of Bernard Bolzano*, trans. and ed. Steve Russ (Oxford: Oxford University Press, 2005).

Bonsiepen, Wolfgang. *Die Begründung einer Naturphilosophie bei Kant, Schelling, Fichte und Hegel. Mathematische versus spekulative Naturphilosophie* (Frankfurt a.

M.: Klostermann, 1996).

Boyer, Carl. *History of Analytic Geometry* (New York: Dover Publications, 2004).

– . *The History of Calculus and Its Conceptual Development* (New York: Dover Publications, 1959).

Brandt, Reinhard. »Materialien zur Entstehung der Kritik der reinen Vernunft (John Locke und Johann Schultz)» in: *Beiträge zur Kritik der reinen Vernunft*, 1781–1981, eds. Ingeborg Heidemann and Wolfgang Ritzel (Berlin: Walter de Gruyter, 1981), pp. 59–66.

Breazeale, Daniel. »Fichte's Aenesidemus Review and the Transformation of German Idealism« *Review of Metaphysics* 34 (1981): 545–568.

– . »Fichte's Nova Methodo Phenomenologica: On the methodological role of ›intellectual intuition‹ in the later Jena Wissenschaftslehre« in: *Revue Internationale de Philosophie* 206 (1998), pp. 587–616.

– . »Die *Neue Bearbeitung der Wissenschaftslehre* (1800): Letzte ›frühere‹ oder erste ›spätere‹ Wissenschaftslehre?» *Fichte-Studien* 17 (2000): 43–67.

– . »Inference, Intuition, and Imagination: On the Methodology and Methods of the First Jena *Wissenschaftslehre*« in: eds. D. Breazeale and T. Rockmore, *New Essays in Fichte's Foundation of the Entire Doctrine of Scientific Knowledge* (Amherst NY: Humanity Books, 2001), pp. 19–36.

– . »Fichte und Spinoza: ›Gemeinschaftlicher Standpunkt – wesentlicher Gegensatz‹» (unpublished manuscript from 2005).

– . »Toward a *Wissenschaftslehre more geometrico* (1800–1801)», in: *After Jena: New Essays on Fichte's Later Philosophy*, eds. D. Breazeale and T. Rockmore (Evanston, Illinois: Northwestern University Press, 2008), pp. 3–40.

– . »›Men at Work‹: Fichte and Schelling on Philosophical Construction« (paper for the Fichte Congress in Brussels in 2009, forthcoming in conference proceedings).

Brittan, Gordon. »Kant's Philosophy of Mathematics« in: ed. Graham Bird, *Blackwell Companion to Kant* (Oxford: Blackwell, 2006), pp. 222–235.

Brouwer, L.E.J. »Intuitionism and Formalism« in: *Philosophy of Mathematics: Selected Readings* (1983), eds. Paul Benacerraf and Hilary Putnam (Cambridge: Cambridge University Press, 2nd edition 1983), pp. 66–77.

Buhl, Günter. »Über eine Bemerkung Fichtes zur Quadratur des Kreises« *Archiv für Geschichte der Philosophie* 59 (1977), 281–288.

Buzaglo, Meir. *Solomon Maimon: Monism, Skepticism and Mathematics* (Pittsburgh: University of Pittsburgh Press, 2002).

Cajori, Florian. *A History of Mathematics* (New York: Macmillan, 1919; Reprint: Muschamp Press, 2007).

Carnap, Rudolf. »Die logizistische Grundlegung der Mathematik«, *Erkenntnis* 2 (1931): 91–121. English translation: »The Logicist Foundations of Mathematics« in: *Philosophy of Mathematics: Selected Readings*, eds. P. Benacerraf and H. Putnam, (Englewood Cliffs, NJ: Prentice-Hall, 1964), pp. 31–41.

Carson, Emily. »Kant on Intuition in Geometry« *Canadian Journal of Philosophy* 27 (1997), 489–512.

– . ed. *Intuition and the Axiomatic Method* (Dordrecht: Springer Verlag, 2006)

– . »Hintikka on Kant's mathematical method« *Revue internationale de Philosophie*

2009: 435–449.

Cattin, Emmanuel. »Le néoplatonisme de Fichte« in: J.-C. Goddard and A. Schnell (eds.), *L'être et le phénomène. Sein und Erscheinung. J. G. Fichte. Die Wissenschaftslehre (1804)* (Paris: Vrin, 2009), pp. 439–451.

Coolidge, J.L. »The Rise and Fall of Projective Geometry«, *American Mathematical Monthly* 41 (1934): 217–228.

Copleston, Frederick. *A History of Philosophy* (Vol. 7: *Eighteenth and Nineteenth German Philosophy*, first published 1963; reprint, New York: Continuum, 2003).

Courtine, Jean-François. *Extase de la raison. Essais sur Schelling* (Paris: Galilée, 1990).

– . »Les métamorphoses schellingiennes de l'intuition intellectuelle dans le *Système de l'idéalisme transcendantal* (intuition transcendantale et intuition esthétique)» in: *Schelling et l'élan du Système de l'idéalisme transcendantal* (Paris: L'Harmattan, 2001), pp. 41–58.

Csech, Werner Volker. »Von der Reichweite mathematischen Denkens bei Fichte und Novalis« *Theologie und Philosophie. Vierteljahrschrift* 70 (1995): 41–61.

– . *Die Raumlehre Johann Gottlieb Fichtes: Mit Berücksichtigung philosophiegeschichtlicher Konstellationen* (Frankfurt: Peter Lang, 1999). [= *Raumlehre*]

Curley, Edwin. *Behind the Geometrical Method: A Reading of Spinoza's Ethics* (Princeton: Princeton University Press, 1988).

DePaul, Michael R., and William Ramsey, eds. *Rethinking Intuition: The Psychology of Intuition and its Role in Philosophical Inquiry* (Lanham Maryland: Rowman & Littlefield Publishers, Inc., 1998).

Fischbein, Efraim. *Intuition in Science und Mathematics: An Educational Approach* (Dordrecht: Springer, 1987).

Förster, Eckart. »Goethe and the ›Auge des Geistes‹», *Deutsche Vierteljahrschrift für Literaturwissenschaft und Geistesgeschichte* 75 (2001): 87–101.

– . »›Da geht der Mann dem wir alles verdanken!‹ – Ein Untersuchung zum Verhältnis Goethe-Fichte« *Deutsche Zeitschrift für Philosophie* 45 (1997): 331–344.

Fowler, David H. *The Mathematics of Plato's Academy: A New Reconstruction* (Oxford University Press, 1987, 2nd ed. 1999).

Frängsmyr, Tore. »Christian Wolff's Mathematical Method and its Impact on the Eighteenth Century«, *Journal of the History of Ideas* 36 (1975): 653–668.

Frank, Manfred. *Unendliche Annäherung* (Frankfurt am Main: Suhrkamp, 1998).

Franks, Paul W. *All or Nothing. Systematicity, Transcendental Arguments, and Skepticism in German Idealism* (Cambridge MA: Harvard University Press, 2005).

– . »Intellectual Intuition, Geometry, and Fichte's Response to Maimon's Naturalistic Challenge« (Jan., 2006 London Talk. PdF File).

Freudenthal, Gideon. *Definition and Construction. Salomon Maimon's Philosophy of Geometry* (Berlin: Preprint 317 of the Max Planck Institute for the History of Science, 2006).

Friedman, Michael. »Kant's Theory of Geometry« *Philosophical Review* 94 (1985): 455–506.

– . *Kant and the Exact Sciences* (Cambridge MA: Harvard University Press, 1992).

– . »Geometry, Construction and Intuition in Kant and his Successors,« in: eds., Sher, G. & Tieszen. *Between Logic and Intuition: Essays in Honor of Charles Parsons*

(Cambridge: Cambridge University Press, 2000), pp. 186–218.

Fuchs. Erich. *Wirklichkeit als Aufgabe. Die doxischen Konstitutiva der theoretischen Konzeption des faktischen Gegenstandes in J. G. Fichtes »Grundlage der gesammten Wissenschaftslehre« von 1794/95* (Dissertation, University of Munich, 1973).

– . ed., *Fichte im Gespräch*, 7 vols. (Stuttgart/Bad Cannstatt: Frommann-Holzboog, 1978 ff). [= FG]

– . »Einleitung« to *Züricher Vorlesungen über den Wissenschaftslehre Februar 1794.* (Neuried/Munich: Ars Una, 1996), pp. 7–51.

– . »Fichte: A System of Freedom? Biographical-Philosophical Reflections« *Idealistic Studies* 41 (2011), pp. 113-123

Gentile, Marino. »Il matematismo spinoziano nella formazione della filosofia di Fichte«, *Giornale di metafisica* 12 (1957): 330–340.

Goddard, Jean-Christophe. »Bergson: une lecture néoplatonicienne de Fichte«, in: *Les études philosophiques* 4 (2001): 465–479.

– . »Dans quelle mesure Fichte est-il spinoziste?» in: C. Bouton (ed.) *Dieu et la nature. La question du panthéisme dans l'idéalisme allemand* (Hildesheim/Zurich/N.Y.: Georg Olms Verlag, 2005) pp. 75–89.

Gueroult, Martial. *L'évolution et la structure de la Doctrine de la Science chez Fichte*, 2 vols. (Paris: Les Belles-Lettres, 1930).

Hadamard, Jacques. *The Psychology of Invention in the Mathematical Field* (Princeton: Princeton University Press, 1949).

Hahn, Hans. »The Crisis in Intuition« in: James R. Newman, ed. *The World of Mathematics, Vol. III, pp. 1956–1976.* (Reprinted in: H. Hahn, *Empiricism, Logic and Mathematics* (Dordrecht: Reidel, 1980), pp. 73–102.

Hammacher, Klaus. »Fichte und Spinoza« in: *Spinoza und der deutsche Idealismus*, ed. M. Walther (Würzburg: Könighausen & Neumann, 1992), pp. 82–107.

– . »Spinoza und Fichte« in: M. Czelinski *et al* (eds.) *Transformation der Metaphysik in die Moderne* (Würzburg: Könighausen & Neumann, 2003), pp. 192–201.

Hart, James G. »Husserl and Fichte: With Special Regard to Husserl's Lectures on ›Fichte's Ideal of Humanity‹» in: *Husserl Studies* (1995): 135–163.

Hatfield, Gary. »Kant on the Perception of Space (and Time)» in: *The Cambridge Companion to Kant and Modern Philosophy*, ed. Paul Geyer (Cambridge: Cambridge University Press, 2006), pp. 61–93.

Hempel, Carl G. »On the Nature of Mathematical Truth« *American Mathematical Monthly* 52 (1945), 543–556.

Henrich, Dieter. »Fichtes Ursprüngliche Einsicht« in: Subjektivität und Metaphysik: Festschrift für Wolfgang Cramer (Frankfurt am Main: Klostermann, 1967).

Hilbert, David. *Grundlagen der Geometrie.* (Stuttgart: Tuebner, 1962, 9th ed.). English Translation: *The Foundations of Geometry*, trans. E. Townsend (Illinois: Open Court, 1950).

– . »Axiomatisches Denken« *Mathematische Annalen* 78 (1918): 405–415.

– . *Lectures on the Foundations of Geometry 1891–1902*, eds. M. Hallet and U. Majer (Berlin: Springer, 2004).

Hintikka, J. »Kant's Theory of Mathematics Revisited« *Philosophical Topics* 12 (1981): 201–215.

– . »Kant's Transcendental Method and his Theory of Mathematics« *Topoi* 3 (1984): 99–108.
– . ed. *Logic of Discovery and Logic Discourse* (New York: Plenum Press, 1985).
Horstmann, Hubert. »Aspekte der Raumlehre Fichtes« *Deutsche Zeitschrift für Philosophie* 36 (1988): 909–918.
– . »Unterschiede zwischen den Raumauffassungen Fichtes und Kants« in: ed. Albert Mues, *Transzendentalphilosophie als System. Die Auseinandersetzung zwischen 1794 und 1806* (Hamburg: Meiner, 1989), pp. 58–67.
Husserl, Edmund. *Gesammelte Werke* (Berlin: Springer, 1978ff).
– . *Fichtes Menschheitsideal*, in: *Husserliana*, XXV (1986): 267–293. English translation: »Fichte's Ideal of Humanity« in: *Husserl Studies* 12 (1995): 111–133.
Ivaldo, Marco. »Transzendentalphilosophie und ›Realistische‹ Metaphysik. Das Fichtesche Spinoza-Verständnis« in: ed. Manfred Walter, *Spinoza und der deutsche Idealismus* (Würzburg: Königshausen & Neumann, 1992), pp. 59–79.
– . »L'intuizione intellettuale come ›punto fermo‹ della filosofia nel Saggio di una nuova esposizione della dottrina della scienza (1797–1798)» in: *Sul trascendentale moderno - Genesi, Struttura, Problemi*, ed. Roberto Perini (Naples: E. S. I., 2004), pp. 91–109.
Jahnke, Hans Niels. *Mathematik und Bildung in der Humboldtschen Reform* (Göttingen: Vandenhoeck & Ruprecht, 1990).
Janke, Wolfgang. *Vom Bilde des Absoluten. Grundzüge der Phänomenologie Fichtes* (Berlin/New York: Walter de Gruyter, 1993).
Jergius, Holger. »Fichtes ›geometrische‹ Semantik«, *Fichte-Studien* 10 (1997): 51–63.
Klein, Felix. *Vorlesungen über die Entwicklung der Mathematik im 19. Jahrhundert* (New York: Chelsea, 1967).
Kline, Morris. *Mathematical Thought from Ancient to Modern Times*, 3 vols. (Oxford: Oxford University Press, 1972).
Koch, Reinhard. *Fichtes Theorie des Selbstbewusstsein: Ihre Entwickelung von den ›Eignen Meditationen über Elementar Philosophie‹ 1793 bis zur ›Neuen Bearbeitung der W.L.' 1800* (Würzburg: Königshausen & Neumann, 1989).
Koriako, Darius. *Kants Philosophie der Mathematik* (Hamburg: Meiner, 1999).
Kueker, D., and Smith, C. *Learning and Geometry* (Boston: Birkhäuser, 1995).
Lahbib, Olivier. *De Husserl à Fichte: Liberté et réflexivité dans le phénomène* (Paris: L'Harmattan, 2009).
– . »Husserl, lecteur de Fichte«, *Archives de Philosophie* 67 (2004): 421–443.
Lask, Emil. *Fichtes Idealismus und die Geschichte* (Tübingen/Leipzig: Mohr, 1902).
Lauth, Reinhard. »J. G. Fichtes Gesamtidee der Philosophie«, *Philosophisches Jahrbuch* 71 (1964/65): 353–384.
– . »Genèse du Fondement de toute la Doctrine de la science de Fichte à partir de ses ›Méditations personnelles sur l'élémentarphilosophie‹, *Archives de philosophie* 34 (1971): 51–79.
– . *Die Konstitution der Zeit im Bewusstsein* (Hamburg, Meiner, 1981).
– . *Die Transzendentale Naturlehre Fichtes nach den Prinzipien der Wissenschaftslehre* (Hamburg: Meiner, 1984).
– . »Fichtes Sicht der Philosophie Spinozas« in: *Transzendentale Entwicklungslinien*

von Descartes bis zu Marx und Dostojewski (Hamburg: Meiner, 1989), 24–43.
– . *Vernünftige Durchdringung der Wirklichkeit. Fichte und sein Umkreis.* (Neureid/Munich: Ars Una, 1994).
Léon, Xavier. *Fichte et son temps*, 2 vols. (Paris: Armand Colin, 1922–1927).
Lindemann, Ferdinand. »Ueber die Zahl π« in: *Mathematische Annalen* 20 (1882): 213–225.
Lorenzen, Paul. »Wie ist Philosophie der Mathematik möglich« *Philosophia Naturalis* 4 (1957): 192–208.
– . »Constructive Mathematics as a Philosophical Problem« *Compositio Mathematica* 20 (1968): 133–142.
Magnani, Lorenzo. *Philosophy and Geometry: Theoretical and Historical Issues* (Dordrecht: Kluwer Academic Publications, 2001).
Majer, Ulrich. »Geometry, intuition and experience: From Kant to Husserl« *Erkenntnis* 42 (1995): 261–285.
Martin, Wayne. »In Defense of Bad Infinity. A Fichtean Response to Hegel's *Differenzschrift*«, *Bulletin of the Hegel Society of Great Britain* 55/56 (2007): 168–187.
Metz, Wilhelm. »Fichtes genetische Deduktion von Raum und Zeit in Differenz zu Kant« *Fichte-Studien* 6 (1994): 71–94.
Moiso, Francesco. »Geometrische Notwendigkeit, Naturgesetz und Wirklichkeit: Ein Weg zur *Freiheitsschrift*« in: *Weg zur Freiheitsschrift. Legende und Wirklichkeit*, eds. H. M. Baumgartner and W. G. Jacobs (Stuttgart-Bad Cannstatt: Frommann-Holzboog, 1996), pp. 132–186.
Neumann, Johann von. »The Formalist Foundations of Mathematics«, in: *Philosophy of Mathematics: Selected Readings*, eds. P. Benacerraf and H. Putnam, (Englewood Cliffs, NJ: Prentice-Hall, 1964), pp. 50–54.
Paimann, Rebecca. *Die Logik und das Absolute. Fichtes Wissenschaftslehre zwischen Wort, Begriff und Unbegreiflichkeit* (Würzburg: Könighausen & Neumann, 2006).
Paret, Peter. *Clausewitz and the State* (Oxford: Oxford University Press, 1976).
Parsons, Charles. »Mathematical Intuition« *Proceedings of the Aristotelian Society* 80 (1980): 145–68.
Philonenko, Alexis. »Die intellektuelle Anschauung bei Fichte« in *Der transzendentale Gedanke: Die gegenwärtige Darstellung der Philosophie Fichtes*, ed. Klaus Hammacher (Hamburg: Meiner, 1981), pp. 91–106.
– . *La liberté humaine dans la philosophie de Fichte* (Paris: Vrin, 1966; 3rd ed., 1999).
Piché, Claude. »Évidence géométrique et certitude morale chez Fichte: le problème de la vérité en philosophie«, book chapter in: *Kant et ses épigones, le jugement critique en appel* (Paris: Vrin, 1995), pp. 132–189.
Pinkard, Terry. »Hegel's Philosophy of Mathematics« *Philosophy and Phenomenological Research* 41 (1981): 452–464.
Poincaré, Henri. *Science et méthode* (Paris: Flammarion, 1908). English translation: *Science and Method* (New York: Dover Publications, 2003).
– . *The Value of Science*, translated by G.B. Halsted (New York: Cosimo, 2007).
Posy, Carl J. (ed.). *Kant's Philosophy of Mathematics: Modern Essays* (Dordrecht: Kluwer Academic Publications, 1992).
Radbruch, Knut. »Die Bedeutung der Mathematik für die Philosophie bei Fichte«

Fichte-Studien 22 (2003): 251–263.
Radrizzani, Ives. *Vers la fondation de l'intersubjectivité chez Fichte. Des Principes à la Nova Methodo* (Paris: Vrin, 1993).
– . »La « première » Doctrine de la science de Fichte. Introduction et traduction«, *Archives de philosophie* 60 (1997): 615–638.
– . »Fichtes ›erste‹ Wissenschaftslehre«, *Fichte-Studien* 16 (1999): 409–431.
Rametta, Gaetano. »La dissolution de la théologie politique chez le dernier Fichte« in: J.-C. Goddard & M. Maesschalck, eds. *La philosophie de la maturité (1804–1814)*, vol. II., (Paris: Vrin, 2003) pp. 31–50.
Rath, Michael. *Selbstbewusstsein und Will: Untersuchungen zu Fichtes Wissenschaftslehre nova methodo* (Heidelberg: PhD manuscript, 1985).
Rohs, Peter. »Über die Zeit als das Mittelglied zwischen dem Intelligiblen und dem Sinnlichen« *Fichte Studien* 6 (1994): 95–116.
Sassen, Brigitte. *Kant's Early Critics*, ed. (Cambridge: Cambridge University Press, 2000).
Schmidt, Andreas. *Der Grund des Wissens: Fichtes Wissenschaftslehre in den Versionen 1794/95, 1804/II und 1812* (Paderborn: Schöningh, 2004).
– . »Johann Gottlieb Fichte (1762–1814)» in: ed. Otfried Höffe, *Klassiker der Philosophie: Von Immanuel Kant bis John Rawls* (Munich: Beck, 2008), pp. 29–42.
Scholz, Erhard. »Hermann Weyl's ›purely infinitesimal‹ geometry«, in: *Proceedings of the International Congress of Mathematicians*, Zurich, Switzerland 1994 (Basel: Birkhäuser, 1995), vol. 1, 2, pp. 1592–1603.
Scholz, Heinrich. »Die Wissenschaftslehre Bolzanos: Eine Jahrhundert-Betrachtung« (1937), in: *Mathesis Universalis: Abhandlungen zur Philosophie als Strenger Wissenschaft*, eds. H. Hermes, F. Kambartel & J. Richter (Basel: Benno Schwabe & Co., 1961), pp. 219–267.
– . »Warum Mathematik« in: *Neue Jahrbücher für Wissenschaft und Jugendbildung* 10 (1934), pp. 458–469 (reprinted: H. Scholz, *Mathesis Universalis: Abhandlungen zur Philosophie als Strenger Wissenschaft*, pp. 312–323).
Schüler, Wolfgang. *Grundlegungen der Mathematik in transzendentaler Kritik. Frege und Hilbert* (Hamburg: Meiner, 1983).
Shabel, Lisa. *Mathematics in Kant's Critical Philosophy* (New York and London: Routledge, 2003).
– . »Kant's Philosophy of Mathematics« in: ed. Paul Geyer, *The Cambridge Companion to Kant and Modern Philosophy* (Cambridge: Cambridge University Press, 2006) pp. 94–128.
Shapiro, Steward. *Thinking about Mathematics: The Philosophy of Mathematics* (Oxford: Oxford University Press, 2000).
– . ed. *Oxford Handbook of the Philosophy of Mathematics and Logic* (Oxford: Oxford University Press, 2005).
Sher, G. & Tieszen R. (eds.). *Between Logic and Intuition: Essays in Honor of Charles Parsons* (New York: Cambridge University Press, 2000).
Sieroka, Norman. »Weyl's ›Agens Theory‹ of Matter and the Zurich Fichte«, *Studies in History and Philosophy of Science* 38 (2007): 84–107.
– . *Umgebungen: Symbolischer Konstruktivismus im Anschluss an Hermann Weyl und*

Fritz Medicus (Zurich: Chronos, 2010).

Smith, David E. *History of Mathematics* (New York: Dover Publications, 1958).

Speiser, Andreas. »Die Grundlagen der Mathematik von Plato bis Fichte«, *Eranos Jahrbuch* 14 (1946): 11–38.

– . *Ein Parmenideskommentar – Studien zur platonischen Dialektik*, 2nd ed. (Leipzig: K. F. Koehler, 1959).

Staël, Germaine de. *De l'Allemagne*, 2 vols. (Paris: Flammarion, 1968).

Steiner, Mark. *Mathematical Knowledge* (Ithaca: Cornell University Press, 1975).

Stolzenberg, Jürgen. *Fichtes Begriff der intellektuellen Anschauung: Die Entwickelung in den Wissenschaftslehre von 1793/94 bis 1801/02* (Stuttgart: Klett-Cotta, 1986).

Thomas-Fogiel, Isabelle. *Critique de la représentation. Étude sur Fichte* (Paris: Vrin, 2000).

Tignol, Jean-Pierre. *Galois' Theory of Algebraic Equations* (London: World Scientific Publishing Co., 2001).

Tilliette, Xavier. *L'intuition intellectuelle de Kant à Hegel* (Paris: Vrin, 1995).

– . *Fichte: la science de la liberté* (Paris: Vrin, 2003).

Tonelli, Giorgio. »Der Streit über die mathematische Methode in der Philosophie in der ersten Hälfte des 18 Jahrhunderts und die Entstehung von Kants Schrift über die ›Deutlichkeit‹,« *Archiv für Philosophie* 9 (1959): 37–66.

Toretti, Roberto. *Philosophy of Geometry from Riemann to Poincaré* (Dordrecht: D. Riedel Publishing Co., 1978).

Tutor, Juan Ignacio Gómez. *Die wissenschaftliche Methode bei Christian Wolff*, in: Christian Wolff, *Gesammelte Werke*, III. Abt.: Materialien und Dokumente, vol. 90 (Hildesheim: Georg Olms Verlag, 2004).

Veblen, Oswald and Young, John Wesley. »A Mathematical Science«, reprinted in: *The World of Mathematics*, ed. J. Newman (New York: Dover Publications, 2000, Volume 3, pp. 1696–1707).

Vuillemin, Jules. *L'héritage kantien et la révolution copernicienne: Fichte, Cohen, Heidegger* (Paris: PUF, 1954).

– . *Physique et métaphysique kantiennes* (Paris: PUF, 1955).

– . »Fichte, 1762–1814« in: *Les philosophes célèbres*, ed. M. Merleau-Ponty (Paris: Edition Mazenod, 1956), pp. 226–233.

– . *La philosophie de l'algèbre* (Paris: PUF, 1962, 2nd ed. 1993).

– . *What are Philosophical Systems?* (Cambridge: Cambridge University Press, 1986).

– . *Mathématiques pythagoriciennes et platoniciennes* (Paris: Éditions Albert Blanchard, 2001).

Waibel, Violetta. *Hölderlin und Fichte: 1794–1800* (Paderborn: Schöningh, 2000).

Weyl, Hermann. *Das Kontinuum* (Leipzig: Veit & Comp., 1918).

– . *Gesammelte Abhandlungen*, edited by K. Chandrasekharan (Berlin: Springer, 1968).

Widmann, Joachim. *Analyse der formalen Strukturen des transzendentalen Wissens in Joh. Gottl. Fichtes 2. Darstellung der »Wissenschaftslehre« aus dem Jahre 1804* (Dissertation, University of Munich, 1961).

– . *Die Grundstruktur des Transzendentalen Wissens nach Joh. Gottl. Fichtes Wissenschaftslehre 1804²* (Felix Meiner Verlag, Hamburg, 1977). [= *Grundstruktur*].

– . »Exact Concepts: Fichte's Contribution on a Problem of Tomorrow« *Idealistic*

Studies 11 (1981): 41–47.
– . »Dialektik more geometrico. Fichtes Prinzip des Systemgenesis«, in: *Kant oder Hegel? Formen der Begründung der Philosophie*, ed. Dieter Henrich, (Stuttgart: Klett-Cotta, 1983), pp. 186–194.
Wood, David W. »Fichte's Conception of Infinity in the *Bestimmung des Menschen*« (forthcoming in the proceedings of the tenth biennial meeting of the North American Fichte Society, Lisbon, Portugal, April 2010).
– . »›*Mathematische Wissenschaftslehre*‹: On a Late Fichtean Reflection of Novalis« (Paper held at Conference: Relevance of Romanticism, Villanova University, USA, April 2010, forthcoming in proceedings, 2011).
Wright, Walter E. »The Shadow of Spinoza in Fichte's WL 1804²« *Idealistic Studies* 33 (2003): 161–174.
Wundt, Max. *Fichte-Forschungen* (Stuttgart: Frommann, 1929).
Ziche, Paul. *Mathematische und naturwissenschaftliche Modelle in der Philosophie Schellings und Hegels* (Stuttgart-Bad Cannstatt: Frommann-Holzboog, 1996).
– . »Raumkonstruktion, Deduktion der Dimensionen und idealistische Prinzipientheorie. Problemlagen im Fichte-Schelling-Briefwechsel vom November 1800« *Fichte-Studien* 25 (2005): 21–42.
Zöller, Günter. »An Eye for an I. Fichte's Transcendental Experiment«, in: eds. David E. Klemm and Günter Zöller, *Figuring the Self: Subject, Absolute, and Others in Classical German Philosophy* (Albany: SUNY Press, 1997) pp. 73–95.
– . *Fichte's Transcendental Philosophy: The Original Duplicity of Intelligence and Will* (Cambridge: Cambridge University Press, 1998).
– . »Fichte als Spinoza, Spinoza als Fichte. Jacobi über den Spinozismus der Wissenschaftslehre« in: W. Jaeschke/B. Sandkaulen (eds.) *Friedrich Heinrich Jacobi. Ein Wendepunkt der geistigen Bildung der Zeit* (Hamburg: Meiner, 2004, 37–52).
– . »From Transcendental Philosophy to *Wissenschaftslehre*: Fichte's Modification of Kant's Idealism«, *European Journal of Philosophy* 15 (2007): 249–269.

List of Abbreviations

GA	(e.g. GA I/4: 307) *J. G. Fichte-Gesamtausgabe der Bayerischen Akademie der Wissenschaften* (series, volume, page)
SW	(e.g. SW X: 254) *J. G. Fichte sämmtliche/nachgelassene Werke* ed. I. H. Fichte, Bonn/Berlin (I–XI)
StA-1/SWV-1	Fichte, Johann Gottlieb: *Die späten wissenschaftlichen Vorlesungen I, 1809–1811.* eds. Hans Georg von Manz, Erich Fuchs, Reinhard Lauth und Ives Radrizzani. Stuttgart-Bad Cannstatt: Frommann-Holzboog, 2000
StA-2/SWV-2	Fichte, Johann Gottlieb: *Die späten wissenschaftlichen Vorlesungen II.* Wissenschaftslehre 1811. Über das Wesen der Philosophie 1811. Von den Thatsachen des Bewußtseyns 1811. Hrsg. von Hans Georg von Manz, Erich Fuchs, Reinhard Lauth und Ives Radrizzani. Stuttgart-Bad Cannstatt: Frommann-Holzboog, 2003
AzsL	*Anweisung zum seeligen Leben* 1806
BdG	*Vorlesungen über die Bestimmung des Gelehrten* 1794
BdG-1811	*Über die Bestimmung des Gelehrten* 1811
BdM	*Die Bestimmung des Menschen* 1800
Beiträge	*Beiträge zur Berichtigung der Urtheile des Publikums* etc. 1793/94
BWL	*Ueber den Begriff der Wissenschaftslehre* 1794
Diarium-I	*Diarium*, March- 1813
Diarium-II	*Diarium* August/September 1813
Diarium-III	*Diarium* Oktober 1813/January 1814
ErE	*Erste Einleitung in die Wissenschaftslehre* 1797
GB	*Über Geist und Buchstab in der Philosophie*
GNR	*Grundlage des Naturrechts* 1796
GdgZ	*Die Grundzüge des gegenwärtigen Zeitalters* 1806
GWL	*Grundlage der gesammten Wissenschaftslehre* 1794/95
GHS	*Der geschlossene Handelsstaat* 1800
Principien	*Die Principien der Gottes-, Sitten- und Rechtslehre* 1805
RL-1812	*Rechtslehre* 1812
Reden	*Reden an die deutsche Nation* 1808
SL	*Das System der Sittenlehre* 1798
SL-1812	*Sittenlehre* 1812
StL	*Die Staatslehre, oder über das Verhältnis des Urstaates zum Vernunftreiche* 1813
TdB	*Die Thatsachen des Bewußtseins*

TL I	*Transzendentale Logik* April to August 1812
TL II	*Transzendentale Logik* October to December 1812
UI	*Ultima Inquirenda. J. G. Fichtes letzte Bearbeitungen der Wissenschaftslehre* 1813/1814, ed. Reinhard Lauth. Stuttgart-Bad Cannstatt, 2001
UM	*Ueber Macchiavell* 1807
VnD	*Versuch einer neuen Darstellung der Wissenschaftslehre* 1798
WdG	*Über das Wesen des Gelehrten, und seine Erscheinungen im Gebiete der Freiheit* 1805
WL	*Wissenschaftslehre*
WL-1801/02	*Wissenschaftslehre* 1801/02
WL-1804-I/II/III	*Wissenschaftslehre* of 1804, first, second, third series of lectures
WLnm	*Wissenschaftslehre nova methodo* 1796–1799
WLnm-K	*Wissenschaftslehre nova methodo* 1798/1799, Nachschrift K. C. F. Krause
ZdDf	*Zurückforderung der Denkfreiheit* 1793
ZV	*Züricher Vorlesungen über den Begriff der Wissenschaftslehre*
ZwE	*Zweite Einleitung in die Wissenschaftslehre* 1797
FG	*Fichte im Gespräch*. ed. E. Fuchs
AA	*Immanuel Kant's gesammelte Schriften, herausgegeben von der Königlich Preußischen Akademie und ihren Nachfolgern*, Berlin 1900ff; *Nachdruck der Druckschriften* (vols. 1–9), Berlin 1968.
KdU	Kant: *Kritik der Urteilskraft*
KpV	Kant: *Kritik der praktischen Vernunft*
KrV	Kant: *Kritik der reinen Vernunft*
TWA	G. W. F. Hegel: *Theorie-Werkausgabe. Werke in zwanzig Bänden. Auf der Grundlage der Werke von 1832–1845 neu edierte Ausgabe.* Eds. Eva Moldenhauer and Karl Markus Michel. Frankfurt am Main: Suhrkamp, 1970

www.ingramcontent.com/pod-product-compliance
Lightning Source LLC
LaVergne TN
LVHW011115300926
849160LV00008B/392